Nanoparticles in Healthcare
Applications in Therapy, Diagnosis, and Drug Delivery

Edited by

Dileep Francis[1], Manoj Balachandran[2], Elcey C Daniel[1], Shinomol George Kunnel[1]

[1]Department of Life Sciences, Kristu Jayanti College, Autonomous, Bengaluru, Karnataka, India

[2]Department of Physics and Electronics, CHRIST (Deemed to be University), Bengaluru, Karnataka, India

Copyright © 2024 by the authors

Published by **Materials Research Forum LLC**
Millersville, PA 17551, USA

Published as part of the book series
Materials Research Foundations
Volume 160 (2024)
ISSN 2471-8890 (Print)
ISSN 2471-8904 (Online)

Print ISBN 978-1-64490-296-7
eBook ISBN 978-1-64490-297-4

Distributed worldwide by

Materials Research Forum LLC
105 Springdale Lane
Millersville, PA 17551
USA
https://www.mrforum.com

Manufactured in the United States of America
10 9 8 7 6 5 4 3 2 1

Table of Contents

Preface

Nanoparticles, with diameters ranging from 1 to 100 nm, represent a captivating realm where the inherent properties of materials undergo significant transformations. When substances are broken down into nano-dimension particles, they exhibit physical, chemical, and electrical characteristics that diverge dramatically from their bulk counterparts. These remarkable attributes have ignited the flourishing domain of nanotechnologies, spanning a multitude of sectors. In healthcare, nanoparticles have taken the spotlight, finding applications in therapeutics, diagnostics, and drug delivery. This edited volume contains 11 chapters contributed by renowned researchers in the field of nanoparticle research. The eleven chapters collectively offer an extensive exploration of the transformative potential of nanoparticles in healthcare, providing invaluable insights into their synthesis, applications, and contributions across diverse medical domains.

Chapter 1 introduces the use of nanomaterials in the healthcare sector. It delves into nanomaterial fabrication, surface modification, and their wide-ranging applications. Examples illustrate the utilization of nanomaterials in optical and electrical-based detection schemes, emphasizing their capacity to offer cost-effective healthcare solutions when tailored to meet specific requirements.

Chapter 2 navigates through diverse strategies employed for synthesizing nanosized functional materials. It elucidates the principles, challenges, and benefits underpinning physical, chemical, and biological approaches. The chapter underscores the apt choice of technique with respect to the desired nanomaterial. Additionally, it spotlights green protocols currently applied for crafting nanostructured materials.

Chapter 3 places herbal extracts in the spotlight as safer avenues for generating nanomaterials. Secondary metabolites within these extracts serve as stabilizing and reducing agents, facilitating nanoparticle formation. The focus is on recent advancements in the biomedical applications of plant extract-derived nanoparticles, particularly in cancer treatment and combating viral infections. The chapter also addresses toxicological testing protocols, vital for efficient biomedical applications.

Chapter 4 explores the realm of nanoantibiotics (nAbts), dissecting their composition, types, modes of action, and the challenges they confront concerning safety, toxicity, manufacturing complexities, and regulatory hurdles. The chapter underscores the advantages and disadvantages of nAbts and delves into the development of resistance among bacterial pathogens against them.

Chapter 5 conducts an extensive examination of the use of nanoparticles in bioimaging. It hones in on the structural characteristics, advantages, and progress of contrast agents relying on nanoparticles for various biomedical imaging techniques. Recent advancements in nanoparticle-based bioimaging are discussed, emphasizing their exceptional qualities. The chapter underscores the contributions of metallic, polymeric, carbonaceous, and lipid-based nanoparticles to bioimaging, highlighting their pivotal role in advancing this critical research field.

Chapter 6 provides insight into nanotechnology's role in clinical diagnostics. It outlines various nanoparticle types, including quantum dots, gold nanoparticles, magnetic nanoparticles, fluorescent nanoparticles, and more, used in nanodiagnostics assays for tumor and infectious disease detection. The chapter elucidates the advantages of nanodiagnostics over conventional diagnostics while addressing associated limitations and concerns. It also explores the potential of nanodevices, like nanorobotics, for combined diagnosis and therapeutics.

Chapter 7 underscores the significance of environmentally friendly nanoparticle (NP) production in cancer therapy through plant-mediated green synthesis. It highlights the benefits of this approach, including simplicity, safety, and biocompatibility, while addressing environmental concerns tied to conventional NP manufacturing. The chapter explores the potential of eco-friendly NPs in cancer diagnosis and treatment, with a focus on their selectivity towards cancer cells. It also delves into the mechanisms by which these NPs operate in cancer applications, offering promising prospects for the field.

Chapter 8 spotlights the potential medical applications of metal doped nanoparticles. Nanoparticles, with their unique physico-chemical properties, offer a myriad of applications. Synthesis methods, encompassing top-down and bottom-up approaches, involve physical, chemical, and biological routes. Doping with elements like silver, copper, cobalt, and rare earth elements enhances their functionality and biological effects. These nanoparticles exhibit remarkable antimicrobial properties and find application in wound dressings loaded with medications for efficient drug delivery, accelerating the healing process.

Chapter 9 initiates a discussion on neurodegenerative diseases, focusing on the challenges of drug delivery across the blood-brain barrier, uptake mechanisms of nanoparticles, and nano stem cell hybrids. It delves into the various nanomaterials used in the treatment of neurodegenerative diseases, offering insights into green extract nanoparticles, biodegradable nanoparticles, and nanoparticles designed for specific neurodegenerative conditions.

Chapter 10 reviews the potential and significant role of theranostic nanoparticles in the realm of lung diseases. It provides insight into the application of theranostic nanoparticles, highlighting their potential for enhanced drug bioavailability at the target site. The chapter underscores the efficacy and promising approach of theranostic nanoparticles in treating lung diseases.

Chapter 11 delves into the realm of gliomas, addressing conventional treatment strategies and the role of nanoparticles in their treatment. It discusses various polymers employed in nanotherapy, quantum dot photodynamic therapy, magnetic and gadolinium particles, and other methodologies for glioma treatment. The chapter also explores identified targets for glioma nanotherapy.

Dr. Dileep Francis
Dr. Manoj Balachandran
Dr. Elcey C Daniel
Dr. Shinomol George Kunnel
Editors

Nanoparticles in Healthcare: Applications in Therapy, Diagnosis and Drug Delivery   Materials Research Forum LLC
Materials Research Foundations 160 (2024) 1-23   https://doi.org/10.21741/9781644902974-1

Chapter 1

Overview of Nanotechnology Applications in Healthcare

Tejaswini Ronur Praful[1]*

[1]St. Joseph's University, Bangalore-560027, Karnataka, India

*tejaswini.praful@sju.edu.in

Abstract

Microorganisms gaining drug resistance, difficulty in disease screening, unavailability of point-of-care diagnostic tools are few challenges in the healthcare industry. Over the past few decades, much research has been directed towards developing new nanomaterials with tunable properties for diverse applications in biomedicine, bioimaging, disease diagnosis, therapeutic formulations, etc. Nanosystems specifically used in the healthcare sector include specially designed and functionalized nanomaterials for suitable applications. Nanotechnology finds important applications in identification of specific types of cancers even before their onset, identification of specific infections, targeted gene and drug delivery systems, etc.

Keywords

Nanoparticle, Nanostructure, Nanomaterial, Nanosystems, Functionalization, Healthcare, Diagnosis, Imaging, Biomedicine, Therapy

Contents

1.   Introduction

Even with enormous scientific developments, access to adequate healthcare, diagnostics tools and therapies still remains a concern in many parts of the world [1, 2]. Advancements in nanotechnology enables delivery of such tools at the doorstep [3]. The word 'nano' derived from the Greek, means 'dwarf'. One nanometer is equal to 10^{-9} meters [4, 5]. Nanotechnology refers to engineering and technological solutions targeting large scale problems from atomic and/or molecular scales of a material [6, 7]. Nanotechnology has stretched its roots into complex areas of biotechnology. Biotechnology in itself is an interdisciplinary science bridging the gaps between various fields of pure biological sciences. Nanobiotechnology is an emerging interdisciplinary area of science that combines the benefits of nanotechnological tools for applications in biotechnology [8, 9]. One can take inspiration from nature to produce nanosystems with advanced capacities for diverse applications in biotechnology. Gene editing [10, 11], development of new vaccines [12, 13], at home diagnostic kits [14], targeted drug and gene delivery systems [15, 16], improved industrial processes [17], bioremediation [17, 18], wastewater treatment [17] are a few examples where nanobiotechnology is applied.

Early detection followed by diagnosis of a specific disease or disorder has always remained an issue in healthcare [19]. Nanotechnology has particularly shown significant potential to develop new systems in healthcare for medicine, therapy and diagnostics enabling new fields of specializations within nanobiotechnology such as nanomedicine [20, 21] and nanotheranostics [22, 23]. Both nanomedicine and nanotheranostics, deal with development of new and innovative nano-based solutions for healthcare applications. Various nanosytems with indefinite options ranging across geometry, texture, and composition are extensively studied. Gold, silver, titanium nitride nanosystems have been choice of materials particularly for healthcare and other biological applications [24, 25].

Also, nanotechnological advancements enable the development of new nanosystems for early detection of potential anthropological materials, capable to cross blood-brain-barrier, leading to various diseases and disorders. Examples of such anthropological materials detected using such nanosystems include: microplastics, pesticides, polyfluoroalkyl substances (PFAS), polyhydroxyalkanoates (PHAs), etc. [17, 18, 26, 27].

Nanomaterials can be used either in the form of a suspension or coated on a solid surface for applications. Top-down and bottom-up approaches [28], employing physical, chemical and biological methods [29] are used for fabrication of nanomaterials. Lately, attention is projected towards green synthesis of nanoparticles utilizing microorganisms and plant-based extracts. Specific ions or enzymes produced by the microbes in a microbial suspension or phytochemicals present in plant extracts enable reduction of precursor molecules to produce nanoparticles.

The size of nanosytems utilized for healthcare range between 5-200 nm with exceptional tunability options for altering optical, electrical, magnetic, mechanical and biological properties [30]. Techniques such as 3D printing, Langmuir-Blodgett techniques for monolayers and solvent-dispersion methods prove to be a blessing for producing nanostructures enabling to mimic complex cellular processes *in vitro* required for healthcare applications [28, 31, 32].

Overall, the goal of this chapter is to discuss materials used for healthcare applications, methods used to produce the same, challenges associated and future perspectives enabling new healthcare solutions.

2.   Nanomaterials used in healthcare

Nanomaterials ranging in composition of metal, metal-oxides, semiconductors, organic and inorganic materials have been explored for their potential applications in healthcare. Cerium [33, 34] is a well-studied lanthanide element for biological applications. Recently, elements such as Europium [35, 36] and Erbium [37, 38] are explored for signal enhancement in opto-electronic devices. Cerium is the most abundant rare earth element existing in two oxidation states. Cerium, in the form of nanoceria and cerium oxide nanoparticles are the two major nano-based forms of cerium used in healthcare sector. It is widely used for topical applications, as it is a good UV absorbing material. The fate of nano-based cerium for in vivo applications is reported to be different, and is dependent on the size and dimension of nanoparticle and presence of any surface modification [39, 40]. While, erbium and europium are currently being explored for fluorescence-based signal amplification mechanism for biosensing. Specifically, erbium is used as a dopant material, while europium has its foothold primarily in addressing quenching mechanisms observed in various optoelectronic devices.

Gold, silver, copper, zinc, cobalt, titanium and many other nanoparticle systems with varying dimensions and surface modifications are used for biosensing, antimicrobial properties, imaging, personalized therapies and many more applications [3-41]. Organic materials employing lipids and polymers are generously used for gene and drug delivery. Some examples of important nanomaterials used in healthcare are discussed below.

Gold and silver are the most extensively studied nanomaterials for biological applications. Their optical and electronic properties are well studied in both thin films and nanoparticle systems. Optical shifts in plasmon resonance of silver and gold thin-films and their respective nanoparticles are efficiently utilized for emerging biosensing applications with and without surface modification [42]. Studies pertaining to antimicrobial properties of these two materials based on their size, shape and dimensions are reported in literature [43]. Due to the nanosize of the nanoparticle systems, the antimicrobial effects are induced through various mechanisms [44]. Disrupting the normal functioning of mitochondria, enabling caspase cascade pathway, DNA and other protein damage, increased rate of reactive oxygen species (ROS) production are few mechanisms to mention as to how nanoparticles can induce anti-microbial effects [45, 46]. Both gold [47] and silver [48] nanoparticles have shown dose-dependent anti-cancer effects upon inducing ROS production, DNA damage and enzyme inactivation [49]. Gold nanoparticles are gaining popularity for various photo-induced therapies, particularly hyperthermia-based cancer therapy and targeted gene and drug delivery [50]. Gold conjugated with active biomolecules such as lipids, enzymes, peptides, antibodies are used as part of immunochemistry-based analysis.

Carbon based materials such as single walled carbon nanotubes (SWCNT), multi walled carbon nanotubes (MWCNT), fullerene, graphene sheet, carbon dots and nanodiamond are utilized in healthcare systems primarily as imaging agents [49]. Various reports indicate nanotubes being particularly toxic. Nonetheless, these materials with optimized dose concentrations are used for anti-cancer therapy and drug delivery. Alternate materials that are less toxic, highly biocompatible, efficient in terms of drug delivery are liposome and their metallic counter-parts metallosomes based systems. Liposomes enable mimicking cellular environment much efficiently with ease of fabrication using simple techniques [51]. Based on the orientation of the polymers and the solution in which they were produced, they can be hydrophobic or hydrophilic in nature. Metallosomes are metallic counterparts of liposomes, where lipid is attached to metal. Metallosomes are gaining popularity due to their exceptional optical properties and ease of fabrication. Metallosomes, not only enable targeted drug/gene delivery, but also can assist in imaging the specific targeted area [52].

3.   Fabrication of nanomaterials for healthcare applications

Various approaches are constantly emerging to improve efficiency, sensitivity and selectivity of nanomaterials intended towards specific applications. Two approaches such as top-down and bottom-up approaches utilize physical, chemical and/or biological methods for fabrication of nanosystems. Most desired methods particularly for biological applications include: sol-gel methods [53, 54, 55], lithography (soft and photo) [56, 57], Langmuir-Blodgett (LB) technique [31] and molecular self-assembly [58, 59].

Sol-gel method shown in Fig. 1, is a bottom-up approach based chemical method used to synthesize nanofilms or nanopowders of metal-alkaloid material. In this technique, the precursor material undergoes hydrolysis followed by condensation, resulting in polymeric gel. The precursor material forms a colloid of solid in liquid phase (sol) upon hydrolysis with subsequent condensation in a continuous liquid phase (gel). The final material produced can be utilized in multiple forms. It can either be spin-coated or dip-coated on to a surface to produce thin or thick films, can be heat treated to form dense films and solvent extracted to produce aerogel/xerogel. Upon solvent extraction, dry powder can be produced.

Nanoparticles in Healthcare: Applications in Therapy, Diagnosis and Drug Delivery   Materials Research Forum LLC

Materials Research Foundations 160 (2024) 1-23   https://doi.org/10.21741/9781644902974-1

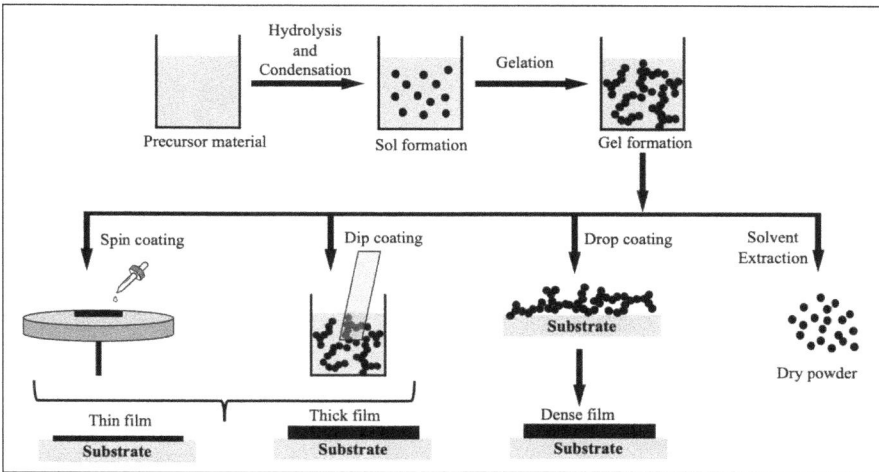

Figure 1. Sol-Gel process for fabrication of nanosystems.

Various optoelectronic sensors and microfluidic devices are heavily used in the healthcare sector. Fabrication of such systems significantly relies on lithography processes. Various forms of lithography are currently available. Photolithography, e-beam lithography, interference lithography, focused ion-beam lithography, are a few variants of lithography to be mentioned here. Lithography refers to writing on a flat surface and it is a top-down approach utilizing physical and chemical methods to create nanostructures onto a solid surface. We will consider photolithography and soft lithography variants for our discussion in this chapter.

Photolithography utilizes materials that are sensitive to specific wavelengths of light. The same technology is used for the fabrication of integrated circuits as part of the semiconductor industry. The photolithography process involves the steps shown in Fig. 2a. In photolithography, light sensitive material polymer (photoresist) is coated onto a substrate (silicon or glass) using spin coating. Photoresists are sensitive to UV light and undergo polymerization upon exposure. Two types of photoresist materials are available for deposition: positive photoresist and negative photoresist.

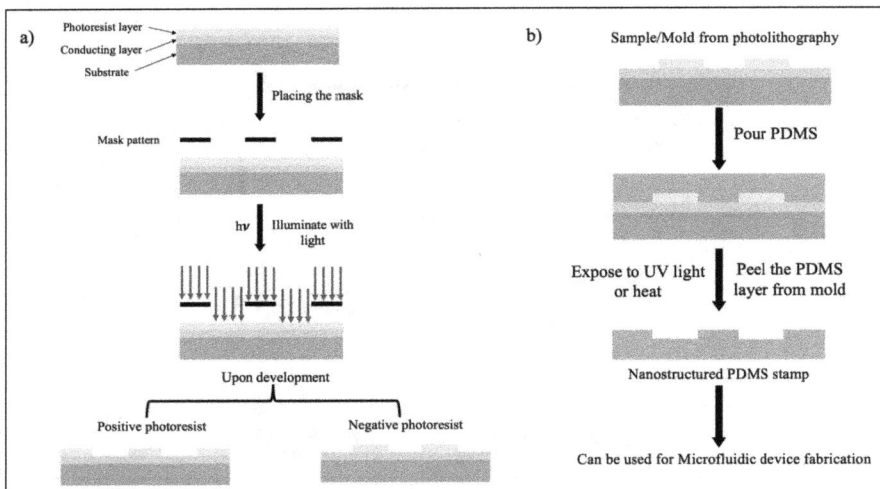

Figure 2. Fabrication of nanostructures using a) Photolithography and b) Soft Lithography.

Mask contains the desirable nanopattern to be transferred onto substrate on interest. The mask is placed onto the polymer layer (in contact mode) or the nanostructure is projected onto the polymer film using an optical system (in projector mode). The film is then exposed to UV light for a defined period of time depending on the thickness of the photoresist polymer on the substrate. This process etches the undesired patterns on the substrate surface. Later, the mask is carefully removed, and the polymer surface is washed using a developer solution. In a photolithography process using positive photoresist, once washed with developer, the regions below the mask that are unexposed to UV light remain. While it is vice versa for a negative photoresist. Any moisture remaining after using the developer can be removed by placing the sample on a hot plate for a few seconds. The patterns generated using photolithography can be used as base for generating patterns using soft lithography [56].

Soft lithography shown in Fig. 2b., also known as micro-contact printing, utilizes polydimethylsiloxane (PDMS) to create soft and flexible devices. Using this technique, one can use PDMS mold produced to print and "ink" onto a surface or shape a polymer. It is a versatile method to mainly fabricate microfluidic devices enabling to mimic various biological processes *in vitro*. If previously made mold is not available, 3D printing of soft lithography stamps using soft polymers is an emerging technology [60]. The PDMS mixed with the curing agent is placed in vacuum to remove air bubbles. This mixture is poured onto mold and allowed to cure in heat or exposed to UV. Once the material is cured, the PDMS stamp is peeled from the mold and used for different purposes.

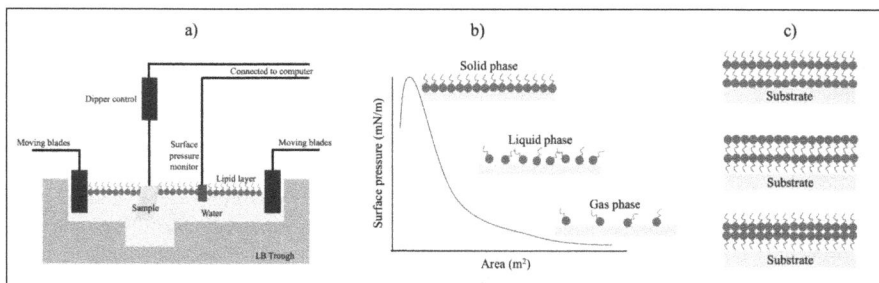

Figure 3. a) Schematics of Langmuir-Blodgett system, b) LB isotherm and c) Various configurations of monolayers produced using LB technique.

In order to efficiently reproduce cellular membrane processes, liposomes may not always provide a solution. Another fascinating technique called Langmuir-Blodgett method (shown in Fig. 3a) enables production of lipid monolayers onto a solid substrate surface. The thickness of each lipid monolayer deposited depends on the chain length of the lipid molecule. To simply increase the thickness of the film, one can add multiple lipid monolayers keeping chemistry of the lipid molecule in mind. The amphiphilic nature of a phospholipid comes handy while developing various configurations of monolayers as shown in Fig. 3c.

A few drops of lipid solution prepared in a non-aqueous volatile solvent are carefully placed onto the water surface inside LB trough. As the solvent evaporates at the surface, lipid molecules disperse randomly in various directions on the water surface (see Fig. 3b). Moving barriers bring these lipid molecules closer resulting in an increase of surface pressure as the distance between barriers reduces. Once a solid phase of packing is achieved, it is observed as sharp increase in surface pressure as shown in Fig. 3b. At this point, barriers are stopped, and the sample dipper is lowered to collect a monolayer of lipids onto the substrate surface. Process has to be repeated to obtain multiple monolayers on the same sample. Monolayers produced using LB techniques find important applications in cellular nanoarchitectonics [31].

Molecular self-assembly acts as a bridge between bottom-up and top-down approaches of nanofabrication. Self-assembly is different when compared to self-organization of molecules. In self-assembly, molecules or part of the molecules tend to form aggregates of ordered systems with no human intervention. DNA based nanotechnology methods such as DNA computing [61] involving DNA origami [62] rely on self-assembly processes. While self-organization requires external energy supply for formation of complex systems. Understanding both play a crucial role while developing nanosytems intended for applications in biology. In line with the above fabrication methods discussed, new and emerging technique refer to the convergence of 3D-printing using bioactive compounds, cells and nanomaterials. This new bioprinting technology provides new avenue of opportunities to create biological materials in lab with unprecedented similarities to those in living systems.

4.   Nanotechnology for healthcare applications

Healthcare refers to the overall well-being of an individual upon inculcating health improvement solutions for prevention, early detection, treatment and cure of a disease. Enhanced accuracies in terms of quantification of analyte plays a crucial role for disease diagnosis followed by treatment. So far in this chapter, we learned about concepts of nanotechnology, nanomaterials used and fabrication of nanosystems intended for healthcare applications. In the healthcare sector, a single disease can be diagnosed using multiple methods, such as CT/MRI/PET scans, blood testing, ultrasound etc. Each diagnosis method will supplement the other for disease identification and its confirmation. The most important question to now ask is, how can nanotechnology assist in addressing healthcare issues? What are the alternate strategies available? Is nanotechnology better? What are the limitations of nanotechnology in healthcare? Let us find answers to these questions using different perspectives discussed further in this chapter.

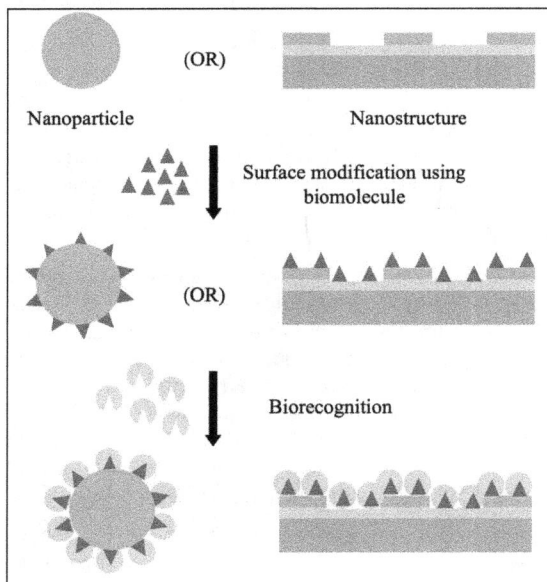

Figure 4: Surface modification of nanoparticle (on left) or nanostructure (on right) with biorecognition elements such as peptides, DNA, RNA, antibody (indicated as red triangles) specific to analyte of interest in sample (indicated as yellow spheres).

Nanotechnology supplements the modern advancements made in the healthcare sector. New wound healing patches, custom hearing aids, blood glucose monitors, cancer detection methods, early infection detection, improved imaging systems are few examples where nanobiotechnology is strongly establishing. Nanotechnology based personalized medicine strategies also look promising for applications. In India, the most pressing challenges in healthcare are pertaining to lack of adequate access to healthcare, affordability of healthcare solutions and accountability of diagnostic methods used [63]. Only when the technology used produces affordable healthcare devices, access to healthcare by the marginalized communities also improves. In order to enable improved detection and imaging capacities, surface modification of nanomaterial plays a crucial role.

Surface modification (shown in Fig. 4) provides the nanomaterial with improved sensitivity and selectivity parameters targeted towards specific analyte of interest. The biorecognition element is attached to the nanosystem either covalently or non-covalently (using hydrogen bonds, Vander Waals force, ionic bonds, electrostatic interaction or adsorption) [64]. Surface modification can be achieved through various processes such as ultrasonication, covalent-attachment of biorecognition elements, radiation or plasma treatment, or even simple changes in pH of the solution. Once the nanomaterial is surface functionalized, it is ready for application, provided detection mechanisms are established. Detection of analyte and its interaction with nanomaterial can be performed using various mechanisms based on optical, electrical, magnetic, thermal and so on. We will discuss optical and electrical detection strategies.

4.1   Detection mechanisms: Optical based

Interaction between analyte and nanomaterial can be studied using various optical detection mechanisms. Small changes in absorbance, reflection, and transmission are observed as either shift in respective peaks, signal enhancement or signal quenching before and after biorecognition as shown in Fig. 5. Various energy band gaps of quantum dots enable the elimination of fluorophore labelling. Various optical mechanisms can be observed appropriately using specific microscopes and/or other imaging systems. In the case of optical detection, additional mechanisms such as diffraction and scattering effects arise do to nanostructures and its interaction can affect the signal output and its quality. Optical detection strategy can be employed for the detection of homogenous or heterogenous analytes in a particular system based on the surface modification employed. The nanosystem used for detection can be in solution form or immobilized onto a surface enabling scanometric based detection. Examples of optical based sensors [65] include brain monitoring devices, various biochemical, plasmonic and photonic sensors.

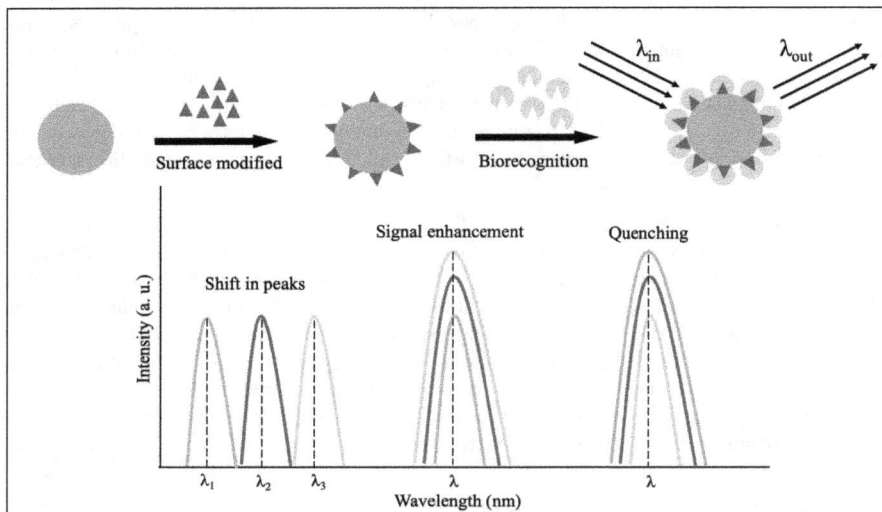

Figure 5: Schematics of changes in optical signal before and after surface modification followed by biorecognition of analyte in a sample.

4.2   Detection mechanisms: Electrical based

Electrical detection is particularly useful for microfluidic, optic-fiber based, surface enhanced Raman scattering (SERS) and microarray-based devices enabling continuous monitoring of samples. Schematics of such devices are shown in Fig. 6. Electrical detection not only enables detection of biochemical changes, but also enables simultaneous detection of other parameters such as degree of hybridization between recognition element and analyte. Small changes in degree of hybridization can result in major potential difference between the measuring electrodes with improved detection capacities. Small changes that are attenuated in optical signal can become more pronounced in electrical detection making it a superior choice for biosensors. Examples of electronic detection-based sensors [66] include glucose monitors, pulse oximeters, pacemakers, etc.

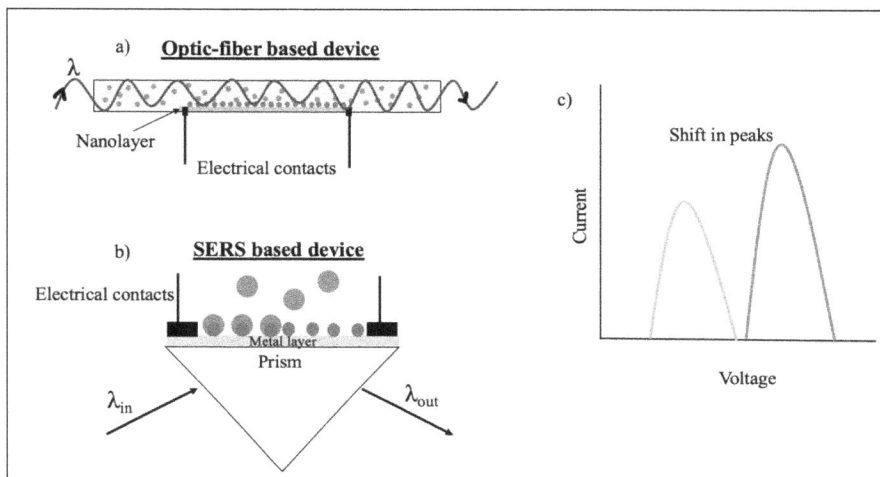

Figure 6. Schematics of a) optic-fiber based sensor, b) SERS based sensor and c) change in signal before and after biorecognition.

4.3 Lab-on-chip systems/nanobiobarcodes/microarray systems

Lab-on-chip systems [67, 68] (schematics shown in Fig. 7) involve miniaturized systems containing mini valves, pumps and integrated circuits all operating together in a microfluidic device. Microarray systems enabled with nanobiobarcodes assist in screening several samples at any given time. Application on an external alternating electric field enables not only detection of molecules, but also its separation [69]. The majority of microarray devices utilize scanometric-based detection scheme of operation, with few working on electrical based detection. The biorecognition element barcoded to a specific analyte is coated onto a nanostructured surface rather than glass slides used in conventional systems. In some cases, biorecognition elements are coated on the surface of nanoparticles. The nanosurface enables the detection of minor changes in degree of hybridization that can play a crucial role in protein expression studies. Currently various diagnostic detection methods come with SERS based plasmonic detection, generating new class of lab-on-chip devices for nanotheranostic applications.

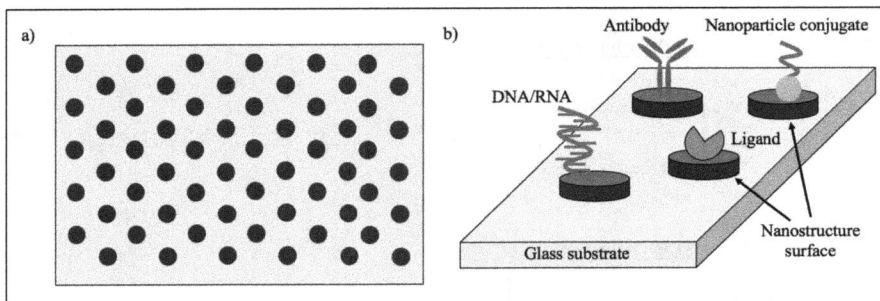

Figure 7. Schematics of a) Lab-on-chip or Microarray device and b) Surface-modification of unit cell with biorecognition elements enabling detection.

4.4   Diagnostic imaging using nanosystems

Early and efficient detection thresholds of abnormalities in tissue samples using CT, MRI and PET scans still remains a concern [70, 71]. Only advanced stages of disease are easily identified using conventional scanning methods. Deep tissue imaging and its analysis is still a challenge. Conventional contrast agents injected to the patient require longer periods of time to reach the tissue before imaging. Nanoformulations and surface modification of imaging agents offer a relief to some extent. Nano-based chelates complexes of gadolinium (3+) [72, 73] and transition metal manganese (2+) [74, 75] have shown promising results with increased signal capacities due to their short relaxation times. While gadolinium is used for brain tumor analysis, manganese is used for neural imaging. Iron based nanoparticles [76, 77], have been shown to be effective for detection of liver cancer in early stages. Main disadvantage of these nanoparticles used as contrast agents is aggregation of nanoparticle systems due to weak Vander Waals forces. Studies have shown that addition of polymer capping to nanoparticle reduces aggregation.

Optical based imaging techniques rely on specialized optical processes called fluorescence and luminescence. For the detection and differentiation of various components within the cell/tissue, labelling with multiple fluorophore molecules is required. Commercially available fluorophores are not photostable and undergo photobleaching and quenching mechanisms. Using nanotechnology, this issue can be addressed easily. Several metal and semiconducting nanoparticles fluoresce differently when size, shape and composition parameters are changed. Quantum dots in particular resist photobleaching and metabolic degradation with high quantum yields enabling superior detection and imaging capacities.

4.5   Nanotheranostics

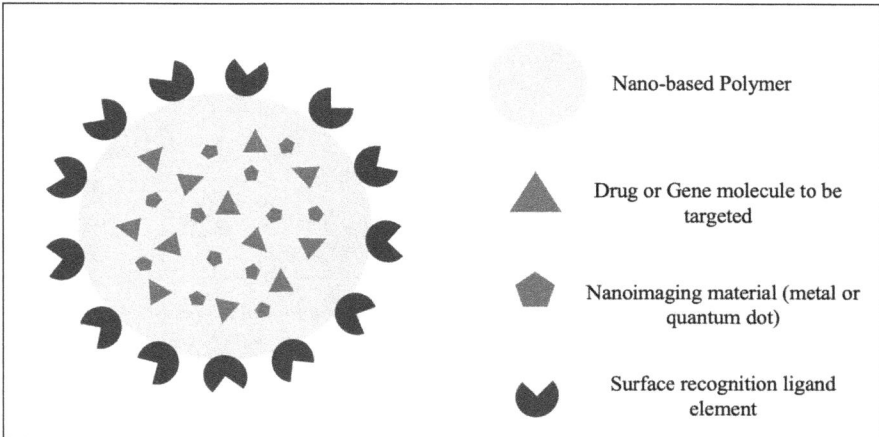

Figure 8. Schematics of surface functionalized polymer based nanotheranostics material encapsulating biorecognition element and imaging nanoparticle within the same system.

Personalized and precision medicine solutions using only biotechnological tools alone is often tedious with little to no promising effects. Nanotheranostics is fairly new in the market and offers a glimmer of hope to those patients suffering from rare diseases and advanced conditions. Nanotheranostics [22, 23] is an emerging personalized medicine tool that enables early detection and therapy simultaneously to patients. Various nanosystems such as metal nanoparticles, magnetic nanoparticles and liposome-based systems are currently being studied for nanotheranostics applications. Nanosystems such as quantum dots and liposomes are studied for biological imaging upon targeted drug delivery, release and studying the disease progress. Due to high surface to volume ratio, nanoparticles can efficiently carry medicine to targeted sites with less time spent in circulation.

Cancer nanotheranostics [78, 79] particularly involves multimodal approaches. Based on the stage of disease progression and drugs available for treatment, conjugates of metal and drug can be designed specifically to meet the needs. Highly pathogenic forms of cancer such as hepatocellular carcinoma (HCC) and other diseases involving metastasis such as endometriosis are providing a strong platform to explore future of nanotheranostic applications [80, 81, 82].

4.6   Virus based nanosystems in healthcare

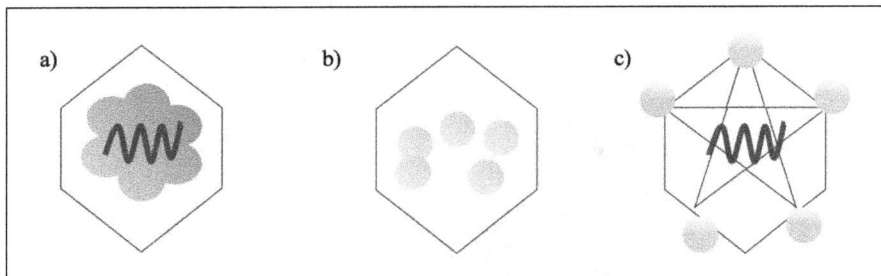

Figure 9. Schematics of virus like particles (VLPs) for biological applications a) VLP encapsulating biomolecule of interest, b) VLP encapsulating drug and c) surface modified VLP for nanotheranostics applications.

Viral nanoparticles are exciting mainly because they exist naturally, they are biocompatible and biodegradable [83. Viral nanoparticles are produced in two forms: viral nanoparticles (VNPs) and virus like particles (VLPs). VNPs [84] include the use of whole viruses such as bacteriophages, plant and animal viruses that can either be infectious or non-infectious. Whereas VLPs [85, 86] include genome free parts of VNPs such as, empty envelope or capsid structures, tail fiber, and certain other viral subunits such as spike proteins. Both VNPs and VLPs find important applications in nanotheranostics. New platforms in medicine such as phage-therapy are already in use in the developed nations. Employing nanotechnology for phage-therapy can ultimately reduce the overall cost of treatment associated which is promising for developing nations [87].

4.7   Nanorobots

Magnetic nanoparticles existing naturally in the form of magnetosomes and those synthesized in lab find important applications as contrasting agents and for nanoparticle induced thermal therapies. Magnetically driven nanorobotic devices, also called nanobots [88, 89] are a new class of engineered nanosystems aiming at nanotheranostic applications. Nanorobots can be self-driven sensing a chemical stimulus in the environment or controlled using an external stimulus through a remote device. Nanorobots are currently used for targeted drug delivery [90]. Since these systems work at nanoscale, a small concentration of drugs is sufficient for the treatment. Although they are gaining popularity in other fields of biotechnology such as water purification and quality analysis, there are a few challenges to be addressed in terms of their design. Challenges in development and use of nanorobots are: technical complexities in terms of their functioning, safety and regulatory issue when applied on human subjects, scalability in terms of their manufacturing process and the overall cost of the device [88, 92].

4.8   Wearable biosensors in healthcare

Frequent hospitalization, multiple sample withdrawals, interval-based biochemical monitoring can be challenging and discomforting to small children and elderly patients. Wearable sensors enable real-time monitoring of biochemical processes with remote monitoring and signal analysis [92]. Wearable sensors refer to those devices that improve the quality of life on a day-to-day basis. These devices are primarily single-use patch systems [93]. Wearable sensors provide a non-invasive method to measure biochemical markers from biofluids such as sweat, urine, blood, tears and saliva [94]. Wearable sensors rely on the fundamental principles of biosensors [95]. Most wearable sensors rely on photolithographic tools for their fabrication followed by surface modification to efficiently determine the presence or absence of an analyte of interest. The measured signal is converted to measurable output using a transducer element such as photo-diodes, electrodes of various forms, thermistors, etc. The measured analog signal is further converted to digital form as part of signalization to generate a measured value [95].

Conclusion and future perspectives

Over the past few decades, research pertaining to nanotechnological advancements has been on a positive slope. Several designs of nanoparticles and nanostructures are exploited for biomedical applications with proven efficiencies. Such positive results offer a chance of hope in various life-threatening conditions. Within a year of COVID-19 pandemic, at home diagnostic kits were readily available on the market as first line of testing. These examples only indicate the true potential of nanotechnology in the healthcare sector but also new revolutionary changes in the biomedical field. Application of nanotechnology in biotechnology is not possible without the fundamental understanding of biological processes.

With new and emerging diseases and corresponding spikes in death rates, development of new diagnostic and therapeutic tools is the need of the hour. Nanotechnology offers various tools to develop, design, fabricate and test new innovative tools addressing such issues. Using top-down and bottom-up approaches, new nano-based solutions are formulated for enhanced disease diagnosis, imaging, treatment and cure. Surface functionalization improves selectivity and sensitivity by several orders of magnitude, enabling detection of definite targets under consideration. New fields of research such as nanotheranostics, personalized precision medicine, nanorobots, wearable sensors offer point-of-care treatments to patients.

Even with tremendous advancements made in the field of nanobiotechnology, metastasis and deep tissue imaging remain to be a challenge. Nanotoxicity of various nanomaterial remains a concern as well. Reliability and reproducibility of nanosytems requires additional research for efficient applications. Despite the fact that several materials and nanosytems are developed, the majority of their applications remain unexplored. Nanoparticles functionalized with biomolecules particularly, virus nanoparticles can be exploited as bioweapons. In order to mitigate the threats associated, the need for the implementation of national and international policies governing the design and use of nanosystems is required. Overall, nanotechnology has proved to be beneficial with diverse applications in healthcare sector.

References

[1] Information on https://www.weforum.org/agenda/2023/04/world-health-day-healthcare-trends/

[2] Information on https://www.who.int/docs/default-source/documents/gs4dhdaa2a9f352b0445bafbc79ca799dce4d.pdf

[3] Syedmoradi, L., Daneshpour, M., Alvandipour, M., Gomez, F. A., Hajghassem, H., & Omidfar, K. (2017). Point of care testing: The impact of nanotechnology. Biosensors & bioelectronics, 87, 373-387. https://doi.org/10.1016/j.bios.2016.08.084 https://doi.org/10.1016/j.bios.2016.08.084

[4] Nouailhat, A. (2010). An introduction to nanoscience and nanotechnology (Vol. 10). John Wiley & Sons.

[5] Joachim, C. (2005). To be nano or not to be nano?. Nature Materials, 4(2), 107-109. https://doi.org/10.1038/nmat1319

[6] Bhattacharyya, S., Sandhu, K., & Chockalingam, S. (2023). Nanotechnology-based healthcare engineering products and recent patents-an update. In Emerging Nanotechnologies for Medical Applications (pp. 273-296). Elsevier. https://doi.org/10.1016/B978-0-323-91182-5.00004-8

[7] Sahoo, S. K., Parveen, S., & Panda, J. J. (2007). The present and future of nanotechnology in human health care. Nanomedicine: Nanotechnology, biology and medicine, 3(1), 20-31. https://doi.org/10.1016/j.nano.2006.11.008

[8] Shahcheraghi, N., Golchin, H., Sadri, Z., Tabari, Y., Borhanifar, F., & Makani, S. (2022). Nano-biotechnology, an applicable approach for sustainable future. 3 Biotech, 12(3), 65. https://doi.org/10.1007/s13205-021-03108-9

[9] Verma, S. K., Suar, M., & Mishra, Y. K. (2022). Green Perspective of Nano-Biotechnology: Nanotoxicity Horizon to Biomedical Applications. Frontiers in Bioengineering and Biotechnology, 10. https://doi.org/10.3389/fbioe.2022.919226

[10] Yan, Y., Zhu, X., Yu, Y., Li, C., Zhang, Z., & Wang, F. (2022). Nanotechnology strategies for plant genetic engineering. Advanced Materials, 34(7), 2106945. https://doi.org/10.1002/adma.202106945

[11] Squire, H. J., Tomatz, S., Voke, E., González-Grandío, E., & Landry, M. (2023). The emerging role of nanotechnology in plant genetic engineering. Nature Reviews Bioengineering, 1-15. https://doi.org/10.1038/s44222-023-00037-5

[12] Huang, X., Kon, E., Han, X., Zhang, X., Kong, N., Mitchell, M. J., Dan, P., & Tao, W. (2022). Nanotechnology-based strategies against SARS-CoV-2 variants. Nature Nanotechnology, 17(10), 1027-1037. https://doi.org/10.1038/s41565-022-01174-5

[13] Gildiz, S., & Minko, T. (2023). Nanotechnology-based nucleic acid vaccines for treatment of ovarian cancer. Pharmaceutical research, 40(1), 123-144. https://doi.org/10.1007/s11095-022-03434-4

[14] Kiremitler, N. B., Kemerli, M. Z., Kayaci, N., Karagoz, S., Pekdemir, S., Sarp, G., ... & Yilmaz, E. (2022). Nanostructures for the prevention, diagnosis, and treatment of SARS-CoV-2: A review. ACS Applied Nano Materials, 5(5), 6029-6054. https://doi.org/10.1021/acsanm.2c00181

[15] Ayub, A., & Wettig, S. (2022). An overview of nanotechnologies for drug delivery to the brain. Pharmaceutics, 14(2), 224. https://doi.org/10.3390/pharmaceutics14020224

[16] Hamimed, S., Jabberi, M., & Chatti, A. (2022). Nanotechnology in drug and gene delivery. Naunyn-schmiedeberg's Archives of Pharmacology, 395(7), 769-787. https://doi.org/10.1007/s00210-022-02245-z

[17] Manikandan, S., Subbaiya, R., Saravanan, M., Ponraj, M., Selvam, M., & Pugazhendhi, A. (2022). A critical review of advanced nanotechnology and hybrid membrane based water recycling, reuse, and wastewater treatment processes. Chemosphere, 289, 132867. https://doi.org/10.1016/j.chemosphere.2021.132867

[18] Jain, S., Gupta, I., Walia, P., & Swami, S. (2022). Application of Actinobacteria in Agriculture, Nanotechnology, and Bioremediation. In Actinobacteria-Diversity, Applications and Medical Aspects. IntechOpen. https://doi.org/10.5772/intechopen.104385

[19] Gowda, G. N., Zhang, S., Gu, H., Asiago, V., Shanaiah, N., & Raftery, D. (2008). Metabolomics-based methods for early disease diagnostics. Expert review of molecular diagnostics, 8(5), 617-633. https://doi.org/10.1586/14737159.8.5.617

[20] Kim, B. Y., Rutka, J. T., & Chan, W. C. (2010). Nanomedicine. New England Journal of Medicine, 363(25), 2434-2443. https://doi.org/10.1056/NEJMra0912273

[21] Riehemann, K., Schneider, S. W., Luger, T. A., Godin, B., Ferrari, M., & Fuchs, H. (2009). Nanomedicine-challenge and perspectives. Angewandte Chemie International Edition, 48(5), 872-897. https://doi.org/10.1002/anie.200802585

[22] Chen, H., Zhang, W., Zhu, G., Xie, J., & Chen, X. (2017). Rethinking cancer nanotheranostics. Nature Reviews Materials, 2(7), 1-18. https://doi.org/10.1038/natrevmats.2017.24

[23] Sharma, A. K. (2023). Current Trends in Nanotheranostics: A Concise Review on Bioimaging and Smart Wearable Technology. Nanotheranostics, 7(3), 258-269. https://doi.org/10.7150/ntno.82886

[24] Pirzada, M., & Altintas, Z. (2019). Nanomaterials for healthcare biosensing applications. Sensors, 19(23), 5311. https://doi.org/10.3390/s19235311

[25] Xie, L., Zhang, Z., Wu, Q., Gao, Z., Mi, G., Wang, R., ... & Du, Y. (2023). Intelligent wearable devices based on nanomaterials and nanostructures for healthcare. Nanoscale. https://doi.org/10.1039/D2NR04551F

[26] Schwab, F., Rothen-Rutishauser, B., & Petri-Fink, A. (2020). When plants and plastic interact. Nature Nanotechnology, 15(9), 729-730. https://doi.org/10.1038/s41565-020-0762-x

[27] Yılmaz, D., Günaydın, B. N., & Yüce, M. (2022). Nanotechnology in food and water security: On-site detection of agricultural pollutants through surface-enhanced Raman spectroscopy. Emergent Materials, 5(1), 105-132. https://doi.org/10.1007/s42247-022-00376-w

[28] Iqbal, P., Preece, J. A., & Mendes, P. M. (2012). Nanotechnology: the "top-down" and "bottom-up" approaches. Supramolecular chemistry: from molecules to nanomaterials. https://doi.org/10.1002/9780470661345.smc195

[29] Srivastava, S., Bhargava, A., Srivastava, S., & Bhargava, A. (2022). Green nanotechnology: an overview. Green Nanoparticles: The Future of Nanobiotechnology, 1-13. https://doi.org/10.1007/978-981-16-7106-7_1

[30] Anjum, S., Ishaque, S., Fatima, H., Farooq, W., Hano, C., Abbasi, B. H., & Anjum, I. (2021). Emerging applications of nanotechnology in healthcare systems: Grand challenges and perspectives. Pharmaceuticals, 14(8), 707. https://doi.org/10.3390/ph14080707

[31] Oliveira Jr, O. N., Caseli, L., & Ariga, K. (2022). The past and the future of Langmuir and Langmuir-Blodgett films. Chemical Reviews, 122(6), 6459-6513. https://doi.org/10.1021/acs.chemrev.1c00754

[32] Capel, A. J., Rimington, R. P., Lewis, M. P., & Christie, S. D. (2018). 3D printing for chemical, pharmaceutical and biological applications. Nature Reviews Chemistry, 2(12), 422-436. https://doi.org/10.1038/s41570-018-0058-y

[33] Charbgoo, F., Ahmad, M. B., & Darroudi, M. (2017). Cerium oxide nanoparticles: green synthesis and biological applications. International journal of nanomedicine, 12, 1401. https://doi.org/10.2147/IJN.S124855

[34] Pansambal, S., Oza, R., Borgave, S., Chauhan, A., Bardapurkar, P., Vyas, S., & Ghotekar, S. (2022). Bioengineered cerium oxide (CeO2) nanoparticles and their diverse applications: a review. Applied Nanoscience, 1-26. https://doi.org/10.1007/s13204-022-02574-8

[35] Lo, W. S., Kwok, W. M., Law, G. L., Yeung, C. T., Chan, C. T. L., Yeung, H. L., Kong, H. K., Chen, C. H., Murphy, M. B., Wong, K. L., & Wong, W. T. (2011). Impressive europium red emission induced by two-photon excitation for biological applications. Inorganic chemistry, 50(12), 5309-5311. https://doi.org/10.1021/ic102465j

[36] Londhe, S., & Patra, C. R. (2022). Biomedical applications of europium hydroxide nanorods. Nanomedicine, 17(1), 5-8. https://doi.org/10.2217/nnm-2021-0351

[37] Ishikawa, I., Aoki, A., & Takasaki, A. A. (2004). Potential applications of Erbium: YAG laser in periodontics. Journal of periodontal research, 39(4), 275-285. https://doi.org/10.1111/j.1600-0765.2004.00738.x

[38] Mondal, S., Park, S., Choi, J., Tran, L. H., Yi, M., Shin, J. H., ... & Oh, J. (2020). Bioactive, luminescent erbium-doped hydroxyapatite nanocrystals for biomedical applications. Ceramics international, 46(10), 16020-16031. https://doi.org/10.1016/j.ceramint.2020.03.152

[39] Rajeshkumar, S., & Naik, P. (2018). Synthesis and biomedical applications of cerium oxide nanoparticles-a review. Biotechnology Reports, 17, 1-5. https://doi.org/10.1016/j.btre.2017.11.008

[40] Barker, E., Shepherd, J., & Asencio, I. O. (2022). The use of cerium compounds as antimicrobials for biomedical applications. Molecules, 27(9), 2678. https://doi.org/10.3390/molecules27092678

[41] De, M., Ghosh, P. S., & Rotello, V. M. (2008). Applications of nanoparticles in biology. Advanced Materials, 20(22), 4225-4241. https://doi.org/10.1002/adma.200703183

[42] Schultz, D. A. (2003). Plasmon resonant particles for biological detection. Current opinion in biotechnology, 14(1), 13-22. https://doi.org/10.1016/S0958-1669(02)00015-0

[43] Nakamura, S., Sato, M., Sato, Y., Ando, N., Takayama, T., Fujita, M., & Ishihara, M. (2019). Synthesis and application of silver nanoparticles (Ag NPs) for the prevention of infection in healthcare workers. International journal of molecular sciences, 20(15), 3620. https://doi.org/10.3390/ijms20153620

[44] Dhayalan, M., Karikalan, P., Riyaz Savaas Umar, M., & Srinivasan, N. (2021). Biomedical Applications of Silver Nanoparticles. IntechOpen. doi: 10.5772/intechopen.99367 https://doi.org/10.5772/intechopen.99367

[45] Guo, C., Sun, L., Chen, X., & Zhang, D. (2013). Oxidative stress, mitochondrial damage and neurodegenerative diseases. Neural regeneration research, 8(21), 2003.

[46] Wang, L., Hu, C., & Shao, L. (2017). The antimicrobial activity of nanoparticles: present situation and prospects for the future. International journal of nanomedicine, 12, 1227. https://doi.org/10.2147/IJN.S121956

[47] Botteon, C. E. A., Silva, L. B., Ccana-Ccapatinta, G. V., Silva, T. S., Ambrosio, S. R., Veneziani, R. C. S., Bastos, J. K., & Marcato, P. D. (2021). Biosynthesis and characterization of gold nanoparticles using Brazilian red propolis and evaluation of its antimicrobial and anticancer activities. Scientific Reports, 11(1), 1974. https://doi.org/10.1038/s41598-021-81281-w

[48] Ratan, Z. A., Haidere, M. F., Nurunnabi, M. D., Shahriar, S. M., Ahammad, A. S., Shim, Y. Y., Martin, J. T. R., & Cho, J. Y. (2020). Green chemistry synthesis of silver nanoparticles and their potential anticancer effects. Cancers, 12(4), 855. https://doi.org/10.3390/cancers12040855

[49] Xu, L., Yi-Yi, W., Huang, J., Chun-Yuan, C., Zhen-Xing, W., & Xie, H. (2020). Silver nanoparticles: Synthesis, medical applications and biosafety. Theranostics, 10(20), 8996. https://doi.org/10.7150/thno.45413

[50] Hu, X., Zhang, Y., Ding, T., Liu, J., & Zhao, H. (2020). Multifunctional gold nanoparticles: a novel nanomaterial for various medical applications and biological activities. Frontiers in Bioengineering and Biotechnology, 8, 990. https://doi.org/10.3389/fbioe.2020.00990

19

[51] Large, D. E., Abdelmessih, R. G., Fink, E. A., & Auguste, D. T. (2021). Liposome composition in drug delivery design, synthesis, characterization, and clinical application. Advanced Drug Delivery Reviews, 176, 113851. https://doi.org/10.1016/j.addr.2021.113851

[52] Hainfeld, J. F., Furuya, F. R., & Powell, R. D. (1999). Metallosomes. Journal of Structural Biology, 127(2), 152-160. doi:10.1006/jsbi.1999.4145 https://doi.org/10.1006/jsbi.1999.4145

[53] Bokov, D., Turki Jalil, A., Chupradit, S., Suksatan, W., Javed Ansari, M., Shewael, I. H., Valiev, G. H., & Kianfar, E. (2021). Nanomaterial by sol-gel method: synthesis and application. Advances in Materials Science and Engineering, 2021, 1-21. https://doi.org/10.1155/2021/5102014

[54] Yilmaz, E., & Soylak, M. (2020). Functionalized nanomaterials for sample preparation methods. In Handbook of Nanomaterials in analytical chemistry (pp. 375-413). Elsevier. https://doi.org/10.1016/B978-0-12-816699-4.00015-3

[55] Aguilar, G. V. (2018). Introductory chapter: a brief semblance of the sol-gel method in research. Sol-Gel Method-Design and Synthesis of New Materials with Interesting Physical, Chemical and Biological Properties.

[56] Thompson, L. F. (1983). An introduction to lithography. https://doi.org/10.1021/bk-1983-0219.ch001

[57] Xia, Y., & Whitesides, G. M. (1998). Soft lithography. Annual review of materials science, 28(1), 153-184. https://doi.org/10.1146/annurev.matsci.28.1.153

[58] Zhang, S. (2003). Fabrication of novel biomaterials through molecular self-assembly. Nature biotechnology, 21(10), 1171-1178. https://doi.org/10.1038/nbt874

[59] Whitesides, G. M., Mathias, J. P., & Seto, C. T. (1991). Molecular self-assembly and nanochemistry: a chemical strategy for the synthesis of nanostructures. Science, 254(5036), 1312-1319. https://doi.org/10.1126/science.1962191

[60] Kajtez, J., Buchmann, S., Vasudevan, S., Birtele, M., Rocchetti, S., Pless, C. J., Heiskanen, A., Barker, R.A., Martínez-Serrano, A., Parmar, M. & Emnéus, J. (2020). 3D-printed soft lithography for complex compartmentalized microfluidic neural devices. Advanced Science, 7(16), 2001150. https://doi.org/10.1002/advs.202001150

[61] Roweis, S., Winfree, E., Burgoyne, R., Chelyapov, N. V., Goodman, M. F., Rothemund, P. W., & Adleman, L. M. (1998). A sticker-based model for DNA computation. Journal of Computational Biology, 5(4), 615-629. https://doi.org/10.1089/cmb.1998.5.615

[62] Dey, S., Fan, C., Gothelf, K. V., Li, J., Lin, C., Liu, L., ... & Zhan, P. (2021). DNA origami. Nature Reviews Methods Primers, 1(1), 13. https://doi.org/10.1038/s43586-020-00009-8

[63] Kasthuri, A. (2018). Challenges to healthcare in India-The five A's. Indian journal of community medicine: official publication of Indian Association of Preventive & Social Medicine, 43(3), 141.

[64] Sperling, R. A., & Parak, W. J. (2010). Surface modification, functionalization and bioconjugation of colloidal inorganic nanoparticles. Philosophical Transactions of the Royal

Society A: Mathematical, Physical and Engineering Sciences, 368(1915), 1333-1383. https://doi.org/10.1098/rsta.2009.0273

[65] Vavrinsky, E., Esfahani, N. E., Hausner, M., Kuzma, A., Rezo, V., Donoval, M., & Kosnacova, H. (2022). The current state of optical sensors in medical wearables. Biosensors, 12(4), 217. https://doi.org/10.3390/bios12040217

[66] Chen, S., Qi, J., Fan, S., Qiao, Z., Yeo, J. C., & Lim, C. T. (2021). Flexible wearable sensors for cardiovascular health monitoring. Advanced Healthcare Materials, 10(17), 2100116. https://doi.org/10.1002/adhm.202100116

[67] Wu, J., Dong, M., Rigatto, C., Liu, Y., & Lin, F. (2018). Lab-on-chip technology for chronic disease diagnosis. NPJ digital medicine, 1(1), 7. https://doi.org/10.1038/s41746-017-0014-0

[68] Hou, T., Chang, H., Jiang, H., Wang, P., Li, N., Song, Y., & Li, D. (2022). Smartphone based microfluidic lab-on-chip device for real-time detection, counting and sizing of living algae. Measurement, 187, 110304. https://doi.org/10.1016/j.measurement.2021.110304

[69] Miller, M. B., & Tang, Y. W. (2009). Basic concepts of microarrays and potential applications in clinical microbiology. Clinical microbiology reviews, 22(4), 611-633. https://doi.org/10.1128/CMR.00019-09 https://doi.org/10.1128/CMR.00019-09

[70] Dang, X., Bardhan, N. M., Qi, J., Gu, L., Eze, N. A., Lin, C. W., Kataria, S. Hammond, P. T., & Belcher, A. M. (2019). Deep-tissue optical imaging of near cellular-sized features. Scientific reports, 9(1), 1-12. https://doi.org/10.1038/s41598-019-39502-w

[71] Appel, A. A., Anastasio, M. A., Larson, J. C., & Brey, E. M. (2013). Imaging challenges in biomaterials and tissue engineering. Biomaterials, 34(28), 6615-6630. https://doi.org/10.1016/j.biomaterials.2013.05.033

[72] Ho, S. L., Yue, H., Tegafaw, T., Ahmad, M. Y., Liu, S., Nam, S. W., Chang, Y., & Lee, G. H. (2022). Gadolinium neutron capture therapy (GdNCT) agents from molecular to nano: Current status and perspectives. ACS omega, 7(3), 2533-2553. https://doi.org/10.1021/acsomega.1c06603

[73] Marasini, R., Thanh Nguyen, T. D., & Aryal, S. (2020). Integration of gadolinium in nanostructure for contrast enhanced-magnetic resonance imaging. Wiley interdisciplinary reviews. Nanomedicine and nanobiotechnology, 12(1), e1580. https://doi.org/10.1002/wnan.1580

[74] Cai, X., Zhu, Q., Zeng, Y., Zeng, Q., Chen, X., & Zhan, Y. (2019). Manganese Oxide Nanoparticles As MRI Contrast Agents In Tumor Multimodal Imaging And Therapy. International journal of nanomedicine, 14, 8321-8344. https://doi.org/10.2147/IJN.S218085

[75] Kunuku, S., Lin, B. R., Chen, C. H., Chang, C. H., Chen, T. Y., Hsiao, T. Y., Yu, H.K., Chang, Y.J., Liao, L.C., Chen, F.H., & Niu, H. (2023). Nanodiamonds Doped with Manganese for Applications in Magnetic Resonance Imaging. ACS omega, 8(4), 4398-4409. https://doi.org/10.1021/acsomega.2c08043

[76] Chen, C., Ge, J., Gao, Y., Chen, L., Cui, J., Zeng, J., & Gao, M. (2022). Ultrasmall superparamagnetic iron oxide nanoparticles: A next generation contrast agent for magnetic resonance imaging. Wiley Interdisciplinary Reviews: Nanomedicine and Nanobiotechnology, 14(1), e1740. https://doi.org/10.1002/wnan.1740

[77] Geppert, M., & Himly, M. (2021). Iron oxide nanoparticles in bioimaging - an immune perspective. Frontiers in Immunology, 12. https://doi.org/10.3389/fimmu.2021.688927

[78] Chen, X., & Wong, S. (Eds.). (2014). Cancer theranostics. Academic Press. https://doi.org/10.1016/B978-0-12-407722-5.00001-3

[79] Iravani, S., & Varma, R. S. (2022). MXenes in cancer nanotheranostics. Nanomaterials, 12(19), 3360. https://doi.org/10.3390/nano12193360

[80] Ladju, R. B., Ulhaq, Z. S., & Soraya, G. V. (2022). Nanotheranostics: A powerful next-generation solution to tackle hepatocellular carcinoma. World journal of gastroenterology, 28(2), 176-187. https://doi.org/10.3748/wjg.v28.i2.176

[81] Chavda, V. P., Balar, P. C., & Patel, S. B. (2023). Interventional nanotheranostics in hepatocellular carcinoma. Nanotheranostics, 7(2), 128-141. https://doi.org/10.7150/ntno.80120

[82] Cao, L., Zhu, Y. Q., Wu, Z. X., Wang, G. X., & Cheng, H. W. (2021). Engineering nanotheranostic strategies for liver cancer. World journal of gastrointestinal oncology, 13(10), 1213-1228. https://doi.org/10.4251/wjgo.v13.i10.1213

[83] Koudelka, K. J., Pitek, A. S., Manchester, M., & Steinmetz, N. F. (2015). Virus-Based Nanoparticles as Versatile Nanomachines. Annual review of virology, 2(1), 379-401. https://doi.org/10.1146/annurev-virology-100114-055141

[84] Guenther, C. M., Kuypers, B. E., Lam, M. T., Robinson, T. M., Zhao, J., & Suh, J. (2014). Synthetic virology: engineering viruses for gene delivery. Wiley Interdisciplinary Reviews: Nanomedicine and Nanobiotechnology, 6(6), 548-558. https://doi.org/10.1002/wnan.1287

[85] Tariq, H., Batool, S., Asif, S., Ali, M., & Abbasi, B. H. (2022). Virus-like particles: Revolutionary platforms for developing vaccines against emerging infectious diseases. Frontiers in microbiology, 12, 4137. https://doi.org/10.3389/fmicb.2021.790121

[86] Nooraei, S., Bahrulolum, H., Hoseini, Z. S., Katalani, C., Hajizade, A., Easton, A. J., & Ahmadian, G. (2021). Virus-like particles: Preparation, immunogenicity and their roles as nanovaccines and Drug Nanocarriers. Journal of Nanobiotechnology, 19(1). https://doi.org/10.1186/s12951-021-00806-7

[87] Peabody, D. S., Peabody, J., Bradfute, S. B., & Chackerian, B. (2021). RNA Phage VLP-Based Vaccine Platforms. Pharmaceuticals (Basel, Switzerland), 14(8), 764. https://doi.org/10.3390/ph14080764

[88] Soto, F., Wang, J., Ahmed, R., & Demirci, U. (2020). Medical micro/nanorobots in precision medicine. Advanced Science, 7(21), 2002203. https://doi.org/10.1002/advs.202002203

[89] Hortelão, A. C., Patiño, T., Perez-Jiménez, A., Blanco, À., & Sánchez, S. (2018). Enzyme-powered nanobots enhance anticancer drug delivery. Advanced Functional Materials, 28(25), 1705086. https://doi.org/10.1002/adfm.201705086

[90] Feng, Y., An, M., Liu, Y., Sarwar, M. T., & Yang, H. (2023). Advances in Chemically Powered Micro/Nanorobots for Biological Applications: A Review. Advanced Functional Materials, 33(1), 2209883. https://doi.org/10.1002/adfm.202209883

[91] Information on https://builtin.com/robotics/nanorobotics

[92] Kim, J., Campbell, A. S., de Ávila, B. E.-F., & Wang, J. (2019). Wearable biosensors for healthcare monitoring. Nature Biotechnology, 37(4), 389-406. https://doi.org/10.1038/s41587-019-0045-y

[93] Information on https://lifesignals.com/wearable-biosensors/

[94] Sharma, A., Badea, M., Tiwari, S., & Marty, J. L. (2021). Wearable Biosensors: An Alternative and Practical Approach in Healthcare and Disease Monitoring. Molecules (Basel, Switzerland), 26(3), 748. https://doi.org/10.3390/molecules26030748

[95] Bhalla, N., Jolly, P., Formisano, N., & Estrela, P. (2016). Introduction to biosensors. Essays in biochemistry, 60(1), 1-8. https://doi.org/10.1042/EBC20150001

Nanoparticles in Healthcare: Applications in Therapy, Diagnosis and Drug Delivery   Materials Research Forum LLC
Materials Research Foundations 160 (2024) 24-51   https://doi.org/10.21741/9781644902974-2

Chapter 2

Contemporary Approaches in the Synthesis and Fabrication of Nanoparticles

Preetha G. Prasad*

Department of Chemistry and Polymer Chemistry, Kumbalathu Sankupillai Memorial Devaswom Board College, Sasthamcotta, Kollam-690 521, Kerala, India

*preethagp@ksmdbc.ac.in

Abstract

Nowadays engineered nanomaterials and nano phases are fabricated through the manipulation of factors during their synthesis. The current chapter encompasses the trends and challenges of the diversified synthetic strategies including the physic chemical and biological techniques. Methods such as solvo-thermal, reduction, microemulsion, microwave-assisted, sonochemical, electrochemical, polyol and radiolytic processes are discussed under chemical approach. The various physical procedures included are milling, high-energy irradiation, ion implantation, laser ablation and spray pyrolysis. The rising trends of green protocols using biological systems like plant extracts, enzymes and microorganisms towards the surface tuning of these are also discussed.

Keywords

Bottom-Up Method, Solvo Thermal, Microwave Assisted Synthesis, Green Synthesis, Sonochemical Technique, Ion Implantation

Contents

1.   Introduction

Engineered nanometarials and nano phases proffer versatile applications in diversified fields related to environment, aerospace, catalytic reactions, drug delivery, energy harvesting, information technology, etc. The cutting edge research work in the field of nanotechnology has emerged as a consequence of the research efforts to improve the properties at the nanodimensions. New vistas have been opened up for revolutionising the existing technologies through fabrication of novel functional materials in view of the commendable electronic, chemical, environmental, biological and physical properties. Recently nano dimensional materials have notable significance as sustainable functional material owing to the possibility of synthesising in diverse shapes and dimensions pertaining to the specific use. In this context, the scaling up procedures for the synthetic strategies are of great significance due to the chances of manipulation and control of their properties through regulation of the size during synthesis [1]. The properties and functionalities of nanomaterials are determined by composition, morphology, architecture, facet, size and dimensionality. The various approaches mentioned here for the preparation of nanomaterials could be broadly categorised under bottom up or top-down procedures based on the starting material. So, the manipulation, fine tuning and consequent regulation of the properties of nanomaterials are possible through the selection of suitable methodology for synthesis whereby size control can be achieved.

2. Chemical methods of preparation

All the chemical methods discussed in this chapter belongs to the top to bottom category. The various methods are:

2.1 Hydrothermal and solvothermal processes

The fundamental principle is performing the required chemical reaction at the critical points of the solvents, ensured through heating in sealed vessels known as hydrothermal autoclave synchronously with autogeneous pressure [2]. When work up is accomplished in aqueous medium, the strategy is known as hydrothermal process. Solvothermal process utilises non aqueous medium for the synthesis. In certain cases the usage of precursors is essential which can be employed as gel or suspension in the solution. Relatively high concentration of organic or inorganic additives as mineralizer control the particle morphology. The nanocrystal formation takes place through crystal nucleation and auxiliary growth.

Hydrothermal methods are versatile for the design of surface enhanced nanomaterials [3]. An over view of very recent application of this strategy are prearranged in Table 1.

Table 1. Hydrothermal Method of Synthesis of nanomaterials

Nanomaterial	Starting Materials	Application	Morphology	Reference
SnO_2 composites modified by 2D graphitic carbon nitride (g-C_3N_4)	$SnCl_4$, urea and 2D g-C_3N_4 and	Sensing of ethanol gas	2D nanosheets	4
ZnO -based platinum screen-printed electrodes	Pt working electrode, zinc nitrate hexahydrate and hexamethylenetetramine	Detection of bacterial pathogens	Nanorod	5
Titanate	TiO_2 nano powder, 10 M NaOH, 0.1 M HCl	Possible use in solar cells	Nanotubes	6
Iron oxide (α-Fe2O3)	Iron nitrate, hexamethylenediamine (HMDA) and sodium hydroxide (NaOH).	Phenyl hydrazine sensor applications	Nanoparticles	7
Ag modified TiO_2	Different concentrations of titanium sulphate and silver nitrate	Acetone sensing	Nano particles	8

iron oxide/C nanoparticles	NH_4OH, $FeCl_3 \cdot 6H_2O$, $FeSO_4 \cdot 7H_2O$, $Fe(NO_3)_3 \cdot 7H_2O$, ammonium acetate and D-glucose (precursor)/ p-phenylenediamine (precursor)	--------	Nano particles	9
Antimony-tin oxide	Tin chloride, antimony chloride, ethanol, PVP, PVA and ethylene glycol	Anti bacterial agent for gram positive and negative starins, High-performance asymmetric supercapacitor, and photocatalyst	Nano particles	10
Silver decorated reduced graphene oxide (rGO)	Graphite, sodium nitrate, H_2SO_4, H_3PO_4, $KMnO_4$, H_2O_2, HCl, Silver	Photocatalytic activity for wastewater treatment	Nano flakes	11
Powder of Tamarind fruit shell modified with *in situ* generated copper nanoparticles	$CuSO_4 \cdot 5H_2O$, dried tamarind fruit shell powder	Solid waste management	Spherical nano particles	12

There are very recent reports available on the application of microwave assisted hydrothermal method of synthesis for nano materials with diversified applications [13,14].

The solvo thermal process is also versatile for the fabrication of materials of the nano dimension. A general application of this methodology is given in Table 2.

Nanoparticles in Healthcare: Applications in Therapy, Diagnosis and Drug Delivery   Materials Research Forum LLC
Materials Research Foundations 160 (2024) 24-51   https://doi.org/10.21741/9781644902974-2

Table 2. Solvo thermal process for the synthesis of nanomaterials

Nanomaterial	Starting Material	Application	Morphology	Reference
WS_2	Ammonium tetrathiotungstate $((NH_4)_2WS_4)$, oleylamine	Detection of uric acid and quercetin in blood serum by Non-enzymatic technique	Nanoflake-nanorod hybrid	15
$NiFe_2O_4$	$Ni(NO_3)_2$, $Fe(NO_3)_3$, ethylene glycol, urea and ethanol	High-performance supercapacitor	Nanoparticle	16
Cadmium selenide (CdSe)	Cadmium chloride, cadmium nitrate, sodium selenite, cadmium chloride as complxing agents,	Dielectric	Nanoparticle	17
Silver	PEG, PVP, $AgNO_3$	Antimicrobial activity	Nanopaticles	18
Nickel cobalt sulfide	Nickel(II) acetate, cobalt(II) acetate, ethylene glycol and thiourea,	Dye-sensitized solar cell	Nanoparticle	19

Recent reports underscore the significant improvement of efficiency of solvothermal synthesis by microwave. Microwave assisted solvo thermal procedure was successfully employed for the catalytic hydrolysis of a Nerve Agent stimulant for the synthesis of UiO-66-NH_2 [20]. These principles are increasingly in application for the fabrication of nano catalysts for commercially potent chemical reactions [21].

The hydro/solvothermal approaches are versatile as the processes are simple and relatively economic. Being a solution phase technique through judicious selection of solvents, reagents, temperature, pressure, pH, aging time the morphology and surface properties can be easily controlled. Highly homogeneous nano phases could be prepared. Moreover, post annealing treatments are generally excluded and lower size crystalline nano particles with well-defined morphology at relatively lower temperatures can be designed which are useful as high performance functional materials. The commercial production of nano engineered materials would be possible by the proper design of the autoclave. The main disadvantages include the long reaction time, non homogeneous heating, low yield, toxicity of solvents and discontinuity of the reaction.

2.2   Reduction method

This is a substantial method which can be beneficial for the fabrication of metal nanoparticles with attractive catalytic, electronic, optical and magnetic properties. Earlier, metal species were reduced with common reducing agents such as Sodiumborohydride and L-ascorbic and further growth of the nano material takes place with the reduced metal as nuclei.. Novel reduction route was suggested for metallic Cu nanoparticles which are of particular interest in view of their low cost and heterogeneous catalytic properties [22]. Fernandez and co-workers successfully prepared anti-microbial Graphene Oxide–Silver Nanoparticle Nanohybrids with L-ascorbic acid as the reducing agent [23]. Silver-doped copper nitride nanostructures useful in the oxygen reduction reaction was easily synthesised by wet chemical reduction route [24]. Citrate anion and tannic acid were reported as effective reducing agents for silver to the nano dimensions [25].

This method is advantageous due to the ease of preparation of thermally stable and surface controlled nanoparticles. The secondary nucleation could be made to insignificant level by the adjustments of pH, temperature and seed particle concentration. Only a very few reactions can be worked out under this category.

2.3   Microemulsion method

Microemulsions, even though homogeneous in macroscale, are heterogeneous in nanoscale. These are formed as a reversible dispersion stabilized by an interfacial film of surface active molecules where two immiscible liquids of nanosized domains of one or both, in the other. They proffer a significant reaction media for the synthesis of nanomaterials in view of their capacity to solubilize both aqueous and oil-soluble compounds, thermodynamic stability, lower interfacial tension and larger interfacial area. In these media the rate of the formation reaction can be well tuned by their interfacial properties. Hydrocarbon, water and a surfactant are essential for microemulsion. Sometimes, a cosurfactant is also used to enhance the fluidity. The important limitations are: limited solubilizing capacity for high- melting substances, necessitate large quantity of surfactants for stabilizing droplets, influence of parameters such as temperature and pH on their stability, The efficient micro emulsions for the formation of nano sized materials are specified in Table 3.

Table 3. Micro emulsion method for the synthesis of nanomaterials

Nanoparticle	Nature of microemulsion	Application	Morphology	Reference
$Mg_{1-x}\,Ca_x\,Ni_y\,Fe_{2-y}O_4$	Cetlytrimethylammonium bromide (CTAB) in water	....	Nanoparticles	26
Rhodium, Palladium and bimetallic Rh/Pd functionalised multi walled carbon nanotubes	Water in hexane	Hydrogenation of olefins	Nanoparticles	27
Ferrite	Emulsion of polyoxyethylen(15)cety lether/cyclohexene system	....	Nanoparticles	28
Cobalt-ferrite coated with silica	Emulsion of polyoxyethylen(15)cety lether/cyclohexene system	....	Nanoparticles	28
Ag	Castor oil in water,	....	Nanoparticles	29
Silica-coated Fe_3O_4	Water-in-oil emulsion of n-heptane.	Antibacterial activity	Nanoparticles	30
Ni	Water-in-oil emulsion of n-heptane n-Hepatane	Use in energy storage devices	Nanoparticles	31

2.4　Microwave (MW) assisted synthesis

Started in 1950s only and became versatile within the last two decades due to its green applications. The energy of the microwave is directly converted to heat energy and consequently requires a shorter reaction time. The reactions in conventional solvents can be completed within a few minutes if performed under microwave assistance. The shorter reaction time enhances the high yield of product with extra purity because this kind of reaction excludes undesirable side products. Reproducibility due to uniform distribution of microwaves during the reaction, controlled reaction conditions due to which the products are formed with very narrow size distribution. This strategy can be well used for the synthesis of colloidal metals. This method is not generally suitable for scale up. Sometimes the monitoring of the reactions is impossible during this strategy. Special reactants known as capping agents are essential in certain reactions.

Microwave assisted synthetic strategy is well-versed owing to the better control over the reaction by adjusting the specific parameters and solvents. An outline of the method and the various nanoparticles prepared by this methodology is included in Table 4.

Table 4. Microwave Assisted Synthesis of nanomaterials

Nanomaterial	Conditions of MW assisted reaction	Morphology	Application	Reference
Tunable photoluminescent carbon nanodots	Poly(ethylenimine) aqueous solution after vigorous stirring at room temperature irradiation under microwave (200 W) at 180 °C for 15 min	Nanodot	Imaging of cells and sensing of ions.	32
Carbon dots with Hydrophilicity	Glycerol/phosphate buffer solution heated under microwave (750 W) for 14 min	Nanodot	Biolabelling and bioimaging	33
Hydrophobic carbon dots	Microwave irradiation of poloxamer, polyoxyethylene– polyoxypropylene– polyoxyethylene block co-polymer pluronic F-68 (PF-68) and phosphoric acid for 4 minutes (450 W).	Spherical nanodot	Fluorescent	34
Cobalt oxide	Cobalt nitrate in ethylene glycol in presence of surfactant trioctyl phosphine oxide (TOPO) irradiated with QWave 1000 for 5 mintes.	nanoparticle	Magnetic application	35

Ag nanoparticle incorporated chitosan-alginate hydrogel	MW assisted synthesis for 8 min. of solution containing AgNO$_3$ ascorbic acid, chitosan alginate at pH 9	Wound dressing	Hydrogel	36
Colloidal zinc oxide	MW at 60 ∘C for 30 minutes on Potassium hydroxide solution and zinc precursor in methanol.	Biological applications	Nanocrystals	37
Carbon quantum dots dopped with N	Solution of urea in 50 mL of orange juice heated under micro wave for 10 minutes.	Quantum dots	Imaging of cancer cells and detection of palbociclib	38

2.5   Sonochemical method

Synthesis of nanostructured materials which are not possible otherwise through conventional methods can be made facile by consumption of high intensity ultrasound. This is a green method with mild experimental conditions in presence of green energy source. Nebulization and acoustic cavitation are the two underlying physical phenomena pertinent to synthesis. Bubbles are formed, grown and collapsed as a result of acoustic cavitation. Meanwhile sonochemical process are initiated by the extreme conditions arising during the collapse of bubbles in a liquid or liquidsolid slurries [39]. Ultra sound mediates nebulization which is the formation of mist from a liquid and liquid-gas interface. The droplets will get heated up and facilitate further reaction. This is utilised as the basis for ultrasonic spray pyrolysis (USP). Sonochemical reactions provide a unique interaction between hot spots inside the bubbles of ~ 5000 K and pressures of ~ 1000 bar [39].

This strategy is advantageous due to its ability to form even nano composites of uniform sizes rapidly with lesser energy. The acoustic pulses help in the production of nano materials such as metals, metal oxides, alloys, semiconductors etc., in high purity. But this technique has the limitations of difficulty in setting up the conditions for a reference reaction. In presence of an efficient agitation system, the sonochemical effect on the reaction would be reduced. Table 5.

Table 5. Sonochemical method of synthesis of nanomaterials

Nanoparticle	Power or Frequency	Size and shape	Media	Reference
Amorphous Iron oxide	40 kHz	38.9 nm, spherical	4M NaOH	40
$NiFe_2O_4$	20 kHz, 500 W	9 to 17 nm	3 M NaOH	41
Ni	26 kHz	7 to 8 nm	Potassium hydroxide and hydrazine monohydrate in absolute ethanol	42
$NiMoO_4$/chitosan	20 kHz	-------	Ammonium molybdate and nickel(II) acetate as precursors for Mo and Ni respectively	43
CoFe2O4/ Multi walled Carbon Nanotube hybrids	20 kHz	--------	Homogenised solution of functionalized MWCNTs, 0.02 M $Co(CH_3COO)_2$. 4H2O and 0.04 M $Fe(CH_3 COO)_2$	44
Gold	20 kHz	18.5 nm	0.03M aqueous solutions of $HAuCl_4$ and sodium citrate	45
TiO_2	500 W	2–8 nm , spherical	1:3 volumtric ratio if solutions of titanium isopropoxide and isopropanol	46

Nanoparticles in Healthcare: Applications in Therapy, Diagnosis and Drug Delivery   Materials Research Forum LLC
Materials Research Foundations 160 (2024) 24-51   https://doi.org/10.21741/9781644902974-2

2.6   Electrochemical method

The electrochemical method offers highly efficient, low cost and low temperature technique for the fabrication of functional nanomaterials with uses as energy materials with variety of morphological structures as wires, tubes, rods, sheets, composites etc. This method is simple as well as eco-friendly. Both aqueous and non aqueous environments can be used, and high purity nano materials are produced even from low quality initial reagents. The high electrical conductance of the molten chlorides or fluorides of alkali metals produce high current density. This would speed up the process. Batch and continuous process can be designed. This method has some limitations when used for large scale production.

A fascinating application of this strategy was the synthesis of carbo nano onion using two ultra clean and identical platinum electrodes as the working as well as the counter electrodes with silver wire coated with silver chloride (Ag/AgCl) as the quasi-reference electrode [47]. Super capacitor with ultra fast charging and discharging was developed with TiN nanoparticles by the electrochemical method [48].

High purity magnetic alloy nanoparticles are essential for applications in magnetic resonance imaging, drug targeting and magnetic refrigeration systems. In this context, the electrochemical method has immense potential. Cobalt-Samarium alloy was prepared in an aprotic ionic liquid [49]. Another application of this method was reported for developing a high sensitive electrochemical sensor for hydrazine employing extremely surface-roughened nanoporous gold electrode containing Pt nanoparticles. A commercial AuSn alloy with application as sensor was fabricated by anodisation in HCl and subsequent electrodeposition with H_2PtCl_6 as precursor in aqueous solution [50]. Nano Silicon for use in powerful lithium-ion batteries was prepared by lectrochemical method from $KCl–K_2SiF_6$ melts [51].

2.7   Polyol method

It is liquid-phase synthetic route. In earlier days, the preparation was accomplished in a diol medium like ethylenevglycol or tetraethyleneglycol containing NaOH. This methodology was later modified by replacement of the diol by very low concentration of polyol like 1,2-hexadecanediol. The modified polyol process proffers polyol as the reducing agent rather than solvent. The method has limitations of high volatility and toxicity. If the reaction requires precursors, compounds with high thermal stability and good solubility in the reaction medium are selected for the purpose. When the reaction temperature had attained, the quick and homogeneous decomposition of precursor happens throughout the solution.

Magnetic nanoparticles with homogeneous phases are functional materials in tumour targeting, magnetic recording, in MRI etc. This method is well defined for the synthesis of such particles. Polyol reduction below 200 °C was suggested for formulating catalytic Pt nanoparticles on F-doped SnO_2/glass substrates [52]. Conclusive evidence was there in literature for the mediation of polyol method whereby long-chain carboxylate Co (II) precursors were reduced to get nano-rods with the hexagonal close-packed structures [53]. The nano rods so prepared had enhanced magnetic behaviour [53].

2.8   Radiolytic process

High energy radiation is utilised for the synthesis. Ruthenium nanoparticles on pristine (MWCNT) and functionalized carbon nanotubes (f-MWCNT) from the aqueous solutions containing the Ru precursor were conveniently generated employing the high energy gamma rays from ^{60}Co source [54]. Ammonium uranyl tricarbonate up on irradiation with high efficiency gamma rays, UO_2 nanoparticles were formed [55]. The prepared nano particles were stable in air [55]. This result underscores the significance of this method for the production of UO_2 nuclear fuels.

There were reports on the single-step production of homogenised Iridium nanoparticles distributed on to multiwalled carbon nanotubes by means of gamma irradiation from a cobalt-60 source. The average particle size was reported to be in the range 4.5 to 3.4 nm [56]. In the presence of polyvinyl alcohol as capping agent, size-controlled and monodispersed silver nanoparticles were synthesized from an aqueous solution containing aqueous silver nitrate. Isopropyl alcohol was helpful in this reaction for scavenging hydrogen and hydroxide ion [57].

2.9   Sol-gel method

The sol-gel method is a low cost and wet-chemical technique extensively in use for the development of nanomaterials. The versatility of this technique relies on the flexibility attainable for the design of a wide variety of surface modified nanomaterials simply by alterations in pH, temperature, pressure, reaction time and nature of precursors. The liquid precursor used in the reaction mixture is getting transformed to a sol which subsequently getting converted to a gel like network structure [58]. Metal alkoxides were the common choice as precursor for the preparation of variety of metal oxides of nanodimesions. This strategy was reported be very effective and successful for the production of various metal oxide nanoparticles [59]. The nano powder containing TiO_2 was conveniently prepared by the sol-gel technique with $TiCl_4$ solution as the starting material [60]. Photodegradative and catalytic TiO_2 nanoparticles were obtained by this strategy [61]. The synthetic utility of sol gel method was well explored for the synthesis of silica nanoparticle from rice husk. The important application of the nanosilica prepared by the afore mentioned method was its use in the detection of the psychic stimulant Methamphetamine [62]. Photoactive MgO nanoflakes fabricated via this strategy were of immense use in the environmentally application of removal of organic contaminants [63].

3.   Physical methods

3.1   Milling

It is a top down method. This is a cost effective and eco friendly scheme. A solid grain is reduced to nano scale via the defect formation during this mechanochemical technique. A high energy mill facilitates the breaking down of the chemical bonds due to the kinetic energy. The steps involved are generation of mechanical stress due to milling, raise in temperature, breaking of lattice structure of the materials, transfer of mass and energy, crystal deformation and higher defect density. In order to tailor specific nanomaterial, a particular type of ball mill is fabricated [64]. The industrial production of one-dimensional nanomaterials such wire, tubes and rods could be better accomplished by milling technique [65]. Successful application of this method for the fabrication of ZnO as nanorod, nano belt and nano wire forms was reported [66]. Anti-bacterial ZnO nanoparticles were synthesised through high energy ball milling [67].

Important photonic material can be prepared in bulk through this strategy. There is literature evidence for the economic and effective application of wet ball milling as a low cost technique for the production of luminescent colloidal nano material from bulk ApbBr$_3$ (A=Cs, Fe) [68].

3.2   Plasma method

The underlying principle is heating the selected metal beyond its evaporation point using high voltage RF coils encased around the evacuated system. When the nucleation of metal vapour occurs, Helium gas is circulated through the system. The nano sized particles are formed around the collector rod through diffusion. Thermal plasmas found application when extremely high temperatures of several thousand degrees would result in more economic process. In addition, a continuous process can be designed with this work up. While using RF plasma more extended plasma flames with longer mean residence time are possible. RF plasma technique ensures the production of nano sized SiC particles [69]. The raw materials like commercially available silica powder along with various carbon sources like carbon black, char, graphite and residue of tyre pyrolysis were reported as precursors for SiC nanoparticle [69]. The construction of Mg-doped ZnO (MZO) sensing film for Electrolyte–Insulator– Semiconductor sensing (EIS) Urea and Glucose had cited in the literature through the successful execution of Plasma technology where Zinc oxide was treated by ammonium plasma for more than 3minutes [70].

3.3   Laser ablation

Magnetic nanopowders can be expediently made by laser ablation. Laser produces steep temperature gradient. The raw metal oxides selected as the initial reactant, has been evaporated. Hence, the nanoparticles are formed outside the evaporation zone by fast condensation and nucleation. Hybrid nanomaterials of the kind TiO$_2$-ZnO were prepared using Laser Ablation in liquid phase [71]. This field is changing with tremendous advancements and research. A newer procedure has been evolved as a result of which femtosecond laser ablation was reported and used for the novel design of nanomaterials [72]. More advantageous methods like pulsed laser ablation was employed for the fabrication of carbon-based nanoparticles by irradiating at wavelength 532 nm with a pulsed Nd:YAG laser of a graphite target [73]. This technique has the limitations that the yield of the nano particles generated is often low and very difficult for industrial scale preparation. To obtain yield to considerable levels, it is necessary to exploit quite expensive high-energy lasers.

3.4   Spray pyrolysis

This technique requires a precursor solution, atomizer and substrate. Solution of the precursor is atomised into small droplets using a nanoporous nebulizer. These droplets are transferred to the heated substrate leading to the decomposition of the precursor to form the desired material. Solvent mixed spray pyrolysis technique is commonly in use for the facile synthesis of metal oxide nanoparticles. There was an interesting study on the synthesis of MgO nanoparticles by the principles of Spra pyrolysis [74]. Spray pyrolysis has developed as a strategy for the preparation of metal clusters. Recent research work conclusively reported synthesis of anodic nanomaterial $ZnS/CoS/CoS_2$@N-doped carbon for lithium ion battery by spray pyrolysis [75]. Another significant application of this technique was in building resistive switching memory. Graphene oxide thin films prepared by this method was suitable for the said application [76].

Recent advancements in this field encompassed newer techniques employing flame and ultrasonic sound for pyrolysis. Flame spray pyrolysis utilises aqueous solution of the metal salt which is sprayed through a capillary into a flame where up on the formation of small droplets followed by the ignition of organic solvents in the flame converting the salt into the metal oxide. Agglomeration of metal oxide particles results in nano scale organisation of particle size. Generally, a part of the combustion energy was obtained from the high enthalpy of combustion of organic solvents mixed as precursors. Some research workers applied this strategy for the fabrication of Ni dopped Fe/Ce oxygen carrier which could catalyse the environmentally important reaction, dry reforming of methane through which two important greenhouse gases, methane and CO_2 can be utilised [77].

Ultrasonic Spray Pyrolysis (USP) exploits ultrasound for dispersing the precursor solution into droplets. Chemical decomposition occurs for the dissolved material inside the droplets where elevated temperature is produced. Within the droplets, solvent is evaporated followed by pyrolysis and chemical decomposition. Nanosized particle formation will proceed due to sintering. The noted advantages of this method include possible selection of wide variety of raw materials and simplicity of setting up a characteristic route in continuous process. The disadvantage is the low efficiency resulting from loss of the dissolved material as deposits on the structural elements of the USP device. Nano cluster Ru on MoO_2 was accomplished by this strategy and was useful as a catalyst in the hydrogen evaluation reaction [78].

3.5   Ion implantation

It is a bottom-up process effective for introducing selected impurity into the surface region of a substrate at the depth of several microns to form hybrid nanomaterials. It is a surface modification technique where ions of very high speed are bombarded on the surface of the substrate. There is a high selectivity for ion species and extraordinary controllability over amount of doped ion and depth which are vital for the formation of size-controlled nanomaterials. It can be broadly classified into two categories, beamline ion implantation and plasma-immersion ion implantation. The beamline method is not suitable for the synthesis of nanomaterials with complex geometries. Meanwhile, the latter is a hybrid process combining ion implantation and deposition. It results only in thick coatings. Here, intermixing occurs at the atomic level leading to coatings with high bond strength.

The surface physicochemical and biological properties of polylactic acid was modified by the implantation of nano particles of Ag as surface layer [79]. This layer was useful for improving the hydrophilicity [79]. Further investigations proved that polylactic acid modified with nano Ag had promising biomedical applications in view of its compatibility with immune system [79]. Nano Fe particles were introduced into a phenolic resin at ambient temperature under vacuum producing catalytic graphitization [80]. Very recent application of plasma immersion ion implantation was the introduction of osteo immune modulatory role to interleukin4 immobilised membrane that was in use for bone regeneration [81].

3.6   Electrospinning

It is the simplest top-down strategy for the development of nanostructured materials. This methodology finds immense application in the production of nanofibers from a wide variety of materials. This procedure relies up on the influence on a droplet polymer by electrostatic forces. When the electric force overcomes surface tension the droplets will fall towards a conductive collector which is kept at high voltage difference from the spinneret. Hence, the collector causes the droplet to be in the form of a jet with electrical charge resulting in the reduction of the diameter and consequently it is drawn into a fibre.

Antibacterial nanoparticles were developed successfully by electrospinning technique where Fe doped ZnO nanoparticles were incorporated into polyvinyl alcohol nanofibres [82]. Porous silica nanofibres containing nano silver in polymethyl methacrylate capable of reduction of methylene blue dye was processed through electrospinning. The precursors employed were Tetraethyl orthosilicate, poly[3-(trimethoxysilyl) propyl methacrylate and silver nitrate [83]. The nanofibres obtained through the electrospinning are currently high through put functional materials for therapeutic delivery [84].

4.   Green synthesis of nano dimensional materials

The synthesis of nano particles by green protocols offers benefits such as low cost, ease of production, lower energy requirements, safe and environmentally benign aspects. Biosynthesis gains special attention on account of the specificity and susceptibility to easy functionalization. It is a reliable process for the fabrication of a wide array of nanostructures with better defined size, morphology and biocompatibility for pharmaceuticals, medical, agronomical and environmental applications. This scheme never utilise toxic solvents, chemical precursors, reducing agents or reductive chemical atmosphere. These techniques collectively known as biological methods explore the potential of microorganisms, plant extracts, biological templates and enzymes for the preparation of nanomaterials.

4.1   Mediated by microbes

The common microbes employed for the preparation of nanomaterials are bacteria, fungi and algae. The details regarding the conditions, size and application of the nanoparticle are included in Table 4. The microorganism mediated synthesis possesses limitations. The process includes complicated steps such as microbial sampling, isolation, culturing and storage. In addition, the retrieval of the nanoparticle requires downstream processing.

Table 4. *Microbial Synthesis of nanomaterials using various biological species*

Microoraganism	Nanoparticle	Size and morphology	Application	Reference
Bacteria				
Proteus vulgaris ATCC-29905	Iron oxide	19.23 -30.51 nm, spherical	Antioxidant, antibacterial, anti cancer	85
Escherichia coli MF754138, *Exiguobacterium aurantiacumm* MF754139, and *Brevundimonas diminuta* MF754140	Ag	5-50 nm, spherical	Antibacterial	86
Raoultella planticola Pantoea agglomerans	Ag and AgCl	10-50 nm	Antibacterial activity	87
Paraclostridium benzoelyticum strain 5610	ZnO	50 nm	Antidiabetic, antibacterial, antiarthritic, and anti-inflammatory	88
Lactobacillus acidophilus	Se	34.13 nm	Antibacterial activity	89
Bacillus cereus SZT1	Ag	18 to 39 nm, sherical	Bateriside for leaf blight pathogen in rice	90
E. coli	CdSAg	15 nm	Bioimaging	91
Lactobacillus casei	Ag	single (25–50 nm) aggregates (100 nm), spherical	--------	92
Fungus				
Cladosporium halotolerans	Ag		Anti cancer	93
Aspergillus hortai	Ag	57nm, spherical	Antibacterial	94
Aspergillus brunneoviolaceus	Ag		Antibacterial	95
Humicola sp	Au$_2$S	Spherical	Ani cancer activity, anti-leishmanial activity	96

Aspergillus niger STA9	Cu	500 nm	Antibacterial and anticancer ag	97
Alga				
Aqueous extract of Sargassum myriocystum	Ag	22 nm, hexagonal	Anticancer and antibacterial activities	98
Gelidium corneum	Ag	20–50 nm	Anti bacterial and anti biofilm activities	99
Oscillatoria princeps	Ag	------	Anti bacterial activity against multi drug strains	100
Pterocladia capillacea	Cu	62nm	Anti cancer activity	101
Spatoglossum asperum	Fe	16mn, spherical	Anti cancer and anti oxidant activities	102
Extracts of *Ulva rigida Cystoseira, myrica , Gracilaria foliifera*	Ag	12, 17 and 24 nm, spherical	Anticancer and anti microbial agents	103

4.2   Mediated by enzymes

Enzyme-mediated synthesis has provided an alternative and safer approach for assembly of nanoparticles. Generally, active enzymes catalyse the formation of nanoparticles. Sometimes, the enzyme itself may act both as reducing and capping agent as well to facilitate the formation of nanoparticles.   Under certain reaction conditions, even amino acids released from denatured enzymes act as reducing and stabilizing agents as well in the synthetic strategy. Various enzymes which would be instrumental for the formation of nanodimensional particles are given in Table 5.

Table 5. Enzyme mediated Synthesis of nanoparticles

Pure Enzyme	Nanoparticle	Size and morphology	Application	Reference
Sulphite reductase	CdS	5-20 nm, quantum dots	Fluorescent biolabels	104
α-NADPH-dependent sulphite reductase	Au	7–20 nm, hydrosols	-----------	105
α-amylase	Au	----------	-------------	106
α-amylase	Ag	22–44 nm, triangular and hexagonal shape	----------	107
Serratiopeptidase	Ag	-------	Anti inflammatory	108
Macerating enzymes			Antibacterial activity, anti microbial surgical thread	109

4.3   Mediated by plant extracts

The various secondary metabolites like flavones, organic acids, quinones are phenolic compounds, tannins bear good reducing power and the nano particles will get reduced. Some of these extracts and their use in the synthesis of nanoparticles are exhibited in table 6.

Table 6. Plant extract mediated synthesis of nanoparticles

Plant	Nanoparticle	Size and Morphology	Application	Reference
Coleus aromaticus leaf	Ag	44nm	Antibacterial activity	110
Crotalaria verrucosa Leaf	ZnO	16 to 38 nm, hexagonal	Antimicrobial and anti cancer activities	111
Vernonia amygdalina leaf extracts	NiO	17.86 nm, octahedral	---------	112

Acalypha fruticosa L. leaf extract	ZnO	---------	Antibacterial activity	113
Eclipta prostrata leaf	TiO$_2$	36 to 68 nm	---------	114
Eucalyptus globulus leaf	Ag	17.5 ± 5.89 nm	Antibacterial	115
Salvia officinalis leaf	Ag	34.3 ± 7.76	Antibacterial	115

Conclusions and future prospects

Nanomaterials are used for wide-ranging applications. There is immense research work focusing on novel synthetic strategies to obtain optimal size and morphology for their use in diversified disciplines. Depending on the specific application there are physical, chemical or biological methods already in use. More efforts are required for green synthesis with economic and energy saving methodologies. More and more nanomaterial applications are developed for targeted drug delivery, bioimaging and nanomedicine. The optimisation of size, morphology and surface characteristics are the prerequisites. This can be accomplished only through the judicious selection of appropriate synthesis methods. Plasmonics, photonics, informatics and data storage, *etc.* are some areas where scalable synthetic strategies are to be explored and optimized.

References

[1] N. Baig, I. Kammakakam, W. Falath, Nanomaterials: a review of synthesis methods, properties, recent progress, and challenges, Mater. Adv. **2** (2021) 1821-1871. https://doi.org/10.1039/D0MA00807A

[2] P.G. Jamkhande, N.W. Ghule, A.H. Bamer, M.G. Kalaskar, Metal nanoparticles synthesis: An overview on methods of preparation, advantages and disadvantages, and applications, J. Drug Deli. Sci. Tech. 53 (2019) 101174. https://doi.org/10.1016/j.jddst.2019.101174

[3] S. Cao , C. Zhao , T. Han, L. Peng , Hydrothermal synthesis, characterization and gas sensing properties of the WO$_3$ nanofibers, *Mater. Lett.* **169** (2016) 17—20. https://doi.org/10.1016/j.matlet.2016.01.053

[4] J. Cao, C. Qin, Y. Wang, H. Zhang, B. Zhang, Y. Gong, X. Wang, G. Sun, H. Bala, Z. Zhang, Synthesis of g-C$_3$N$_4$ nanosheet modified SnO$_2$ composites with improved performance for ethanol gas sensing, RSC Adv. 7 (2017) 25504-25511.c. https://doi.org/10.1039/C7RA01901G

[5] T.H. Phuoc Nguyen, et.al, Stable electochemical measurements of platinum screen printed electrodes modified with vertical ZnO nanorods for bacterial detection, J. nanomat. 2019 (2019) 1-9. https://doi.org/10.1155/2019/2341268

[6] S. Muniyappana, T. Solaiyammala, K. Sudhakara, A. Karthigeyan, P. Murugakoothan, Conventional hydrothermal synthesis of titanate nanotubes: Systematic discussions on structural, optical, thermal and morphological properties, Modern Elec. Mat. 3(4) (2017) 174-178. https://doi.org/10.1016/j.moem.2017.10.002

[7] S.W. Hwang, A. Umar, G.N. Dar, S.H. Kim, R.I. Badran, Synthesis and characterization of iron oxide nanoparticles for phenyl hydrazine sensor applications, Sens. Lett. 12 (2014) 1-5. https://doi.org/10.1166/sl.2014.3224

[8] Z. Wang, A.A. Haidry, L. Xie, A. Zavabeti, Z. Li, W. Yin, R.L. Fomekong, B. Saruhan, Acetone sensing applications of Ag modified TiO2 porous nanoparticles synthesized via facile hydrothermal method, Appl. Surf. Sci. 533 (2020) 147383. https://doi.org/10.1016/j.apsusc.2020.147383

[9] A.O. da Silva, A.F.C. Campos, M.O. Rodrigues, M.H. Sousa, Tuning magnetic and luminescent properties of iron oxide@C nanoparticles from hydrothermal synthesis: Influence of precursor reagents, Surfaces and Interfaces 36 (2023) 102624. https://doi.org/10.1016/j.surfin.2022.102624

[10] E. Amutha, S. Rajaduraipandian, M. Sivakavinesan, G. Annadurai, Hydrothermal synthesis and characterization of the antimony–tin oxide nanomaterial and its application as a high-performance asymmetric supercapacitor, photocatalyst and antibacterial agent, Nanoscale Adv. 5 (2023) 255-268. https://doi.org/10.1039/D2NA00666A

[11] M. Ikram, A. Raza, M. Imran, A. Ul-Hamid, A. Shahbaz & S. Ali, Hydrothermal synthesis of silver decorated reduced graphene oxide (rGO) nanoflakes with effective photocatalytic activity for wastewater treatment, Nanoscale Res. Lett. 15 (2020) 95-106. https://doi.org/10.1016/j.jallcom.2022.167389

[12] B. Ashok, N. Hariram, S. Siengchin , A. Varada Rajulu, Modification of tamarind fruit shell powder with in situ generated copper nanoparticles by single step hydrothermal method, J. Biores. and Bioproducts 5 (2020) 180-185. https://doi.org/10.1016/j.jobab.2020.07.003

[13] Y. Chen, T. Chen, Z. Qin, Z. Xie, M. Liang, Y. Li, J. Lin, Rapid synthesis of AgInS$_2$ quantum dots by microwave assisted-hydrothermal method and its application in white light emitting diodes, J. Alloys and Comp. 930 (2023) 167389-167403. https://doi.org/10.1016/j.jallcom.2022.167389

[14] T.T. Hoang, H.P. Pham, Q.T. Tran, A facile microwave-assisted hydrothermal synthesis of graphene quantum dots for organic solar cell efficiency improvement, J. Nano. 2020 (2020) 1-8. https://doi.org/10.1155/2020/3207909

[15] L. Durai, C. Yi Kong, S. Badhulika, One-step solvothermal synthesis of nanoflake-nanorod WS$_2$ hybrid for non-enzymatic detection of uric acid and quercetin in blood serum, Mat. Sci. Eng. C 107 (2020) 110217-228. https://doi.org/10.1016/j.msec.2019.110217

[16] M. Sethi, U.S. Shenoy, S. Muthu, D. K. Bhat, Facile solvothermal synthesis of NiFe$_2$O$_4$ nanoparticles for high-performance supercapacitor applications. Frront Mater. Sci. 14 (2020) 120-132. https://doi.org/10.1007/s11706-020-0499-3

[17] S. Suresh, C. Arunseshan, Dielectric properties of cadmium delenide (CdSe) Nanoparticles synthesized by solvothermal method, Appl. Nanosci. 4 (2014) 179–184. https://doi.org/10.1007/s13204-012-0186-5

[18] L. Marinescu, D. Ficai, A. Ficai, O. Oprea, A.I. Nicoara, B.S. Vasile, L.Boanta, A. Marin, E. Andronescu, A.M. Holban, Comparative antimicrobial activity of silver nanoparticles obtained by wet chemical reduction and solvothermal methods, Int. J. Mol. Sci. 23 (2022) 5982. https://doi.org/ 10.3390/ijms23115982

[19] K.M. Aiswarya, T. Raguram, K.S. Rajni, Synthesis and characterisation of nickel cobalt sulfide nanoparticles by the solvothermal method for dye-sensitized solar cell applications, Polyhedron 176 (2020) 114267. https://doi.org/10.1016/j.poly.2019.114267

[20] Z. Zhang, T. Cheng-An, J. Zhao, F. Wang, J. Huang, J. Wang, Microwave-assisted solvothermal synthesis of UiO-66-NH2 and Its catalytic performance toward the hydrolysis of a nerve agent simulant, Catalysts 10 (2020) 1086-97. https://doi.org/10.3390/catal10091086

[21] A. Rizzuti, et.al., Microwave-assisted solvothermal synthesis of Fe3O4/CeO2 nanocomposites and their catalytic activity in the imine formation from benzyl alcohol and aniline, *Catalysts 10*(11) (2020) 1325-47. https://doi.org/10.3390/catal10111325

[22] A. Khan, A. Rashid, R. Younas, R. Chong, A chemical reduction approach to the synthesis of copper nanoparticles. Int. Nano. Lett 6 (2016) 21-26. https://doi.org/10.1007/s40089-015-0163-6

[23] M. Cobos, I. De-La-Pinta, G. Quindós, M.J. Fernández, M.D. Fernández, Graphene oxide–silver nanoparticle nanohybrids: synthesis, characterization, and antimicrobial properties, *Nanomaterials 10*(2) 2020 376-398. https://doi.org/10.3390/nano10020376

[24] A. Ścigała, R. Szczesny, P. Kamedulski, M. Trzcinski, E. Szłyk, Copper nitride/silver nanostructures synthesized via wet chemical reduction method for the oxygen reduction reaction, Biomaterials Research 23 (2019) 27-42. https://doi.org/10.1186/s40824-019-0173-y

[25] K.R. Ranoszek-Soliwoda, E.Tomaszewska, E. Socha, P. Krzyczmonik, A. Ignaczak, P. Orlowski, M. Krzyzowska, G. Celichowski, J. Grobelny, The role of tannic acid and sodium citrate in the synthesis of silver nanoparticles, J. Nanopart. Res. 19 (2017) 273-288. https://doi.org/10.1007/s11051-017-3973-9

[26] R. Ali , M.A. Khan, A. Mahmooda, A.H. Chughtai, A. Sultand, M. Shahide , M. Ishaqf, M.F. Warsia, Structural, magnetic and dielectric behavior of Mg $_{1-x}$Ca$_x$Ni$_y$Fe$_{2-y}$O$_4$ nano-ferrites synthesized by the micro-emulsion method, Ceramics Int. 40 (2014) 3841–3846. https://doi.org/10.1016/j.ceramint.2013.08.024

[27] B. Yoon, C.M. Wai, Microemulsion-templated synthesis of carbon nanotube-supported Pd and Rh nanoparticles for catalytic applications, J. Am. Chem. Soc. 127(49) (2005) 17174–17175. https://doi.org/10.1021/ja055530f

[28] T.Tago, T.Hatsuta, K. Miyajima, M. Kishida, S. Tashiro, K. Wakabayashi, Novel synthesis of silica-coated ferrite nanoparticles prepared using water-in-oil microemulsion, J. Am. Ceram. Soc. 85(9) (2002) 2188-94. https://doi.org/10.1111/j.1151-2916.2002.tb00433.x

[29] R.D. Rivera-Rangela , M.P. González-Muñoza, M. Avila-Rodriguez , T.A. Razo-Lazcanoa, C. Solans, Green synthesis of silver nanoparticles in oil-in-water microemulsion and nano-emulsion using geranium leaf aqueous extract as a reducing agent Colloids and Surfaces A, 536 (2018), 60-67. https://doi.org/10.1016/j.colsurfa.2017.07.051

[30] G. Asab, E.A. Zereffa, T.A. Seghne, Synthesis of silica-coated Fe_3O_4 nanoparticles by microemulsion method: characterization and evaluation of antimicrobial activity, Int. J. Biomat.. 2020 (2020) 4783612-23. https://doi.org/10.1016/j.colsurfa.2017.07.051

[31] Z. Ur Rehman, M. Nawaz, H. Ullah, I. Uddin, S. Shad, et.al, Synthesis and characterization of Ni nanoparticles via the microemulsion technique and its applications for energy storage devices., Materials 16 (2023) 325-336. https://doi.org/10.3390/ma16010325

[32] H. Liu, Z. He, J. Li-Ping, Z. Jun-Jie, Microwave-assisted synthesis of wavelength- tunable photoluminescent carbon nanodots and their potential applications, ACS Appl. Mater. Interfaces 7 (2015) 4913−4920. https://doi.org/10.1021/am508994w

[33] X. Wang, K. Qu, B. Xu, J. Ren, X. Qu, Microwave assisted one-step green synthesis of cell-permeable multicolor photoluminescent carbon dots without surface passivation reagents, J. Mater. Chem. 21 (2011) 2445−2450. https://doi.org/10.1039/C0JM02963G

[34] S. Mitra, S. Chandra, T. Kundu, R. Banerjee, P. Pramanik, A. Goswami, Rapid microwave synthesis of fluorescent hydrophobic carbon dots, RSC Adv. 2 (2012) 12129−12131. https://doi.org/10.1039/C2RA21048G

[35] A.S. Bhatt, D.K. Bhat, T. Cheuk-Wai Tai, M.S. Santosh, Microwave-assisted synthesis and magnetic studies of cobalt oxide nanoparticles, Mater. Chem. Phys. 125 (3) (2011) 347-35. https://doi.org/10.1016/j.matchemphys.2010.11.003

[36] T.Oe, D. Dechojarassri, S. Kakinoki, H. Kawasaki, T. Furuike, H. Tamura, Microwave-assisted incorporation of AgNP into chitosan–alginate hydrogels for antimicrobial applications. J. Funct. Biomater. 199 (2023) 14-30. https:// doi.org/10.3390/jfb14040199

[37] N. Garino, T. Limongi, B. Dumontel, M. Canta, L. Racca, M. Laurenti, M. Castellino, A. Casu, A. Falqui, V. Cauda, A microwave-assisted synthesis of zinc oxide nanocrystals finely tuned for biological applications, Nanomaterials 9 (2019) 212-229. https://doi.org/10.3390/nano9020212

[38] G.Magdy, F. Belal, H. Elmansi, Rapid microwave-assisted synthesis of nitrogen-doped carbon quantum dots as fluorescent nanosensors for the spectrofluorimetric determination of palbociclib: application for cellular imaging and selective probing in living cancer cells, RSC Adv. 13 (2023) 4156-416. https://doi.org/10.1039/D2RA05759J

[39] K.S. Suslick, Sonochemistry, Science, 247 (1990) 1439-1445. https://doi.org/10.1126/science.247.4949.1439

[40] V.K. Yadav, D. Ali, S.H. Khan, G.Gnanamoorthy, N. Choudhary, K.K. Yadav, V.N. Thai, S.A. Hussain, S. Manhrdas, Synthesis and characterization of amorphous iron oxide nanoparticles by the sonochemical method and their application for the remediation of heavy metals from wastewater, Nanomaterials 10 (2020) 1551-1568. https://doi.org/10.3390/nano10081551

[41] M.A.S. Amulya, H.P. Nagaswarupa, M.R.A. Kumar, C.R. Ravikumar, S.C. Prashantha, K.B. Kusuma, Sonochemical synthesis of $NiFe_2O_4$ nanoparticles: characterization and their photocatalytic and electrochemical applications, Appl. Surf. Sci. Advances 1 (2020) 100023-100033. https://doi.org/10.1016/j.apsadv.2020.100023

[42] A.A. Ádám, M. Szabados. G. Varga, Á. Papp, K. Musza, Z. Kónya, Á. Kukovecz, P. Sipos, I. Pálinkó, Ultrasound-assisted hydrazine reduction method for the preparation of nickel nanoparticles, physicochemical characterization and catalytic application in Suzuki-Miyaura cross-coupling reaction, Nanomaterials 10 (2020) 632-650. https://doi.org/10.3390/nano10040632

[43] B.S.Lou, U. Rajaji, C. Shen-MingRajaji, C. Tse-Wei, A simple sonochemical assisted synthesis of NiMoO4/chitosan nanocomposite for electrochemical sensing of amlodipine in pharmaceutical and serum samples, Ultrasonics Sonochemistry, 64 (2020) 104827-36. https://doi.org/10.1016/j.ultsonch.2019.104827

[44] J. Acharya, B. Gnana, B.G.S. Raj, T.H. Ko, M.S. Khil, H.Y. Kim, Byoung-Suhk Kim, Facile one pot sonochemical synthesis of CoFe2O4/ MWCNTs hybrids with well-dispersed MWCNTs for asymmetric hybrid supercapacitor applications, Int. J. Hydrogen energy 45 (2020) 3073-85. https://doi.org/10.1016/j.ijhydene.2019.11.169

[45] M.A. Dheyab, A.A. Aziz, M.S. Jameel, P.M. Khaniabadi, A.A. Oglat, Rapid sonochemically-assisted synthesis of highly stable gold nanoparticles as computed tomography contrast agents, Appl. Sci. 10 (2020) 7020-34. https://doi.org/10.3390/app10207020

[46] A. Qayyum, A. Dimitrios Giannakoudakis, D.Łomot, R.F.C. Quintero, A.P. LaGrow, K. Nikiforow, D. Lisovytskiy, J.C. Colmenares, Tuning the physicochemical features of titanium oxide nanomaterials by ultrasound: elevating photocatalytic selective partial oxidation of lignin-inspired aromatic alcohols, Ultrasonics Sonochemistry 94 (2023) 106306-316. https://doi.org/10.1016/j.ultsonch.2023.106306

[47] Y. Bian, L. Liu, D. Liu, Z. Zhu, Y. Shao,L. Meixian, Electrochemical synthesis of carbon nano onions, : Inorg. Chem. Front. 7 (2020) 4404-12

[48] S.A. Ansari, N.A. Khan, Z. Hasan, A. A. Shaikh, K. Farhana, Ferdousi, H.R. Barai, N.S. Lopa, Md. M. Rahman, Electrochemical synthesis of titanium nitride nanoparticles onto titanium foil for electrochemical supercapacitors with ultrafast charge/discharge, Sustainable Energy and Fuels (2020) 1-11. https://doi.org/10.1039/D0SE00049C

[49] M. Manjum, N. Serizawa, A.Ispas, A. Bund, Y. Katayama, Electrochemical preparation of cobalt-samarium Nanoparticles in an aprotic ionic liquid, J. Electrochem. Soc., 167 (2020) 042505-514. https://doi.org/10.1149/1945-7111/ab7a83

[50] Y. Pei, M.Hu, Y. Xia, W. Huang, Z. Li, S. Chen, Electrochemical preparation of Pt nanoparticles modified nanoporous gold electrode with highly rough surface for efficient determination of hydrazine, Sensors and Actuators B: Chemical 304 (2020) 127416. https://doi.org/10.1016/j.snb.2019.127416

[51] T. Gevel, S. Zhuk, N. Leonova, A. Leonova, A. Trofimov, A. Suzdaltsev, Y. Zaikov, Electrochemical synthesis of nano-sSized silicon from $KCl–K_2SiF_6$ melts for powerful lithium-ion batteries. Appl. Sci. 11 (2021) 10927-39. https://doi.org/10.3390/ app112210927

[52] A. Ghifari, D.X. Long, S. Kim, B. Ma, J. Hong, Transparent platinum counter electrode prepared by polyol reduction for bifacial, dye-sensitized solar cells, Nanomaterials 10 (2020) 502-512. https://doi.org/10.3390/nano10030502

[53] M.A. Bousnina, A.D. Omrani, F. Schoenstein,Y. Soumare, A.H. Barry, J, Y. Piquemal, G.Viau, S.Mercone, N. Jouin, Enhanced magnetic behavior of cobalt nano-rods elaborated by the polyol process assisted with an external magnetic field, Nanomaterials 10 (2020) 334-48. https://doi.org/10.3390/nano10020334

[54] J.V. Rojas, M.T.Gonzalez, M.C.M. Higgins, C.E. Castano, Facile radiolytic synthesis of ruthenium nanoparticles on graphene oxide and carbon nanotubes, Materials Sci. Engg. B 205 (2016) 28–35. https://doi.org/10.1016/j.mseb.2015.12.005

[55] Y. Wang, Q. Chen, X. Shen, Preparation of low-temperature sintered UO_2 nanomaterials by radiolytic reduction of ammonium uranyl tricarbonate, J. of Nuclear Mater. 479 (2016) 162e166. https://doi.org/10.1016/j.jnucmat.2016.07.003

[56] J.V. Rojas, C.H. Castano, Radiolytic synthesis of iridium nanoparticles onto carbon nanotubes, J. Nanopart. Res. 16 (2014) 2567-72. https://doi.org/10.1002/anie.201201726

[57] E. Saion, E. Gharibshahi, K. Naghavi , Size-controlled and optical properties of monodispersed silver nanoparticles synthesized by the radiolytic reduction method, Int. J. Mol. Sci. 14 (2013) 7880-7896. https://doi.org/10.3390/ijms14047880

[58] J. L.Howard,.C. Cao, D.L. Browne, 2018. Mechanochemistry as an emerging tool   for molecular synthesis: what can it offer? Chem. Sci. (2018) 3080-95. https://doi.org/10.1039/C7SC05371A

[59] Y. Chen, C.P. Li, H. Chen, Y. Chen, One-dimensional nanomaterials synthesized using high-energy ball milling and annealing process, Sci. Tech. Adv. Mater. 7 (2006) 839–846. https://doi.org/10.1016/j.stam.2006.11.014

[60] J.S.Lee, K. Park, K. Myung-IL, P.IL-Woo, S.W. Kim, W.K. Cho, H.S. Han, S. Kim, ZnO nanomaterials synthesized from thermal evaporation of ball-milled ZnO powders, J.Crys. Growth, 254(3-4) (2003) 423-431. https://doi.org/10.1016/S0022-0248(03)01197-7

[61] N. Salah, S.S. Habib, Z.H. Khan, A. Memic, A. Azam, E. Alarfaj, N. Zahed, S. Al-Hamedi, High-energy ball milling technique for ZnO nanoparticles as antibacterial material, Int. J.Nanomedicine 6 (2011) 863–869. https://doi.org/10.2147/IJN.S18267

[62] L. Protesescu, S. Yakunin, O. Nazarenko, D.N.Dirin, M.V. Kovalenko, Low-cost synthesis of highly luminescent colloidal lead halide perovskite nanocrystals by wet ball milling, ACS Appl. Nano Mater. 1 (3) (2018) 1300–1308. https://doi.org/10.1021/acsanm.8b00038

[63] Z. Károly, I. Mohai, Sz. Klébert, A. Keszler, I.E. Sajó, J. Szépvölgyi, Synthesis of SiC powder by RF plasma technique, Powder Technology 214 (2011) 300–305. https://doi.org/10.1016/j.powtec.2011.08.027

[64] C.F. Lin, C.H.Kao, C.Y.Lin, K.L. Chen, Y.H. Lin, NH_3 plasma-treated magnesium doped zinc oxide in biomedical sensors with electrolyte–insulator– semiconductor (EIS) structure for urea and glucose applications, Nanomaterials 10 (2020) 583-599. https://doi.org/10.3390/nano10030583

[65] N. Mintcheva, S. Yamaguchi, S.A. Kulinich, Hybrid TiO_2-ZnO nanomaterials prepared using laser ablation in liquid, Materials 13 (2020) 719-733. https://doi.org/10.3390/ma13030719

[66] D. Tan, S. Zhou, J. Qiu, N. Khusro, Preparation of functional nanomaterials with femtosecond laser ablation in solution, J. Photochemistry and Photobiology C: Photochemistry Reviews 17 (2013) 50-68. https://doi.org/10.1016/j.jphotochemrev.2013.08.002

[67] E.A. Ganash, G.A. Al-Jabarti, R.M. Altuwirqi, The synthesis of carbon-based nanomaterials by pulsed laser ablation in water, *Mater. Res. Express* 7 (2020) 015002-13.

[68] K.R. Nemade, S.A.Waghuley, Synthesis of MgO nanoparticles by solvent mixed spray pyrolysis technique for optical investigation, Int. J. Metals 2014 (2014) 1-4. http://dx.doi.org/10.1155/2014/389416

[69] W. Cheng, H. Di, Z. Shi, D. Zhang, Synthesis of ZnS/CoS/CoS2@N-doped carbon nanoparticles derived from metal-organic frameworks via spray pyrolysis as anode for lithium-ion battery, J. Alloys. Compounds 831 (2020) 154607. https://doi.org/10.1016/j.jallcom.2020.154607

[70] A. Moazzeni, H.R. Madvar, S. Hamedi, Z. Kordrostami, Fabrication of graphene oxide-based resistive switching memory by the spray pyrolysis technique for neuromorphic computing, ACS Appl. Nano Mater. 6(3) 2023 2236–2248. https://doi.org/10.1021/acsanm.2c05497

[71] X. Chen, G. Zou, Y. Yuan, Z. Xu, H. Zhao, Flame spray pyrolysis synthesized Ni-doped Fe/Ce oxygen carriers for chemical looping dry reforming of methane, Fuel 343 (2023) 127913. https://doi.org/10.1016/j.fuel.2023.127913

[72] Y. Koo, S. Oh, K. Im, J. Kim, Ultrasonic spray pyrolysis synthesis of nano-cluster ruthenium on molybdenum dioxide for hydrogen evolution reaction, Appl. Surf. Sci., 611, Part B (2023) 155774. https://doi.org/10.1016/j.apsusc.2022.155774

[73] I.A. Kurzina, O.A. Laput , D.A. Zuzaa, I.V. Vaseninaa, M.C. Salvadoria, K.P. Savkin, D.N. Lytkinaa,V.V. Botvina, M.P. Kalashnikov, Surface property modification of biocompatible material based on polylactic acid by ion implantation, Surface coatings and Tech. 388 (2020) 125529-37. https://doi.org/10.1016/j.surfcoat.2020.125529

[74] A. Idesaki, S. Yamamoto, M. Sugimoto, T. Yamaki, Y. Maekawa, Formation of Fe nanoparticles by ion implantation technique for catalytic graphitization of a phenolic resin, Quantum Beam Sci. 4 (2020) 11-22. https://doi.org/10.3390/qubs4010011

[75] F. Wei, Y. Mu, R. P. Tan, S.G. Wise, M.M. Bilek, Y. Zhou,Y. Xiao, Osteo-immunomodulatory role of interleukin-4-immobilized plasma immersion ion implantation membranes for bone regeneration, ACS Appl. Mater. Interfaces 15 (2023) 2590–2601. https://doi.org/10.1021/acsami.2c17005

[76] S. Majeed, M. Danish, M.N.M. Ibrahim, S. H. Sekeri, M. T. Ansari, A. Nanda, G. Ahmad, Bacteria mediated synthesis of iron oxide nanoparticles and their antibacterial, antioxidant, cytocompatibility properties. J Cluster Sci 32 (2021) 1083-1094. https://doi.org/10.1007/s10876-020-01876-7

[77] S. Saeed, A. Iqbal, M.A. Ashraf, Bacterial-mediated synthesis of silver nanoparticles and their significant effect against pathogens. Environ Sci Pollut Res 27 (2020) 37347-37356. https://doi.org/10.1007/s11356-020-07610-0

[78] I. Ghiuta, C. Croitoru, J. Kost, R. Wenkert, D. Munteanu, Bacteria-mediated synthesis of silver and silver chloride nanoparticles and their antimicrobial activity, Appl. Sci. 11 (2021) 3134-47. https://doi.org/10.3390/app11073134

[79] S.Faisal, Abdullah, M.Rizwan, R.Ullah, A.Alotaibi, A.Khattak, N.Bibi, M.Idrees, *Paraclostridium benzoelyticum* bacterium-mediated zinc oxide nanoparticles and their *in vivo* multiple biological applications 2022 (2022) 1-15. https://doi.org/10.1155/2022/5994033

[80] H. Alam, N, Khatoon, M. A. Khan, S. A. Husain, M. Saravanan, M. Sardar, Synthesis of selenium nanoparticles using probiotic bacteria Lactobacillus acidophilus and their enhanced antimicrobial activity against resistant bacteria. J Clust Sci 31 (2020) 1003-1011. https://doi.org/10.1007/s10876-019-01705-6

[81] T. Ahmed, M. Shahid, M. Noman, M.B.K. Niazi, F. Mahmood, I. Manzoor, Y. Zhang, B. Li, Y. Yang, C. Yan, J. Chen, Silver nanoparticles synthesized by using *Bacillus cereus* SZT1 ameliorated the damage of bacterial leaf blight pathogen in rice, Pathogens 9 (2020) 160-177. https://doi.org/10.3390/pathogens9030160

[82] N. Órdenes-Aenishanslins, G. Anziani-Ostuni, J.P. Monrás, A. Tello, D. Bravo, D. Toro-R., S. Soto-Rifo, R. Ricardo, P.N. Prasad, J.M. Pérez-Donoso, Bacterial synthesis of ternary CdSAg quantum dots through cation exchange: tuning the composition and properties of

biological nanoparticles for bioimaging and photovoltaic applications, Microorganisms 8 (2020) 631-50. https://doi.org/10.3390/microorganisms8050631

[83] H. Korbekandi, S. Iravani, S. Abbasi, Optimization of biological synthesis of silver nanoparticles using *Lactobacillus casei* subsp. *Casei,* J.Chemical Tech and Biotechnology. 87(7) (2012) 932-937. https://doi.org/10.1002/jctb.3702

[84] F. Ameen, A. A. Al-Homaidan, A. Al-Sabri, A. Almansob, S. AlNAdhari, Anti-oxidant, anti-fungal and cytotoxic effects of silver nanoparticles synthesized using marine fungus Cladosporium halotolerans. Appl Nanosci 13 (2023) 623-631. https://doi.org/10.1007/s13204-021-01874-9

[85] R. Rai, A. S. Vishwanathan, B. S. Vijayakumar, Antibacterial potential of silver nanoparticles synthesized using *Aspergillus hortai.* Bio Nano Sci. 13 (2023) 203-21. https://doi.org/10.1007/s12668-022-01043-4

[86] H. Mistry, R. Thakor, C. Patil, J. Trivedi, H. Bariya, Biogenically proficient synthesis and characterization of silver nanoparticles employing marine procured fungi Aspergillus brunneoviolaceus along with their antibacterial and antioxidative potency. Biotechnology Lett. 43 (2021) 3077-316. https://doi.org/10.1007/s10529-020-03008-7

[87] A. Syeda, M.H. Al Saedia , A.H. Bahkalia , A.M. Elgorgana, M. Kharatb, K.Pai , J. Pichtel, A. Ahmad, α-Au$_2$S nanoparticles: fungal-mediated synthesis, structural characterization and bioassay, Green Chem. Lett. Rev. 15(1) (2022) 61–70. https://doi.org/10.1080/17518253.2021.1999509

[88] S. Noor, Z. Shah, A.Javed, A. Ali, S.B. Hussain, S.Z.H. Ali, S.A. Muhammad, A fungal based synthesis method for copper nanoparticles with the determination of anticancer, antidiabetic and antibacterial activities, J. Micro. Methods 174 (2020) 105966, https://doi.org/10.1016/j.mimet.2020.105966

[89] P. Balaraman, B. Balasubramanian, D. Kaliannan, M. Durai, H. Kamyab, · S. Park, S. Chelliapan, C.T. Lee, V. Maluventhen, A. Maruthupandian, Phyco-synthesis of silver nanoparticles mediated from marine algae *Sargassum myriocystum* and its potential biological and environmental applications, Waste and Biomass Valorization 11 (2020) 5255-5271. https://doi.org/10.1007/s12649-020-01083-5

[90] B.Y. Öztürk, B.Y. Gürsu, İ. Dağ, Antibiofilm and antimicrobial activities of green synthesized silver nanoparticles using marine red algae *Gelidium corneum*, Process Biochemistry 89 (2020) 208-219. https://doi.org/10.1016/j.procbio.2019.10.027

[91] A. K. Bishoyi, C. R. Sahoo, A. P. Sahoo, R. N. Padhy, Bio-synthesis of silver nanoparticles with the brackish water blue-green alga Oscillatoria princeps and antibacterial assessment. *Applied Nanoscience.* Appl. Nanosci. 11 (2021) 389-398). https://doi.org/10.1007/s13204-020-01593-7

[92] N.M. Aboeita, S.A. Fahmy, M.M.H. El-Sayed, H.M. El-Said Azzazy, T. Shoeib, Enhanced anticancer activity of nedaplatin loaded onto copper nanoparticles synthesized using red algae, Pharmaceutics 14 (2022) 418-32. https://doi.org/10.3390/pharmaceutics14020418

[93] T. Palaniyandi, G. Baskar, V. Bhagyalakshmi, S. Viswanathan, M.R.A. Wahab, M. K.Govindaraj, A. Sivaji, B.K. Rajendran, S. Kaliamoorthy, Biosynthesis of iron nanoparticles using brown algae *Spatoglossum asperum* and its antioxidant and anticancer activities through *in vitro* and *in silico* studies, Particulate Science and Technology.

[94] R. Algotiml, AliGab-Alla, R. Seoudi, H.H. Abulreesh, M.Z. El-Readi, K. Elbanna, Scientifc Reports 12 (2022) 2421-39. https://doi.org/10.1038/s41598-022-06412-3

[95] A.A. Ayoobul; K.S. Anil, M.V. Krishnasastry, M.K. Abyaneh, K.K.Sulabha, A. Absar, M.I. Khan, CdS quantum dots: enzyme mediated *in vitro* synthesis, characterization and conjugation with plant lectins, J.Biomedical Nanotech. 3(4) (2007) 406-413. https://doi.org/10.1166/jbn.2007.045

[96] S.A. Kumar, M.K. Abyaneh, S.W. Gosavi, S.K. Kulkarni, A. Ahmad, M.I. Khan, Sulfite reductase-mediated synthesis of gold nanoparticles capped with phytochelatin, Biotech. Appl.Biochem. 47(4) (2007) 191-195. https://doi.org/10.1042/BA20060205

[97] M. Vanaja, G. Annadurai, *Coleus aromaticus* leaf extract mediated synthesis of silver nanoparticles and its bactericidal activity,   Appl. Nanosci. 3 (2013) 217-223. https://doi.org/10.1007/s13204-012-0121-9

[98] S.S. Sana, D.V. Kumbhakar, A. Pasha, S.C. Pawar, A.N. Grace, R.P. Singh, V.H. Nguyen, Q.V.Le, W. Peng, *Crotalaria verrucosa* leaf extract mediated synthesis of zinc oxide nanoparticles: assessment of antimicrobial and anticancer activity, Molecules 25 (2020) 4896-4917. https://doi.org/10.3390/molecules25214896

[99] A.B. Habtemariam, M. Oumer, Plant extract mediated synthesis of nickel oxide nanoparticles, Mat. Int. 2 (2020) 0205-0209. https://doi.org/10.33263/Materials22.205209

[100] S. Vijayakumar, P. Arulmozhi, N. Kumar, B. Sakthivel, S.P. Kumar, P.K.   Praseetha, *Acalypha fruticosa L.* leaf extract mediated synthesis of ZnO nanoparticles: characterization and antimicrobial activities, Materials Today: Proceedings 23 Part 1 (2020) 73-80. https://doi.org/10.1016/j.matpr.2019.06.660

[101] G. Rajakumar, A.A. Rahuman, B. Priyamvada, V.G. Khanna, D.K. Kumar, P.J. Sujin, *Eclipta prostrata* leaf aqueous extract mediated synthesis of titanium dioxide nanoparticles, Materials Letters 68 (2012) 115-117. https://doi.org/10.1016/j.matlet.2011.10.038

[102] A. Bal`ciunaitiene, M. Liaudanskas, V. Puzeryte, J. Viškelis, V. Janulis, P. Viškelis, E. Griškonis, V. Jankauskaite, *Eucalyptus globulus* and *Salvia officinalis* extracts mediated green synthesis of silver nanoparticles and their application as an antioxidant and antimicrobial agent, Plants 11 (2022) 1085-1101. https://doi.org/10.3390/plants11081085

Nanoparticles in Healthcare: Applications in Therapy, Diagnosis and Drug Delivery   Materials Research Forum LLC
Materials Research Foundations 160 (2024) 52-82   https://doi.org/10.21741/9781644902974-3

Chapter 3

Approaches in Synthesis, Fabrication, and Toxicity of Plant-based Nanomaterials

V. Gayathri [1,a*], S. Ajikumaran Nair [1,b], B. Greeshma Nair [1,c], S.M. Surayya Muhammed [1,d]

[1]Phytochemistry and Phytopharmacology Division, KSCSTE-Jawaharlal Nehru Tropical Botanic Garden and Research Institute (JNTBGRI), Kerala, India

[a] gayathrieaswer@gmail.com, [b] saknair@rediffmail.com
[c]exploregreemist@gmail.com, [d] surayyamohdsm@gmail.com
*gayathrieaswer@gmail.com

Abstract

Plants contain a wide range of phytochemicals, including flavonoids, alkaloids, tannins, saponins, and other metabolites. The nanomaterials have significant implications in current therapeutic approaches for various diseases, including cancer. Herbal extracts offer a safer pathway for the generation of nanomaterials, as the secondary metabolites present in these extracts act as stabilizing and reducing agents, leading to the formation of nanoparticles (NPs). These plant extract-derived NPs find diverse applications in various fields, particularly in nanomedicine. This review focuses on the recent advancements in the biomedical applications of such nanomaterials, specifically for cancer treatment and combating viral infections.

Keywords

Synthesis of Nanoparticles, Toxicity, Plant-Based Nanomaterials

Contents

1.   Introduction

Nanotechnology primarily focuses on the design, synthesis, analysis of structures, and applications of materials smaller than 100 nm. In recent times, there has been a significant exploration of nanomaterials due to their exceptional properties and capabilities, leading to the advancement of interdisciplinary research and practical problem-solving. These materials have brought numerous benefits and promising prospects to various aspects of human life, including medicine, pharmaceuticals, agriculture, environment, catalysis, food, cosmetics, and electronics. This can be attributed to their small size, diverse structures, and a wide range of biochemical and physicochemical properties, making them suitable for diverse fields. Particularly, metallic/metal oxide nanoparticles are highly regarded for their advantageous features such as a large surface area to volume ratio, high biocompatibility, adjustable synthesis, and remarkable stability. As a result, researchers are greatly interested in the synthesis and development of nanoparticles.

In order to achieve the objectives of green synthesis and sustainable development, it is crucial to explore alternative approaches. Physical and chemical synthesis methods may not be the most optimal solutions due to their drawbacks, such as high energy consumption, environmental pollution, and potential risks to human health [1]. The twelve principles of green chemistry serve as guidelines for researchers to adopt innovative approaches in the development of environmentally friendly synthesis methods. [2]. The primary objective of green synthesis methods is to offer greater environmental sustainability, improved safety for human health, and enhanced performance advantages compared to physical or chemical methods [3]. Green synthesis is widely favoured due to its utilization of locally available, cost-effective, and easily accessible raw materials. This approach significantly reduces the production costs of nanoparticles by saving energy and employing simple routes that do not require complex equipment or machinery [4]. The biosynthesis of nanoparticles involves three primary pathways, namely microorganisms, biomolecules, and plants. In the biogenic synthesis process, naturally occurring compounds present in plant and microbial extracts serve as both reducing and stabilizing agents. These functional components facilitate the transformation of metal sources into nanoparticles. [5]. While green synthesis methods utilizing microorganisms and biomolecules have achieved some level of success, they still face numerous limitations and challenges in nanoparticle production. These methods often require a series of technically complex and stringent conditions, posing safety concerns. Additionally, the overall process tends to be slow, making it unsuitable for large-scale nanoparticle production [6].

The utilization of plant extracts as a method for biosynthesizing nanoparticles offers several advantages, including the ability to produce nanoparticles in large quantities, within a short period of time, with high efficiency, and at a low production cost [7]. Plants are abundant in ecosystems and readily available for collection. They contain a diverse array of phytochemicals that can function as substitutes for sodium citrate, sodium borohydride ($NaBH_4$), and ascorbate, which are chemical reducing agents known for their high toxicity, expense, and detrimental environmental impact [3]. Numerous studies have provided evidence that phytochemicals found in plant extracts, including polysaccharides, flavonoids, phenolic acids, and quercetins, possess exceptional capabilities in reducing metal ions such as Ag^+, Cu^{2+}, and Au^{3+} [8,9,10]. Additionally, these phytochemicals exhibit various functions, such as capping, stabilizing, and chelating, during the nanoparticle formation process. Furthermore, these biocompounds can be easily extracted from different parts of plants, such as leaves, flowers, stems, roots, and other plant components, highlighting the advantageous nature of plants in the biosynthesis of nanoparticles [11].

Plant-synthesized nanoparticles not only exhibit remarkable efficacy in green synthesis but also play a significant role in various domains. In the context of water treatment, these nanoparticles possess the ability to facilitate reactions that break down harmful contaminants found in aquatic environments [12,13,14]. Due to their diminutive size, exceptional biocompatibility, substantial surface area, remarkable stability, remarkable versatility, and a plethora of remarkable attributes, environmentally-friendly nanoparticles are well-suited for diverse medical applications. These applications encompass antibacterial and antifungal treatments, combating cancer, and addressing a wide range of diseases [15, 16,17,18]. Nanoparticles have revealed their immense potential in biomedical diagnostics and drug delivery, unveiling a new realm of possibilities for therapeutic applications [19,20]. The utilization of biosynthesized nanoparticles has made a substantial impact on the progress of biomedical technology and environmental remediation.

2.   Synthesis of nanoparticles

2.1   General strategy

The distinctive characteristics of nanoparticles have sparked extensive research, leading to a diverse array of practical applications across various fields, including chemical sensing, heterogeneous catalysis, environmental remediation, nanotechnology, biomedical engineering, and agriculture. The need for customizing and creating nanoparticles through the most straightforward approach is growing. Currently, nanoparticles are produced using two distinct methods: the "top-down" and "bottom-up" approaches. The former involves reducing the size of bulk materials to produce new nano-sized particles using common techniques like sputtering and chemical etching [21]. The latter procedure involves utilizing atoms and molecules to assemble or join small particles together, resulting in the formation of nano-sized materials. This process is achieved through common methods like the sol-gel process, green synthesis, and other similar techniques [22]. The production of nanoparticles can primarily be accomplished through chemical, physical, and biological methods. In the following sections, a comprehensive analysis will be presented, detailing the merits and drawbacks of each approach.

2.2   Physical methods

There are various physical techniques commonly employed for synthesizing nanoparticles, including high-energy ball milling, electrospraying, laser ablation, physical vapor deposition, melt mixing, inert gas condensation, laser pyrolysis, and flash spray pyrolysis [23,24]. These methods primarily fall under the top-down approach, utilizing mechanical or electrical energy to reduce materials into small-sized nanoparticles [23]. Notably, evaporation and condensation conducted in a tube furnace are among the prominent methods for producing metallic nanoparticles at atmospheric pressure [25]. While physical methods are environmentally friendly as they don't involve toxic chemicals, the grinding of materials in these processes consumes a significant amount of energy and necessitates sophisticated equipment, resulting in high production costs and challenges in achieving systematic scalability [26].

2.3   Chemical methods

Chemical approaches are extensively employed for nanoparticle synthesis, encompassing methods such as sol-gel, hydrothermal, microemulsion, chemical reduction, and precipitation [27]. Among these, sol-gel is widely favored for its adjustable implementation and high-yield production. This technique involves the use of metal precursors and chemical reducing agents for synthesis [28]. However, chemical methods not only raise production costs but also pose environmental and health risks through the generation of hazardous waste [29]. To address this drawback, the green synthesis of nanoparticles through biological methods has garnered significant attention in recent decades.

2.4   Biological methods

Biological methods employ various bioreducing agents, such as bacteria, fungi, algae, protozoa, biomolecules (proteins, enzymes, nucleic acids, carbohydrates), and plant extracts, for nanoparticle synthesis. These biological substrates are utilized as alternatives to toxic chemical reductants and stabilizers [28]. Compared to chemical methods, biological approaches offer several advantages as they are more environmentally friendly and highly biocompatible. Microorganisms, for instance, can secrete enzymes that aid in reducing and stabilizing metal nanoparticles. However, microbial synthesis of nanoparticles presents challenges related to culturing and maintaining microbial growth [7]. Biomolecules, particularly enzymes, are capable of binding with metal ions and reducing them to form nanoparticles. Nonetheless, the main limitation of biomolecule-mediated synthesis lies in the nanoparticles' poor stability and extended synthesis duration [30]. On the other hand, plant extract-mediated synthesis offers significant benefits as various phytochemicals found in plant extracts (such as polyphenols, alkaloids, flavonoids, and alcohols) can act as biocapping and bioreducing agents, enhancing nanoparticle stability [31,32]. For instance, Rafique et al. demonstrated that using plant extracts for nanoparticle synthesis brings higher efficiency, ease of handling, safety, and faster results compared to other biological methods [33]. Phytochemicals derived from various plant components such as leaves, flowers, roots, seeds, stems, bark, peel, and latex, are renowned for their exceptional antioxidant properties. These phytochemicals are effectively employed in the reduction process to synthesize diverse nanoparticles. Singh et al., for example, showed that phytochemicals as antioxidants facilitate the formation of MgO nanoparticles [34,35,36]. In other words, these phytochemicals possess the capability to transform metallic precursors into corresponding nanoparticles. Furthermore, a study by Kesharwani et al. showcased the involvement of amino acids, alkaloids, polysaccharides, and reducing sugar compounds in the reduction of Ag+ to Ag0 [37].

In addition to their role as reducing agents, the chemical components present in plant extracts also contribute to the prolonged lifespan of nanoparticles by providing a protective coverage, thereby enhancing their stability. For instance, Dubey et al. utilized flavonoid and terpenoid compounds found in *E. hybrida* extract to ensure the long-term stability of silver nanoparticles [38]. Mittal et al. identified quinol and chlorophyll pigments in plant extracts as responsible for stabilizing such nanoparticles [39]. Interestingly, these chemical compositions also prevent the agglomeration of nanoparticles during synthesis, resulting in improved dispersion and increased active sites [40]. Furthermore, the concentration of the extract, along with other fabrication conditions such as temperature, pH, and time, directly influences the size of the nanoparticles. Elemike et al. explored the potential of *L. africanum* extracts in synthesizing Ag nanoparticles [41,42]. Their findings revealed that Ag nanoparticles achieved smaller sizes (8–35 nm) within the optimal pH range of 6.8 to 7 and at a temperature of 65 °C. The fastest nanoparticle formation rate was observed at a concentration ratio of leaf extract to ionic salt solution of 1:10. Based on the aforementioned points, plant extracts serve as excellent reducing agents for synthesizing nanoparticles with high stability, reduced clustering, and good dispersibility.

2.5   Green conception of nanoparticles

Despite the existing literature on the green synthesis of nanoparticles using plant extracts and their antimicrobial properties, the comprehensive understanding of their formation and antimicrobial mechanisms remains limited. Additionally, only a few previous studies have delved deeply into the biomedical applications of green nanoparticles, such as drug delivery, medical diagnostics, and anti-aging effects. These unexplored potentials are intriguing and deserve further exploration to expand the scope of green nanoparticles. Therefore, the objective of this study is to provide an overview of the green methods for nanoparticle synthesis, with a specific focus on the botanical approach due to its notable advantages, including safety, cost-effectiveness, efficiency, and environmental friendliness. We aim to systematically elucidate the synthesis of botanically derived nanoparticles and highlight their exceptional applications in antibacterial, antifungal, anti-cancer, and biomedical fields. Moreover, we aim to provide a detailed understanding of the specific mechanisms involved in each application and engage in in-depth discussions. Finally, we identify the knowledge gaps, limitations, challenges, and prospects in order to guide future research endeavours.

3.   Phytomolecules

3.1   Chemical composition of plants

As previously mentioned, the utilization of plant extracts in nanoparticle synthesis offers numerous benefits and effectiveness, which can be attributed to the presence of intrinsic phytochemicals such as polyphenols, flavonoids, sugars, terpenes, and more, derived from various parts of plants [43]. Table 1 provides an overview of the different types of phytochemicals found in different plant tissues. Flavonoids, phenolics, quercetin, and terpenoids are commonly found in most plants and play a vital role in the reduction, capping, and stabilization of nanoparticles during synthesis. For instance, Pansambal et al. demonstrated that polyphenols, saponins, flavonoids, coumarins, volatile oils, tannins, and sterols extracted from *A. hispidum* act as capping and chelating agents in the synthesis of CuO nanoparticles [44]. Sundararajan and Ranjitha Kumari highlighted the significant role of polyphenols and flavonoids present in *A. vulgaris* leaves as bio-capping agents in the synthesis of Au nanoparticles [45]. Khan et al. conducted a separate investigation and found that kaempferol, quercetin, caffeic acid, and dihydrokaempferol, present in *P. undulata* extract, displayed effective metal-reducing properties in the production of Au and Ag nanoparticles [46]. Chandraker et al. reported that *A. conyzoides* leaves contain alkaloids, flavonoids, terpenoids, saponins, and tannins, all of which contribute to the reduction, capping, and stabilization of Ag nanoparticles [47]. Tannin and tannic acid, in particular, stand out as noteworthy phytochemicals for efficient reduction processes. These studies underscore the significance of phytochemicals present in plant extracts for nanoparticle synthesis.

3.2 Role of phytochemicals for nanoparticles biosynthesis

The presence of phytochemicals in plant extracts confers advantages to the adsorptive, catalytic, biomedical, and biocompatible properties of nanoparticles. In terms of adsorptive activity, phytochemicals contribute to the enrichment of nanoparticle surfaces by introducing new functional groups [48]. This process enhances surface functionalization and fosters various physicochemical interactions, such as electrostatic, π–π, and hydrogen bonding. A recent study by Hammad and Asaad compared the sorption affinity of methylene blue dye between FeO nanoparticles synthesized biologically and chemically [49]. The iron nanoparticles derived from *C. vulgaris* extract exhibited a higher surface area (85.7 m2/g) compared to those synthesized by the chemical precipitation method (71.6 m2/g). The authors attributed the reduction in particle size (4.47 nm) of the biologically synthesized FeO nanoparticles, as opposed to the chemically synthesized ones (9.07 nm), to the presence of phytochemicals from C. vulgaris. Moreover, the maximum dye adsorption capacity of the biologically synthesized FeO nanoparticles (29.14 mg/g) was significantly greater than that of the chemically synthesized nanoparticles (19.08 mg/g). These findings validate the crucial role of phytochemicals in enhancing surface functionalization and improving adsorption performance.

In terms of catalytic activity, the presence of phytochemicals plays a beneficial role in reducing the band gap energy of nanomaterials [50]. This reduction leads to a decrease in the activation energy required for electron splitting, promoting efficient electron separation and facilitating the generation of reactive oxygen species (•OH, •O2–, H2O2) [51]. Additionally, the phytochemicals present on the surface of nanoparticles impede electron recombination processes [52]. For instance, ZnO nanoparticles derived from S. nigrum exhibited a high efficiency of 98.89% in the photocatalytic degradation of methylene blue, whereas chemically synthesized ZnO nanoparticles achieved only 81.94% efficiency [53]. In another study, Abdullah et al. demonstrated the excellent photocatalytic ability of ZnO nanoparticles biofabricated from *M. acuminata* peel, achieving complete degradation of basic blue (100%), whereas chemically synthesized ZnO nanoparticles achieved only 87.71% degradation [48].

In relation to biomedical activity, phytochemicals such as phenolic and flavonoid compounds act as a protective layer surrounding nanoparticles, enhancing their durability when in contact with bacterial surfaces [50]. These phytochemicals also contribute to the diversification of the nanoparticles' surface chemistry, promoting their binding to bacterial cells through electrostatic interactions [54]. Furthermore, they exhibit superior capability compared to chemically synthesized nanoparticles in supporting the generation of reactive oxygen species [55]. Additionally, Muthuvel et al. highlighted the high bactericidal abilities of certain phytochemicals present in plant extracts, including terpenoids, tannins, flavonoids, alkaloids, carbohydrates, and saponins [50 For instance, CuO nanoparticles derived from *S. nigrum* leaves exhibited inhibition zones of 11 mm and 12 mm against B. subtilis and E. coli, respectively, whereas chemical CuO nanoparticles had significantly lower inhibition zones of 2 mm and 3 mm [54]. Similarly, ZnO nanoparticles synthesized from *C. halicacabum* demonstrated inhibition zones of approximately 20 mm and 19 mm for S. aureus and P. aeruginosa, respectively, while chemically synthesized ZnO nanoparticles showed smaller inhibition zones of 13 mm and 12 mm, respectively [54]. This difference in antibacterial activity could be attributed to the presence of alkaloids, which act as capping agents facilitating the binding of nanoparticles to bacterial surfaces. Another study revealed that Ni nanoparticles obtained from *D. gangeticum* exhibited an inhibition zone of 5.67 mm against *S. aureus*, whereas chemically synthesized ZnO nanoparticles showed a smaller inhibition zone of 5.14 mm [56]. Regarding anticancer activity, Virmani et al. demonstrated the involvement of phytochemicals in the bioactivity of Au nanoparticles. Their study showed 50% cell viability for green Au nanoparticles and 80% for chemical Au nanoparticles when tested on HeLa cells [55]. These findings open up new possibilities for the application of nanoparticles derived from plant extracts in various biomedical fields.

In terms of biocompatibility, Amooaghaie et al. demonstrated the use of non-toxic phytochemical coatings around nanoparticles to reduce their toxicity and enhance their biocompatibility [57]. Their study focused on Ag nanoparticles produced from *N. sativa*, which showed an 11-fold lower rate of apoptosis in bone stem cells of mice compared to chemical Ag nanoparticles. This result indicates reduced cytotoxicity and improved biocompatibility of the green nanoparticles. Similarly, Dowlath et al. investigated the biocompatibility of FeO nanoparticles synthesized from *C. halicacabum* in comparison to chemically synthesized FeO nanoparticles, using peripheral blood mononuclear cells. The study revealed that the cell viability of green FeO nanoparticles was significantly higher at 84.04% compared to the chemical FeO nanoparticles [58]. These promising findings regarding biocompatibility make nanoparticles suitable for various biomedical applications, including drug delivery, disease diagnosis, and treatment.

Green nanoparticles are recognized for their stability, which can be attributed to the presence of phytochemicals in their coating. This characteristic sets them apart from nanoparticles that are synthesized chemically. Assessing the stability of these nanoparticles involves measuring their zeta potential. If the zeta potential displays a higher negative or positive value, the nanoparticles tend to repel each other more strongly, leading to reduced clustering and increased stability. Typically, nanoparticles with a zeta potential below -30 or above +30 demonstrate favorable stability [59]. Research has shown that the zeta potential of Pt nanoparticles derived from P. farcta was more negatively charged (-34.6 mV) compared to chemically produced nanoparticles (-15.6 mV). Similarly, Nithya and Kalyanasundharam discovered zeta potentials of -32.06 and -17.89 mV for ZnO nanoparticles obtained from *C. halicacabum* and the chemical method, respectively [54]. These findings indicate that green nanoparticles exhibit higher stability than their chemically produced counterparts. Mousavi-Khattat et al. examined zeta potential histograms of green and chemical Ag nanoparticles and observed significant stability in the green Ag nanoparticles even after 2 months [60]. Therefore, the presence of phytochemicals enhances the stability and prolongs the shelf life of nanoparticles, making them suitable for various applications. In summary, phytochemicals offer numerous benefits to nanoparticles in terms of adsorption, catalysis, biomedical applications, stability, and biocompatibility.

The objective of green chemistry is to develop safer chemical products and procedures that minimize or eliminate the use of hazardous components during manufacturing and usage. This approach incorporates several key principles of green chemistry, including reducing environmental pollution, utilizing renewable feedstocks, employing non-toxic or safer solvents and auxiliaries, minimizing the use of derivatives, and preventing or reducing waste generation. Green synthesis involves the synthesis of nanomaterials using plant extracts, microorganisms, and other biological materials. This process, known as biological synthesis, is both environmentally friendly and cost-effective. Microorganisms such as yeast, viruses, fungi, and bacteria serve as nanofactories for the production of nanomaterials by reducing metallic salts in the presence of reductase enzymes. Furthermore, scientists are increasingly exploring the field of phytonanotechnology, which involves synthesizing nanomaterials using various parts of plants such as fruits, seeds, roots, stems, and leaves. However, there are certain challenges associated with the biological green synthesis of nanomaterials, including difficulties in recovering metallic nanoparticles, storing microorganisms, and performing isolation, culturing, and sampling processes. These complexities pose obstacles to the green synthesis of nanomaterials. The following subsections will delve into different methods of green synthesis for nanomaterials.

4. Mechanisms for fabricating nanoparticles using plant extracts

4.1 Ag, Au, and Pt nanoparticles

The overall procedure for producing Ag, Au, and Pt nanoparticles using plant extracts entails bioreduction facilitated by biomolecules found within the plants. Plant extracts commonly consist of functional groups such as carbonyl, hydroxyl, and amine, which can interact with metal ions, leading to the reduction of these ions into nanoparticles with diverse shapes and sizes. For example, specific compounds like flavonoids, proteins, sugars, terpenoids, and other bioactive substances have been identified as potential agents involved in the reduction of Ag+ to Ag0 [61]. This was confirmed by observing significant changes in the levels of sugars and flavonoids before and after the reduction reaction. Ghotekar et al. analyzed the components of *L. leucocephala* leaf extract and identified tannins, saponins, carbohydrates, coumarins, steroids, flavonoids, phenols, and amino acids [62]. The presence of polyphenols with antioxidant properties in the *L. leucocephala* extract facilitated the reduction of Ag+ ions, leading to the formation of Ag nanoparticles. Furthermore, the authors found that various substances like gallic acid, β–sitosterol, mimosine, caffeic acid, and chrysoenol played a crucial role in stabilizing the Ag nanoparticles. Similarly, Zheng et al. discovered that the reduction of Pt(II) to Pt(0) nanoparticles depended on factors such as temperature, reducing sugars, and flavonoid content [63]. Another study indicated that plant extracts potentially have a mechanism for the reduction and stabilization of chloroaurate ions (AuCl4–) to Au nanoparticles [64].

In a study conducted by Murthy et al., the synthesis of Cu nanoparticles using *H. abyssinica* plant extracts was documented. The presence of phenolic compounds, anthraquinone glycosides, and tannins, known for their high antioxidant properties, facilitated the binding of Cu2+ ions in the precursor salt solution Cu(NO3)2, leading to their reduction to Cu0 form [65]. These biomolecules, after the reduction process, played a crucial role in enveloping and stabilizing the Cu nanoparticles. Similarly, in another study by Naghdi et al., the involvement of quercetin, a polyphenol present in C. reflexa leaf extract, was hypothesized in the concise mechanism for the formation of Cu nanoparticles [66]. It was proposed that Cu2+ would oxidize quercetin, resulting in the formation of a Cu(I)-quinone complex as an intermediate. The Cu(I) in this complex would then further oxidize quercetin, leading to the formation of Cu nanoparticles and quinone compounds. The presence of phytochemicals in the extract played a significant role in the stabilization of Cu nanoparticles. Consequently, the ratio between Cu2+ concentration and plant extract concentration was found to influence the formation of Cu nanoparticles. Additionally, Nagar and Devra suggested that the concentration of the precursor salt had a notable impact on the shape and particle size of Cu nanoparticles synthesized from *A. indica* leaf extract [67]. Specifically, an increase in the concentration of CuCl2 from $6 \times 10{-}3$ mol/L to $7.5 \times 10{-}3$ mol/L resulted in an increase in particle size from 48.01 to 78.51 nm, respectively. The authors explained that a higher concentration of CuCl2 provided a larger number of nuclei, leading to a higher degree of agglomeration of Cu nanoparticles.

4.2   ZnO nanoparticles

The presence of phytochemicals in plant extracts plays a significant role in the formation of green oxide metal nanoparticles, including ZnO nanoparticles. In a study conducted by Król et al., the fusion of ZnO nanoparticles was investigated using biomolecules derived from *M. sativa* extract [68]. The coordination chemistry of zinc enables the zinc aqua complex to undergo exchange with water molecules when binding to protein ligands. Additionally, flavonoids such as quercetin, rutin, and galangin possess specific metal ion binding sites and can chelate with Zn2+ ions (Fig. 5). The formation of the Zn-flavonoid complex occurs, which is subsequently calcined at high temperatures to yield ZnO nanoparticles. In another study by Kumar et al., it was demonstrated that phytochemicals such as flavonoids, limonoids, and carotenoids, which contain free -OH and -COOH groups, can react with ZnSO4 to form the Zn-(flavonoid, limonoid, carotenoid) complex [69]. This complex is then subjected to a furnace at a temperature of 150 °C, leading to the production of ZnO nanoparticles.

Various operational factors significantly influence the morphology and particle size of nanoparticles. For instance, the synthesis temperature plays a crucial role, as demonstrated by Hassan Basri et al. in their study on ZnO nanoparticles synthesized from pineapple peel extract [70]. At a synthesis temperature of 60 °C, the resulting ZnO nanoparticles exhibited a combination of spherical and rod-shaped structures. In contrast, a synthesis temperature of 28 °C resulted in ZnO nanoparticles with a spherical flower-shaped structure. Extraction concentration is another influential factor. In the case of green ZnO particles, an increase in extract concentration led to a decrease in particle size. Soto-Robles et al. showed that using an 8% concentration of *H. sabdariffa* extract resulted in ZnO nanoparticles with a particle size ranging from 5 to 12 nm, whereas a 1% concentration condition yielded particles with a size range of 20 to 40 nm [71]. To create green nanoparticles with the necessary qualities, careful control of operating parameters including temperature and extraction concentration is essential.

4.3   MgO nanoparticles

The biocapping and chelating properties of the biomolecules found in plant extracts are also essential to the production mechanism of MgO nanoparticles. In more detail, the precursor's Mg2+

ions have the capacity to engage in chelation with biomolecules, leading to the creation of metal complexes [72]. MgO nanoparticles with excellent stability are formed with the involvement of isoleucine acid, a phytochemical found in *L. acidissima* fruit extract [73]. In this particular study, the formation of the isoleucine-MgO complex occurred through the binding of magnesium nitrate to isoleucine acid. Subsequently, the complex underwent high-temperature calcination (500-800 °C), resulting in the formation of MgO nanoparticles characterized by good dispersion and high stability.

In the case of MgO nanoparticle synthesis from insulin plant extracts, Suresh et al. proposed a possible mechanism [74]. According to their findings, the following steps took place: (i) diosgenin present in the extract reacted with a magnesium nitrate salt solution, forming complexes held together by weak hydrogen bonds; (ii) the resulting complex was subjected to a temperature of approximately 80 °C, leading to the precipitation of hydroxide; (iii) the final product, MgO nanoparticles, was obtained through calcination at 450 °C.

The pH of the extract solution plays a crucial role in determining the structure of MgO nanoparticles. Recent research by Jeevanandam et al. [75] revealed a significant correlation between pH and the zeta potential of MgO nanoparticles. Specifically, higher pH values resulted in more negative zeta potential values, indicating increased stabilization of the green MgO nanoparticles. Conversely, unfavorable pH conditions could promote agglomeration of the MgO nanoparticles, leading to larger particle sizes. The study provided noteworthy findings regarding the characteristic properties of MgO nanoparticles at different pH levels. At pH 3, hexagonal MgO nanoparticles exhibited a size smaller than 44 nm, while at pH 5 and pH 8, their sizes increased to 50.75 nm and 58.77 nm, respectively. Therefore, pH adjustment serves as a crucial step in controlling the structure of green MgO nanoparticles.

4.4   CuO nanoparticles

CuO nanoparticles can be synthesized by incorporating a high-temperature calcination step under an oxygen atmosphere, which facilitates the transformation of zero-valent Cu into CuO nanoparticles. For instance, Nagore et al. demonstrated the potential of natural biomolecules found in P. longifolia leaf extract to chelate with Cu^{2+} ions at 85°C, leading to the bonding of chemical constituents with Cu^{2+} ions [76]. The authors identified various compounds, including saponins, flavonoids, tannins, polyphenols, steroids, and alkaloids, involved in the chelation process. Following the chelation, the resulting complex can undergo calcination at 400°C in the presence of airflow, resulting in the formation of CuO crystals. This study emphasizes the significant role of phytochemicals in the synthesis of CuO nanoparticles.

5.   Characterization

5.1   UV–Visible Spectroscopic Analysis

UV-visible spectrophotometry is commonly used to characterize the formation of metallic nanoparticles from their respective metallic salts. Noble metals like Ag and Au exhibit strong absorption in the visible region, with maximum absorption at 400-450 nm and 500-550 nm, respectively, due to surface plasmon resonance (SPR) [78]. The resonant collective oscillations of the conduction electrons along the transverse direction of the electromagnetic field give rise to SPR in the UV-Vis region [79]. The intensity and width of the SPR band are affected by factors such as particle shape, dielectric constant of the medium, and temperature [80]. Therefore, UV-vis spectrophotometry is typically the first technique employed to verify the formation of metallic nanoparticles, and several studies have used this method to confirm their formation [78].

When noble metal salts are mixed with plant biomass/extract, a color change in the suspension occurs due to the surface plasmon vibrations in the metal nanoparticles. UV-vis spectrophotometry is a common method used to determine the adsorption peak and confirm the presence of metallic nanoparticles. The characteristic peak increases with time and concentration of the plant biomass/extract and confirms nanoparticle formation. The colour change from brown to ruby red indicates the formation of Au nanoparticles, as seen in the work of Narayanan and Sakthivel [81], while the change from pale green to deep brown confirms the formation of Ag nanoparticles, as seen in the work of Gurunathan et al. The color of the solution turns from brownish yellow to dark brown in the presence of Pd nanoparticles, as observed by Yang et al. The color change of Pt reduced solutions from yellow to black confirms the presence of Pt nanostructures [82].

5.2   Fourier Transform Infrared Spectroscopic Analysis (FTIR)

FTIR spectroscopy is a technique used for surface chemical analysis which can be used to identify functional groups involved in bioreduction. By comparing the FTIR spectra of plant biomass/extract and synthesized nanoparticles, information about the functional groups responsible for bioreduction can be gathered. The biomolecules involved in reducing and capping agents are polysaccharides, proteins, terpenoids, and flavonoids. Significant absorption bands have been observed in the 1000-1800 cm-1 range, confirming the role of these biomolecules in bioreduction. The stretching vibrations of N-H and O-H groups appear in the 3200-3500 cm-1 range.

An FTIR pattern for gold nanoparticles produced using plant extracts was reported with bands at 617 cm−1, 1125 cm−1, 1376 cm−1, 1658 cm−1, and 3278 cm−1 [7,8,9,10,11,12]. Bands at 3402 cm−1, 1606 cm−1, and 1518 cm−1, which indicate the presence of phenols, were assigned to the aromatic hydroxyl and benzene rings. Additionally, bands at 2931 cm−1 and 1402 cm−1 were associated with methylene stretching and methyl deformation vibrations, respectively. The sugar content was identified by bands at 1260 cm−1, 1113 cm−1, and 1076 cm−1, which corresponded to an epoxy bond, semi-acetal, and primary alcohol [83].

5.3   Transmission Electron Microscope (TEM)

To investigate the surface morphology and shape of nanoparticles, transmission electron microscopy (TEM) is commonly used by researchers due to its high magnification and resolution compared to SEM. TEM images provide accurate information about the size, shape, and crystallography of the nanoparticles. For example, Philip [20] observed Ag and Au nanoparticles with various shapes such as triangular, hexagonal, dodecahedral, and spherical, with an average size of 13-14 nm using Hibiscus rosa-sinensis leaf extract. Similarly, Dhayananthaprabhu et al. used Cassia auriculata flower extract for the biosynthesis of Au nanoparticles and observed

Nanoparticles in Healthcare: Applications in Therapy, Diagnosis and Drug Delivery   Materials Research Forum LLC
Materials Research Foundations 160 (2024) 52-82   https://doi.org/10.21741/9781644902974-3

spherical, hexagonal, and triangular-shaped nanoparticles ranging in size from 10-55 nm using TEM images. In addition, TEM analysis can distinguish amorphous structures from crystalline structures using the selected area electron diffraction technique, which is not possible using SEM images [84].

The analysis of nanoparticle morphology is often performed using three types of microscopic techniques: SEM, TEM, and AFM. These methods have been widely used for this purpose, as previously discussed. Compared to SEM, TEM provides higher magnification and resolution. Additionally, TEM uses selected area electron diffraction (SAED) to differentiate between crystalline and amorphous structures. [85,86]. Atomic force microscopy (AFM) has also been utilized by some researchers to study the morphology of nanoparticles. For example, Das et al. used AFM to examine the surface morphology of Ag nanoparticles synthesized from Sesbania grandiflora. The AFM images revealed that the Ag nanoparticles were spherical in shape and had a size ranging from 10 to 25 nm. [87].

5.4   X-ray Diffraction (XRD)

The X-ray diffraction (XRD) method is employed to investigate the structural characteristics of crystalline metallic nanoparticles. As X-rays carry a considerable amount of energy, they are capable of penetrating deep into the materials, providing information on the bulk structure [88]. The Analytical Expert MRD instrument is commonly used to analyze synthetic Au nanoparticles. The nanoparticles are loaded onto the XRD sample holder, and Debye-Scherer's equation is used to calculate the average crystallite size. XRD patterns are used to determine the intensity, position, width, and FWHM of the diffraction peaks, and the XRD data shows that the nanoparticles are crystalline and face-centered cubic (fcc). The Debye-Scherrer equation is used to compute crystallite sizes, and the high-energy X-rays can reveal important details about the bulk structure. The use of XRD patterns/peaks in gold nanoparticle production has been reported [89,90,91,92,93,94,95].

6.   Factors affecting synthesis of nanoparticles

Various factors can contribute to the size and stability of nanoparticles which can greatly influence the efficiency of nanoparticles. The synthesis, size, and shape of nanoparticles can be influenced by various factors including pH, temperature, plant extract concentration, metal solution concentration, and incubation/reaction time.

6.1   pH

The size of nanoparticles is influenced by the pH of the medium. According to various reports, pH plays a crucial role in the bio-formation of metallic nanoparticles [96]. Additionally, the study suggests that the size of gold nanoparticles can be manipulated by adjusting the pH of the medium [97]. Furthermore, it has been observed that plant extracts from different sources and different parts of the same plant may exhibit varying pH values, leading to nanoparticles with different sizes when synthesized using these extracts [98]. Notably, it has been documented that lower pH levels (ranging from 2 to 4) result in the formation of larger nanoparticles compared to higher pH levels [99].

6.2   Temperature

Temperature plays a significant role in nanoparticle biosynthesis, adding to its intriguing aspects. The temperature of the reaction medium serves as a crucial factor that governs the characteristics of the resulting nanoparticles [99]. In the case of utilizing plant extract for metallic nanoparticle synthesis, it was found that elevated temperatures led to a notable reduction in the percentage of gold nano-triangles compared to spherical particles, while lower temperatures predominantly promoted the formation of nano-triangles [100]. Moreover, it was discovered that higher temperatures facilitated a higher rate of gold nanoparticle formation. At elevated temperatures, nano-rod and platelet-shaped gold nanoparticles were synthesized, whereas lower temperatures favored the formation of spherical nanoparticles [101]. Some researchers have also documented an accelerated synthesis rate of silver nanoparticles at higher temperatures [102]. Therefore, these studies collectively suggest that temperature plays a critical role in controlling the size and shape of nanoparticles.

6.3   Incubation time

The incubation time refers to the duration necessary for the completion of all reaction steps, and it has been found to have an impact on nanoparticle synthesis. It has been observed that the plant-mediated synthesis of silver and gold nanoparticles initiates within 5 minutes of the reaction, but an increase in contact time contributes to the sharpening of peaks in both silver and gold nanoparticles [99]. When using chenopodium leaf extract, an augmentation in the sharpness of UV absorption spectra peaks was reported with an increase in contact time [101]. Furthermore, a slight variation was noted between nanoparticles synthesized within 15 minutes of the reaction, and this variation increased up to 2 hours. Optimal incubation duration is crucial for the stability of silver and gold nanoparticles as it allows for complete nucleation and subsequent nanoparticle stability [103].

7.   Applications of nanoparticles

Nanotechnology is a scientific discipline that focuses on the utilization of matter and devices with dimensions smaller than 100 nm. It finds extensive applications in various fields such as applied physics, materials science, chemistry, biology, surface science, biomedical engineering, electrical engineering, and robotics. In the realm of drug delivery systems, nanotechnology has demonstrated its ability to enhance the biodistribution and bioavailability of drugs compared to their parent compounds, primarily due to their reduced size. Additionally, it can improve the solubility of hydrophobic compounds, reduce the frequency of drug administration, facilitate targeted drug delivery, and mitigate potential toxicity [104]. Within the domain of nanotechnology-based drug delivery, different types of nanocarrier systems are employed, including liposomes, micelles, dendrimers, nanovesicles, nanogels, and nanoemulsions, among others [105, 106].

The ability to manipulate the size of materials through nanotechnology has opened up a wide range of potential applications and sparked extensive research in the field. Nanotechnology, as defined by the National Nanotechnology Initiative in the USA, involves the manipulation of matter with dimensions ranging from 1 to 100 nanometers. It encompasses diverse scientific disciplines, including surface science, organic chemistry, molecular biology, semiconductor physics, and microfabrication. The field of nanoscience holds immense potential for revolutionary advancements in various research areas and applications. In recent years, nanotechnology has emerged as a significant player in fields such as medicine, healthcare, food, agriculture, and more. One prominent research focus within this field is the synthesis of nanoparticles with precise chemical compositions, sizes, shapes, and controlled variations. In the past decade, there has been a notable emphasis on the biosynthesis of noble metal nanoparticles, including silver, gold, platinum, palladium, and others, driven by the increasing demand for environmentally friendly material synthesis technologies.

7.1   Plant- based silver nanoparticles for cancer

Cancer cells have the ability to evade apoptosis, which is programmed cell death, leading to uncontrolled proliferation. To address this, the use of phytomolecules incorporating silver nanoparticles (AgNPs) is emerging as an effective approach for tackling cancer. Two signaling pathways, namely the intrinsic pathway and extrinsic pathway, are involved in the activation of programmed cell death or apoptosis. However, cancerous cells often lack this apoptotic response due to DNA damage or severe cellular stress. Greenly synthesized AgNPs using a bioactive fraction of Pinus roxburghii have shown cytotoxic activity against lung and prostate cancer cells. The induction of apoptosis was observed through the intrinsic pathway, involving mitochondrial depolarization and DNA damage. Additionally, an increase in reactive oxygen species (ROS), cell cycle arrest, and caspase-3 activation were found to contribute to the apoptosis of cancer cells [107]. AgNPs synthesized using Phyllanthus emblica leaf extract also exhibited anticancer activity against hepatocellular carcinoma (HCC) [108]. AgNP-dipalmitoyl-phosphatidylcholine composites in the form of liposomes (Lipo-AgNP) were found to be cytotoxic by inducing ROS formation and DNA damage. Activation of the proapoptotic protein Bax and inhibition of the antiapoptotic protein Bcl-2 led to the release of cytochrome C, subsequently activating caspase and causing apoptosis in macrophages [109]. Biosynthesized AgNPs using phycocyanin showed antimicrobial and anticancer activity. Cytotoxic effects were observed against breast cancer cell lines and Ehrlich ascites carcinoma-bearing mice [110]. AgNPs of two sizes, 2 nm and 15 nm, induced endoplasmic reticulum stress via the unfolded protein response (UPR) and enhanced activation of caspase 9 and caspase 7, leading to cell death in MCF-7 and T-47D cells [111]. AgNPs have also been shown to exhibit strong cytotoxicity by arresting the cell cycle at the G2/M phase. In A549 lung epithelial cells, AgNPs downregulated protein kinase-C (PKCζ), resulting in cell cycle arrest. AgNPs also upregulated P-53 protein, Bax and Bid, caspase-3, generation of ROS, and downregulated antiapoptotic proteins Bcl-2 and Bcl-w [112]. AgNPs synthesized using Cynara scolymus (Artichoke) were found to modulate mitochondrial apoptosis by generating ROS, regulating apoptotic proteins, and causing cell death in MCF7 breast cancer cells when combined with photodynamic therapy [113]. Similarly, AgNPs synthesized using *Moringa oleifera* [114], *Tropaeolum majus* [115], *Punica granatum* [116], *Gloriosa superba* [117], and *Teucrium polium*

[118] plant extracts have demonstrated cytotoxicity against cancer cell lines. There is abundant evidence from numerous related investigations and research that supports the potential effectiveness of AgNPs as candidates for cancer therapy, offering the possibility of targeted cancer treatments.

7.2   Plant-based silver nanoparticles for viral-infection

Treating viral infections is a challenging task due to the rapid multiplication and spread of viruses. Humans are currently facing numerous emerging and life-threatening viruses, including coronavirus, Ebola virus, dengue virus, HIV, and influenza virus, which pose a significant threat to the human health. In response to this, there has been a surge in research focusing on the potential of silver nanoparticles (AgNPs) as effective antiviral agents. Various studies have shed light on the modes of antiviral action exhibited by AgNPs. These actions can occur intracellularly, where AgNPs block viral replication, or extracellularly, by interacting with specific viral proteins such as gp120, thereby blocking viral entry. It is important to note that the specific mode of action may vary depending on the type of virus being targeted. AgNPs are considered to the potent and novel pharmacological agent possessing effective antiviral activity against feline coronavirus (FCoV) [119], Influenza virus [120], HIV [121], Adenovirus [122], Herpes simplex virus [123], Dengue virus [124,125], Chikungunya virus [126], Norovirus [127], bovine Herpesvirus [128], Human parainfluenza virus type 3 [129].

8.   Toxicity evaluation of nanoparticles for biomedical applications

Lack of standardized protocols for nanomaterial toxicity assessment has made the testing a challenging aspect.

All biomaterial necessities to be evaluated for environment and scientific safety which must meet the prerequisite of the bodies like ISO, Good Manufacturing Practices (cGMP), Organisation for Economic Co-operation and Development (OECD) and International Council for Harmonisation (ICH), US EPA: United States Environmental Protection Agency, US- FDA, European Environment Agency (EEA) is the agency of the European Union (EU).

Foremost five guidelines for toxicity valuation of the biomaterial is itemized below

(a) Cyto- Compatibility: bio material must be cell compatible, i,e cellular inert, repress severe inflammatory response.

(b) Storage properties: biomaterial needs to be economical, follow good manufacturing practice (GMP) and also non-hazardous to the environment. Leaching and degradation should be minimised.

(c) Physical properties: biomaterial mechanically strong thus should possess greater tensile strength.

(d) Raw materials: materials used in fabrication of a biomaterial need to be assessed for toxicity before the fabrication.
Regular tests used in the biomaterial toxicity evaluation are listed below

a. ISO Standard 10993-5- The cytotoxicity and cytocompartibility toxicity direct or indirect contact assay is tested in an in vitro cell line culture system by MTT and LDH assays.

b. ISO 10993-10 - Sensitisation test like Adjuvant mediated guinea pig maximization test (GPMT) triggers the skin sensitization. Necrosis in epidermis, dermis ulceration, erythema, oedema are observations noticeable. Irritation like Patch assay elicits non-specific inflammation of the skin. Intracutaneous reactivity test invovles irritant chemicals in biomaterial which appear as erythema.

c. ISO 10993-11, OECD and WHO guidelines - Systemic circulation of leachants and their assimilation are observed. Tests for pyrogenicity are also a type of ISO 10993-11 test.

d. OECD 423, 407, 452, OECD and WHO guidelines - Acute, subacute and chronic toxicity tests serum biochemical and haematological parameters of the biocomposites material toxic nature is assessed where no other documented toxicology information is available.

e. Genotoxicity and Carcinogenicity test (ISO 10993-3) - Genotoxicity test determines whether gene mutations, changes in chromosome structure, or other DNA or gene changes are caused by the test biocomposites materials. This test is carried out in mammalian or non-mammalian cells, bacteria, yeasts, fungi or whole animal. Carcinogenicity determines carcinogenic potential of biocomposite materials or their by-products using multiple exposures to test animal for its major portion of its life span.

f. Developmental and Reproductive toxicity test (OECD 421)- Teratogencity involves the development of the foetus after the exposure to the biocomposite material and reproductive test describes the effects of male and female reproductive performance when exposed to the scaffold materials. Gross pathology of necropsied tissues and hispathological are evaluated.

g. Haemocompatibility test (ISO 10993-4) - Blood compatibility of biocomposite material or its by-products for the possible thrombosis, coagulation, platelet function is evaluated.

h. Immunotoxicology test (ISO 10993-20) - Nano composites and biocomposite material are tested for hypersensitivity, photosensitivity, induced autoimmunity, developmental immunotoxicity, immune suppression and stimulation on cultured cells [130,131]. Figure 1 gives a schematic flow chart suitable for nanomaterial testing.

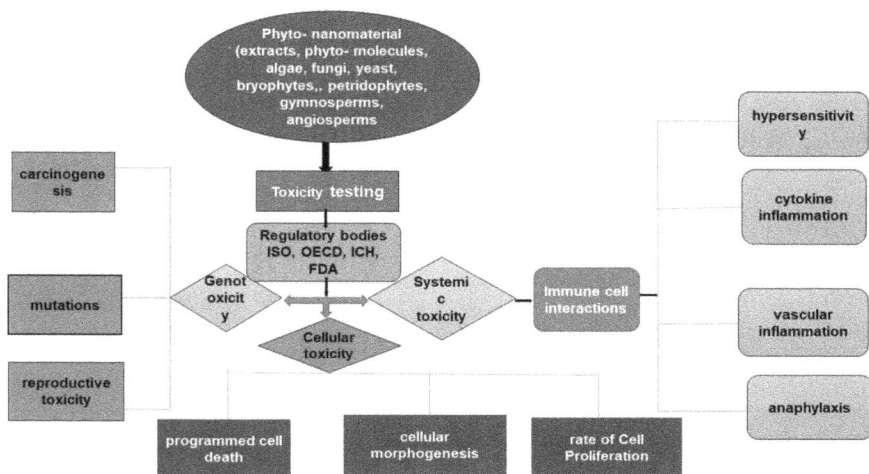

Figure 1: Schematic representation of the toxicological evaluation of nanomaterials and the regulatory tests before the usage.

Conclusion

Biosynthesis methods offer significant advantages in the field of nanomaterials, particularly in medical applications, due to the presence of biomolecule entities or a lipid layer that enhances stability and physiological solubility. This characteristic sets them apart from other synthesis techniques. However, the large-scale use of biogenic nanomaterials poses challenges that need to be addressed. These challenges include issues such as lack of uniformity, batch-to-batch variability, low production rates, and labor-intensive procedures. To tackle these challenges, this chapter explores environmentally friendly nanomaterials derived from plants. Unlike traditional chemical and synthetic approaches, plant-based nanofactories utilize various parts of plants such as fruits, flowers, peels, leaves, shoots, and roots. These plant components serve as capping and reducing agents during synthesis and contribute to the therapeutic effects of the nanomaterials. The accessibility and ease of use of plant extracts further enhance their advantages. Additionally, microorganisms such as actinomycetes, algae, yeasts, fungi, and bacteria can be utilized for the production of nanomaterials. Studies have shown that microbial synthesis techniques incorporate events such as trapping, bio-reduction, and capping during the synthesis process, both extracellularly and intracellularly. These biogenic nanomaterials have found wide-ranging applications in industries such as food, water treatment, textiles, and biomedicine. Therefore, this chapter not only discusses the use of various biogenic nanomaterials in different applications but also outlines a roadmap for advancing the field of green synthesis. It serves as a valuable resource for scientists looking to embark on research in the green synthesis of nanomaterials, building upon the knowledge and progress made by previous researchers.

Future perspective

The increasing demand for green chemistry and nanotechnology has spurred the development of environmentally friendly methods for synthesizing nanomaterials using natural resources such as plants and microbes. Researchers have been actively exploring green synthesis techniques for nanoparticles, which offer a cost-effective, non-toxic, and readily available approach with eco-friendly characteristics. Plant extracts possess unique compounds that facilitate the synthesis process and enhance the kinetics of nanoparticle formation. The utilization of plants in green nanoparticle synthesis represents an intriguing and emerging aspect of nanotechnology that has profound implications for environmental impact and the long-term sustainability and advancement of nanoscience. These plant-based nanoparticles hold significant potential for diverse applications in fields such as catalysis, medicine, cosmetics, agriculture, food packaging, water treatment, dye degradation, textile engineering, bioengineering sciences, sensors, imaging, biotechnology, electronics, optics, and other biological sectors. Particularly, they show promise as a driving force in the biomedical field, particularly for drug delivery systems. Furthermore, these green nanoparticles can find applications in phytopathogen treatment in agriculture or water disinfection for environmental remediation. The popularity of this green approach to nanoparticle synthesis is expected to grow exponentially in the future. However, it is essential to address the long-term effects on animals and humans, as well as the accumulation and impact of these nanoparticles on the environment, in order to ensure their safe and responsible use.

The field of green nanomaterial synthesis is continually evolving, offering promising and environmentally friendly solutions across multiple sectors. However, there are specific areas that necessitate further exploration and advancement to fully harness and maximize the advantages presented by biogenic nanoparticles: • The field of green nanomaterial synthesis is still in its infancy, and there is vast untapped potential in exploring natural materials, including biopolymers, microbes, and waste materials, for the synthesis of biogenic nanoparticles. Consequently, there is a pressing need to expand research efforts in green synthesis, considering its unique advantages that cannot be replicated by chemical or physical synthesis methods. • While progress has been made in manipulating the morphologies of nanomaterials, there is still a lag in effectively controlling their dimensions, especially when it comes to size. The benefits associated with the sizes of nanomaterials are confined within a certain range. Therefore, it is essential to enhance the control of nanomaterial morphologies in order to unlock new possibilities for application by leveraging their unique properties. • The limited understanding of the underlying mechanisms involved in biosynthesis processes, particularly microbial synthesis, poses challenges in harnessing the full potential of biological agents for nanomaterials synthesis. This lack of understanding leads to unpredictable and inconsistent synthesis methods, even for producing the same material. The synthesis process can become complex and time-consuming, requiring significant resources. Therefore, it is crucial to develop predictable synthesis methods based on precise and theoretically anticipated synthesis mechanisms. This requires the establishment of agreed-upon theorems and laws, which, in turn, necessitates further advancements in nuanced and accurate nano-analysis and detection methods to study the kinetics of the reactions. • To achieve successful synthesis, it is imperative to have a comprehensive understanding of the biological components, chemical agents, and molecular mechanisms involved. This includes recognizing and separating the relevant compounds and comprehending the structure and function of the capping and reducing agents. Such knowledge is essential for effectively controlling the synthesis process and obtaining desired nanomaterial properties. • The practical application of green synthesis methods is currently limited to laboratory settings, with challenges in scaling up the production of consistent and homogeneous nanomaterials. Batch-to-batch variations hinder large-scale production, necessitating the development of scalable, continuous flow-based synthesis techniques that are stable, efficient, and environmentally friendly. The key obstacles to overcome include controlling flow parameters such as velocity, viscosity, surface tension, and mixing time, as well as optimizing reactant placement, solvent selection, and preventing contamination. By addressing these challenges, it is possible to achieve high-throughput commercial fabrication of nanomaterials at reduced costs. Further investigations and techno-economic feasibility studies are crucial for advancing the production and utilization of nanomaterials on an industrial scale. • Studies that examine the optimisation of parameters impacting the growth of the microorganisms and the syntheses of nanomaterials need to be conducted in order to overcome longer incubation times and high maintenance of microbial synthesis. • The majority of commercially accessible consumer nanoproducts have not been designed with environmental preservation and safety as their top priorities. Additionally, the negative impacts of nanomaterials are not well understood. To reduce the dangers of toxicity, the use of nanomaterials calls for analysis and new understanding of their movement and mode of action. • It is strongly advised to carefully review the entire raw material life cycle, from manufacture to disposal. Risk management throughout production, processing, storage, and disposal also needs more attention.

Reference

[1] Mohamad NAN, Arham NA, Jai J, Hadi A (2013) Plant extract as reducing agent in synthesis of metallic nanoparticles: a review. Adv Mater Res 832:350–355. https://doi.org/10.4028/www.scientific.net/AMR.832.350

[2] Ivanković A (2017) Review of 12 Principles of green chemistry in practice. Int J Sustain Green Energy 6:39. https://doi.org/10.11648/j.ijrse.20170603.12

[3] Ahmed S, Ahmad M, Swami BL, Ikram S (2016) A review on plants extract mediated synthesis of silver nanoparticles for antimicrobial applications: a green expertise. J Adv Res 7:17–28. https://doi.org/10.1016/j.jare.2015.02.007

[4] Cuong HN, Pansambal S, Ghotekar S et al (2022) New frontiers in the plant extract mediated biosynthesis of copper oxide (CuO) nanoparticles and their potential applications:a review. Environ Res 203:111858. https://doi.org/10.1016/j.envres.2021.111858

[5] Dabhane H, Ghotekar S, Tambade P et al (2021) A review on environmentally benevolent synthesis of CdS nanoparticle and their applications. Environ Chem Ecotoxicol 3:209–219. https://doi.org/10.1016/j.enceco.2021.06.002

[6] Ghotekar S, Pansambal S, Bilal M et al (2021) Environmentally friendly synthesis of Cr2O3 nanoparticles: characterization, applications and future perspective—a review. Case Stud Chem Environ Eng 3:100089. https://doi.org/10.1016/j.cscee.2021.100089

[7] Srikar SK, Giri DD, Pal DB et al (2016) Green synthesis of silver nanoparticles: a review. Green Sustain Chem 06:34–56. https://doi.org/10.4236/gsc.2016.61004

[8] Agarwal H, Venkat Kumar S, Rajeshkumar S (2017) A review on green synthesis of zinc oxide nanoparticles—an eco-friendly approach. Resour Technol 3:406–413. https://doi.org/10.1016/j.reffit.2017.03.002

[9] Ong CB, Ng LY, Mohammad AW (2018) A review of ZnO nanoparticles as solar photocatalysts: Synthesis, mechanisms and applications. Renew Sustain Energy Rev 81:536–551. https://doi.org/10.1016/j.rser.2017.08.020

[10] Jadoun S, Arif R, Jangid NK, Meena RK (2021) Green synthesis of nanoparticles using plant extracts: a review. Environ Chem Lett 19:355–374. https://doi.org/10.1007/s10311-020-01074-x

[11] Beyene HD, Werkneh AA, Bezabh HK, Ambaye TG (2017) Synthesis paradigm and applications of silver nanoparticles (AgNPs), a review. Sustain Mater Technol 13:18–23. https://doi.org/10.1016/j.susmat.2017.08.001

[12] Veisi H, Rashtiani A, Barjasteh V (2016) Biosynthesis of palladium nanoparticles using Rosa canina fruit extract and their use as a heterogeneous and recyclable catalyst for Suzuki-Miyaura coupling reactions in water. Appl Organomet Chem 30:231–235. https://doi.org/10.1002/aoc.3421

[13] Rasheed T, Nabeel F, Bilal M, Iqbal HMN (2019) Biogenic synthesis and characterization of cobalt oxide nanoparticles for catalytic reduction of direct yellow-142 and methyl orange dyes. Biocatal Agric Biotechnol 19:101154. https://doi.org/10.1016/j.bcab.2019.101154

[14] Pakzad K, Alinezhad H, Nasrollahzadeh M (2020) <scp> Euphorbia polygonifolia </scp> extract assisted biosynthesis of Fe $_3$ O $_4$ @CuO nanoparticles: Applications in the removal of metronidazole, ciprofloxacin and cephalexin antibiotics from aqueous solutions under UV irradiat. Appl Organomet Chem 34:e5910. https://doi.org/10.1002/aoc.5910

[15] Narendhran S, Sivaraj R (2016) Biogenic ZnO nanoparticles synthesized using L. aculeata leaf extract and their antifungal activity against plant fungal pathogens. Bull Mater Sci 39:1–5. https://doi.org/10.1007/s12034-015-1136-0

[16] Pansambal S, Deshmukh K, Savale A, et al (2017) Phytosynthesis and biological activities of fluorescent CuO nanoparticles using Acanthospermum hispidum L. extract. J Nanostructures 7:165–174. https://doi.org/10.22052/JNS.2017.03.001

[17] Qasim Nasar M, Zohra T, Khalil AT et al (2019) Seripheidium quettense mediated green synthesis of biogenic silver nanoparticles and their theranostic applications. Green Chem Lett Rev 12:310–322. https://doi.org/10.1080/17518253.2019.1643929

[18] Youssif KA, Haggag EG, Elshamy AM et al (2019) Anti-Alzheimer potential, metabolomic profiling and molecular docking of green synthesized silver nanoparticles of Lampranthus coccineus and Malephora lutea aqueous extracts. PLoS ONE 14:e0223781. https://doi.org/10.1371/journal.pone.0223781

[19] Fazal S, Jayasree A, Sasidharan S et al (2014) Green Synthesis of anisotropic gold nanoparticles for photothermal therapy of cancer. ACS Appl Mater Interfaces 6:8080–8089. https://doi.org/10.1021/am500302t

[20] Sriramulu M, Shukla D, Sumathi S (2018) Aegle marmelos leaves extract mediated synthesis of zinc ferrite: Antibacterial activity and drug delivery. Mater Res Express 5:115404. https://doi.org/10.1088/2053-1591/aadd88

[21] Jadoun S, Arif R, Jangid NK, Meena RK (2021) Green synthesis of nanoparticles using plant extracts: a review. Environ Chem Lett 19:355–374. https://doi.org/10.1007/s10311-020-01074-x

[22] Rath M, Panda SS, Dhal NK (2014) Synthesis of silver nano particles from plant extract and its application in cancer treatment: a review. Int J Plant Anim Env Sci 4:137–145

[23] Dhand C, Dwivedi N, Loh XJ et al (2015) Methods and strategies for the synthesis of diverse nanoparticles and their applications: a comprehensive overview. RSC Adv 5:105003–105037. https://doi.org/10.1039/C5RA19388E

[24] Vishnukumar P, Vivekanandhan S, Misra M, Mohanty AK (2018) Recent advances and emerging opportunities in phytochemical synthesis of ZnO nanostructures. Mater Sci Semicond Process 80:143–161. https://doi.org/10.1016/j.mssp.2018.01.026

[25] Abbasi E, Milani M, Fekri Aval S et al (2014) Silver nanoparticles: synthesis methods, bio-applications and properties. Crit Rev Microbiol 42:1–8. https://doi.org/10.3109/1040841X.2014.912200

[26] Iravani S, Korbekandi H, Mirmohammadi SV, Zolfaghari B (2014) Synthesis of silver nanoparticles: chemical, physical and biological methods. Res Pharm Sci 9:385

[27] Jamkhande PG, Ghule NW, Bamer AH, Kalaskar MG (2019) Metal nanoparticles synthesis: an overview on methods of preparation, advantages and disadvantages, and applications. J Drug Deliv Sci Technol 53:101174. https://doi.org/10.1016/j.jddst.2019.101174

[28] Abinaya S, Kavitha HP, Prakash M, Muthukrishnaraj A (2021) Green synthesis of magnesium oxide nanoparticles and its applications: a review. Sustain Chem Pharm 19:100368. https://doi.org/10.1016/j.scp.2020.100368

[29]   Bandeira M, Giovanela M, Roesch-Ely M et al (2020) Green synthesis of zinc oxide nanoparticles: a review of the synthesis methodology and mechanism of formation. Sustain Chem Pharm 15:100223. https://doi.org/10.1016/j.scp.2020.100223

[30]   Palomo JM (2019) Nanobiohybrids: a new concept for metal nanoparticles synthesis. Chem Commun 55:9583–9589. https://doi.org/10.1039/C9CC04944D

[31]   Lee J, Park EY, Lee J (2014) Non-toxic nanoparticles from phytochemicals: preparation and biomedical application. Bioprocess Biosyst Eng 37:983–989. https://doi.org/10.1007/s00449-013-1091-3

[32]   Ikram M, Javed B, Raja NI, Mashwani Z-R (2021) Biomedical potential of plant-based selenium nanoparticles: a comprehensive review on therapeutic and mechanistic aspects. Int J Nanomedicine 16:249–268. https://doi.org/10.2147/IJN.S295053

[33]   Rafique M, Sadaf I, Rafique MS, Tahir MB (2017) A review on green synthesis of silver nanoparticles and their applications. Artif Cells Nanomedicine Biotechnol 45:1272–1291. https://doi.org/10.1080/21691401.2016.1241792

[34]   Singh A, Joshi NC, Ramola M (2019a) Magnesium oxide Nanoparticles (MgONPs): green synthesis, characterizations and antimicrobial activity. Res J Pharm Technol 12:4644. https://doi.org/10.5958/0974-360X.2019.00799.6

[35]   Singh J, Kukkar P, Sammi H et al (2019b) Enhanced catalytic reduction of 4-nitrophenol and congo red dye By silver nanoparticles prepared from Azadirachta indica leaf extract under direct sunlight exposure. Part Sci Technol 37:434–443. https://doi.org/10.1080/02726351.2017.1390512

[36]   Singh J, Kumar V, Kim K-H, Rawat M (2019c) Biogenic synthesis of copper oxide nanoparticles using plant extract and its prodigious potential for photocatalytic degradation of dyes. Environ Res 177:108569. https://doi.org/10.1016/j.envres.2019.108569

[37]   Jadoun S, Arif R, Jangid NK, Meena RK (2021) Green synthesis of nanoparticles using plant extracts: a review. Environ Chem Lett 19:355–374. https://doi.org/10.1007/s10311-020-01074-x

[38]   Dubey M, Bhadauria S, Kushwah BS (2009) Green synthesis of nanosilver particles from extract of Eucalyptus hybrida (safeda) leaf. Dig J Nanomater Biostruct 4:537–543

[39]   Mittal AK, Chisti Y, Banerjee UC (2013) Synthesis of metallic nanoparticles using plant extracts. Biotechnol Adv 31:346–356. https://doi.org/10.1016/j.biotechadv.2013.01.003

[40]   Abinaya S, Kavitha HP, Prakash M, Muthukrishnaraj A (2021) Green synthesis of magnesium oxide nanoparticles and its applications: a review. Sustain Chem Pharm 19:100368. https://doi.org/10.1016/j.scp.2020.100368

[41]   Elemike E, Onwudiwe D, Ekennia A et al (2017a) Green synthesis of Ag/Ag2O nanoparticles using aqueous leaf extract of eupatorium odoratum and its antimicrobial and Mosquito Larvicidal activities. Molecules 22:674. https://doi.org/10.3390/molecules22050674

[42]   Elemike EE, Onwudiwe DC, Arijeh O, Nwankwo HU (2017b) Plant-mediated biosynthesis of silver nanoparticles by leaf extracts of Lasienthra africanum and a study of the influence of kinetic parameters. Bull Mater Sci 40:129–137. https://doi.org/10.1007/s12034-017-1362-8

[43]   Vishnukumar P, Vivekanandhan S, Misra M, Mohanty AK (2018) Recent advances and emerging opportunities in phytochemical synthesis of ZnO nanostructures. Mater Sci Semicond Process 80:143–161. https://doi.org/10.1016/j.mssp.2018.01.026

[44]   Pansambal S, Deshmukh K, Savale A, et al (2017) Phytosynthesis and biological activities of fluorescent CuO nanoparticles using Acanthospermum hispidum L. extract. J Nanostructures 7:165–174. https://doi.org/10.22052/JNS.2017.03.001

[45]   Sundararajan B, Ranjitha Kumari BD (2017) Novel synthesis of gold nanoparticles using Artemisia vulgaris L. leaf extract and their efficacy of larvicidal activity against dengue fever vector Aedes aegypti L. J Trace Elem Med Biol 43:187–196. https://doi.org/10.1016/j.jtemb.2017.03.008

[46]   Khan M, Al-hamoud K, Liaqat Z et al (2020) Synthesis of Au, Ag, and Au–Ag bimetallic nanoparticles using pulicaria undulata extract and their catalytic activity for the reduction of 4-Nitrophenol. Nanomaterials 10:1885. https://doi.org/10.3390/nano10091885

[47]   Chandraker SK, Lal M, Shukla R (2019) DNA-binding, antioxidant, H2O2 sensing and photocatalytic properties of biogenic silver nanoparticles using Ageratum conyzoides L. leaf extract. RSC Adv 9:23408–23417. https://doi.org/10.1039/C9RA03590G

[48]   Abdullah FH, Bakar NHHA, Bakar MA (2021) Comparative study of chemically synthesized and low temperature bio-inspired Musa acuminata peel extract mediated zinc oxide nanoparticles for enhanced visible-photocatalytic degradation of organic contaminants in wastewater treatment. J Hazard Mater 406:124779. https://doi.org/10.1016/j.jhazmat.2020.124779

[49]   Hammad DM, Asaad AA (2021) A comparative study: biological and chemical synthesis of iron oxide nanoparticles and their affinity towards adsorption of methylene blue dye. Desalin WATER Treat 224:354–366. https://doi.org/10.5004/dwt.2021.27053

[50]   Muthuvel A, Jothibas M, Manoharan C (2020a) Synthesis of copper oxide nanoparticles by chemical and biogenic methods: photocatalytic degradation and in vitro antioxidant activity. Nanotechnol Environ Eng 5:14. https://doi.org/10.1007/s41204-020-00078-w

[51]   Khan MM, Saadah NH, Khan ME et al (2019) Potentials of Costus woodsonii leaf extract in producing narrow band gap ZnO nanoparticles. Mater Sci Semicond Process 91:194–200. https://doi.org/10.1016/j.mssp.2018.11.030

[52]   Ganesan K, Jothi VK, Natarajan A et al (2020) Green synthesis of copper oxide nanoparticles decorated with graphene oxide for anticancer activity and catalytic applications. Arab J Chem 13:6802–6814. https://doi.org/10.1016/j.arabjc.2020.06.033

[53]   Muthuvel A, Jothibas M, Manoharan C (2020b) Effect of chemically synthesis compared to biosynthesized ZnO-NPs using Solanum nigrum leaf extract and their photocatalytic, antibacterial and in-vitro antioxidant activity. J Environ Chem Eng 8:103705. https://doi.org/10.1016/j.jece.2020.103705

[54]   Nithya K, Kalyanasundharam S (2019) Effect of chemically synthesis compared to biosynthesized ZnO nanoparticles using aqueous extract of C. halicacabum and their antibacterial activity. OpenNano 4:100024. https://doi.org/10.1016/j.onano.2018.10.001

[55]   Virmani I, Sasi C, Priyadarshini E et al (2020) Comparative Anticancer Potential of Biologically and Chemically Synthesized Gold Nanoparticles. J Clust Sci 31:867–876. https://doi.org/10.1007/s10876-019-01695-5

[56]   Sudhasree S, Shakila Banu A, Brindha P, Kurian GA (2014) Synthesis of nickel nanoparticles by chemical and green route and their comparison in respect to biological effect and toxicity. Toxicol Environ Chem 96:743–754. https://doi.org/10.1080/02772248.2014.923148

[57]   Amooaghaie R, Saeri MR, Azizi M (2015) Synthesis, characterization and biocompatibility of silver nanoparticles synthesized from Nigella sativa leaf extract in comparison with chemical silver nanoparticles. Ecotoxicol Environ Saf 120:400–408. https://doi.org/10.1016/j.ecoenv.2015.06.025

[58]   Dowlath MJH, Musthafa SA, Mohamed Khalith SB et al (2021) Comparison of characteristics and biocompatibility of green synthesized iron oxide nanoparticles with chemical synthesized nanoparticles. Environ Res 201:111585. https://doi.org/10.1016/j.envres.2021.111585

[59]   Jameel MS, Aziz AA, Dheyab MA (2020) Comparative analysis of platinum nanoparticles synthesized using sonochemical-assisted and conventional green methods. Nano-Struct Nano-Objects 23:100484. https://doi.org/10.1016/j.nanoso.2020.100484

[60]   Mousavi-Khattat M, Keyhanfar M, Razmjou A (2018) A comparative study of stability, antioxidant, DNA cleavage and antibacterial activities of green and chemically synthesized silver nanoparticles. Artif Cells Nanomedicine Biotechnol 46:S1022–S1031. https://doi.org/10.1080/21691401.2018.1527346

[61]   Borase HP, Salunke BK, Salunkhe RB et al (2014) Plant extract: a promising biomatrix for ecofriendly, controlled synthesis of silver nanoparticles. Appl Biochem Biotechnol 173:1–29. https://doi.org/10.1007/s12010-014-0831-4

[62]   Ghotekar S, Savale A, Pansambal S (2018) Phytofabrication of fluorescent silver nanoparticles from Leucaena leucocephala L. leaves and their biological activities. J Water Environ Nanotechnol 3:95–105. https://doi.org/10.22090/JWENT.2018.02.001

[63]   Zheng B, Kong T, Jing X et al (2013) Plant-mediated synthesis of platinum nanoparticles and its bioreductive mechanism. J Colloid Interface Sci 396:138–145. https://doi.org/10.1016/j.jcis.2013.01.021

[64]   Huang J, Zhan G, Zheng B et al (2011) Biogenic silver nanoparticles by cacumen platycladi extract: synthesis, formation mechanism, and antibacterial activity. Ind Eng Chem Res 50:9095–9106. https://doi.org/10.1021/ie200858y

[65]   Murthy HCA, Desalegn T, Kassa M et al (2020) Synthesis of green copper nanoparticles using Medicinal Plant Hagenia abyssinica (Brace) JF. Gmel. leaf extract: antimicrobial properties. J Nanomater 2020:1–12. https://doi.org/10.1155/2020/3924081

[66]   Naghdi S, Sajjadi M, Nasrollahzadeh M et al (2018) Cuscuta reflexa leaf extract mediated green synthesis of the Cu nanoparticles on graphene oxide/manganese dioxide nanocomposite and its catalytic activity toward reduction of nitroarenes and organic dyes. J Taiwan Inst Chem Eng 86:158–173. https://doi.org/10.1016/j.jtice.2017.12.017

[67]   Nagar, Niharika; Devra, Vijay (2018). Green synthesis and characterization of copper nanoparticles using Azadirachta indica leaves. Materials Chemistry and Physics, (), S0254058418302694–. doi:10.1016/j.matchemphys.2018.04.007

[68]   Król A, Railean-Plugaru V, Pomastowski P, Buszewski B (2019) Phytochemical investigation of Medicago sativa L. extract and its potential as a safe source for the synthesis of ZnO nanoparticles: the proposed mechanism of formation and antimicrobial activity. Phytochem Lett 31:170–180. https://doi.org/10.1016/j.phytol.2019.04.009

[69]   Kumar B, Smita K, Cumbal L, Debut A (2014) Green approach for fabrication and applications of zinc oxide nanoparticles. Bioinorg Chem Appl 2014:1–7. https://doi.org/10.1155/2014/523869

[70]   Hassan Basri H, Talib RA, Sukor R, Othman SH, Ariffin H. (2020) Effect of Synthesis Temperature on the Size of ZnO Nanoparticles Derived from Pineapple Peel Extract and Antibacterial Activity of ZnO–Starch Nanocomposite Films. Nanomaterials.; 10(6):1061. https://doi.org/10.3390/nano10061061

[71]   Soto-Robles, C.A.; Luque, P.A.; Gómez-Gutiérrez, C.M.; Nava, O.; Vilchis-Nestor, A.R.; Lugo-Medina, E.; Ranjithkumar, R.; Castro-Beltrán, A. (2019). Study on the effect of the concentration of Hibiscus sabdariffa extract on the green synthesis of ZnO nanoparticles. Results in Physics, 15(), 102807–. doi:10.1016/j.rinp.2019.102807

[72]   Singh A, Joshi NC, Ramola M (2019a) Magnesium oxide Nanoparticles (MgONPs): green synthesis, characterizations and antimicrobial activity. Res J Pharm Technol 12:4644. https://doi.org/10.5958/0974-360X.2019.00799.6

[73]   Nijalingappa TB, Veeraiah MK, Basavaraj RB et al (2019) Antimicrobial properties of green synthesis of MgO micro architectures via Limonia acidissima fruit extract. Biocatal Agric Biotechnol 18:100991. https://doi.org/10.1016/j.bcab.2019.01.029

[74]   Suresh J, Pradheesh G, Alexramani V et al (2018) Green synthesis and characterization of hexagonal shaped MgO nanoparticles using insulin plant (Costus pictus D. Don) leave extract and its antimicrobial as well as anticancer activity. Adv Powder Technol 29:1685–1694. https://doi.org/10.1016/j.apt.2018.04.003

[75]   Jeevanandam, Jaison & Pan, Sharadwata & Akula, Harini & Danquah, Michael. (2021). Challenges in the Risk Assessment of Nanomaterial Toxicity Towards Microbes. 10.1201/9780429321269-4

[76]   Nagore P, Ghotekar S, Mane K et al (2021) Structural properties and antimicrobial activities of Polyalthia longifolia leaf extract-mediated CuO nanoparticles. Bionanoscience 11:579–589. https://doi.org/10.1007/s12668-021-00851-4

[77]   Alsaiari NS, Alzahrani FM, Amari A, Osman H, Harharah HN, Elboughdiri N, Tahoon MA. (2023) Plant and Microbial Approaches as Green Methods for the Synthesis of Nanomaterials: Synthesis, Applications, and Future Perspectives. Molecules. 28(1):463. https://doi.org/10.3390/molecules28010463

[78]   M. Noruzi, (2015) Biosynthesis of gold nanoparticles using plant extracts, Bioprocess and biosystems engineering 38,1-14.

[79]   C.F. Bohren, D.R. Huffman, (1983) Absorption and Scattering of Light by Small Particles;Wiley: New York, NY, USA.

[80]   A.C. Templeton, J.J. Pietron, R.W. Murray, P. Mulvaney, (2000) Solvent refractive index and core charge influences on the surface plasmon adsorbance of alkanethiolate monolayerprotected gold clusters, Journal of Physical Chemistry B 104,564-570.

[81]   K.B. Narayanan, N. Sakthivel, (2008) Coriander leaf mediated biosynthesis of gold nanoparticles, Mater. Lett. 62, 4588-4590.

[82]   R. Venu, T.S. Ramulu, S. Anandakumar, V.S. Rani, C.G. Kim, (2011) Bio-directed synthesis of platinum nanoparticles using aqueous honey solutions and their catalytic applications, Colloids and Surfaces A: Physicochemcal Engineering Aspects 384, 733-738.

[83]   Vijayaraghavan, K., & Ashokkumar, T. (2017). Plant-mediated biosynthesis of metallic nanoparticles: A review of literature, factors affecting synthesis, characterization techniques and applications. Journal of environmental chemical engineering, 5(5), 4866-4883.

[84]   J. Dhayananthaprabhu, R. Lakshmi Narayanan, K. Thiyagarajan, (2013) Facile synthesis of gold (Au) nanoparticles using Cassia auriculata flower extract, Adv. Mater. Res. 678,12-16.

[85]   Pasca, R.-D.; Mocanu, A.; Cobzac, S.-C.; Petean, I.; Horovitz, O.; Tomoaia-Cotisel, M. Biogenic Syntheses of Gold Nanoparticles Using Plant Extracts. *Part. Sci. Technol.* **2014**, *32*, 131–137.

[86]   Jun, S.H.; Kim, H.-S.; Koo, Y.K.; Park, Y.; Kim, J.; Cho, S.; Park, Y. (2014) Root Extracts of *Polygala tenuifolia* for the Green Synthesis of Gold Nanoparticles. *J. Nanosci. Nanotechnol. 14*, 6202–6208.

[87]   J. Das, M. Paul Das, P. Velusamy, (2013) Sesbania grandiflora leaf extract mediated green synthesis of antibacterial silver nanoparticles against selected human pathogens, Spectrochimica Acta - Part A: Molecular and Biomolecular Spectroscopy 104 ,265-270.

[88]   J. Huang, Q. Li, D. Sun, Y. Lu, Y. Su, X. Yang, (2007) Biosynthesis of silver and gold nanoparticles by novel sundried Cinnamomum camphora leaf, Nanotechnology 18 105104-105114.

[89]   Rajasekharreddy, P.; Rani, P.U.; Sreedhar, B. (2010) Qualitative Assessment of Silver and Goldnanoparticle Synthesis in Various Plants: A Photobiological Approach. J. Nanopart. Res., 12, 1711–1721.

[90]   Yu, J.; Xu, D.; Guan, H.N.; Wang, C.; Huang, L.K.; Chi, D.F. (2016) Facile One-Step Green Synthesis of Gold Nanoparticles Using Citrus maxima Aqueous Extracts and Its Catalytic Activity. Mater. Lett., 166, 110–112.

[91]   Pasca, R.-D.; Mocanu, A.; Cobzac, S.-C.; Petean, I.; Horovitz, O.; Tomoaia-Cotisel, M. (2014) Biogenic Syntheses of Gold Nanoparticles Using Plant Extracts. Part. Sci. Technol., 32, 131–137.

[92]   Jun, S.H.; Kim, H.-S.; Koo, Y.K.; Park, Y.; Kim, J.; Cho, S.; Park, Y. (2014) Root Extracts of Polygala tenuifolia for the Green Synthesis of Gold Nanoparticles. J. Nanosci. Nanotechnol., 14, 6202–6208.

[93]   Putnam, C.D.; Hammel, M.; Hura, G.L.; Tainer, J.A. (2007) X-Ray Solution Scattering (SAXS.) Combined with Crystallography and Computation: Defining Accurate Macromolecular Structures, Conformations and Assemblies in Solution. Q. Rev. Biophys., 40, 191–285.

[94]   Dubey, S.P.; Lahtinen, M.; Sillanpää, M. Tansy (2010) Fruit Mediated Greener Synthesis of Silver and Gold Nanoparticles. Process Biochem., 45, 1065–1071.

[95]   Geng, G.; Chen, P.; Guan, B.; Liu, Y.; Yang, C.; Wang, N.; Liu, M. Sheetlike (2017) Gold Nanostructures/Graphene Oxide Composites via a One-Pot Green Fabrication Protocol and Their Interesting Two-Stage Catalytic Behaviors. RSC Adv., 7, 51838–51846.

[96]   Gardea-Torresdey J L, Tiemann K J, Gamez G, Dokken K, Tehuacamanero S and Jose-Yacaman M 1999 J. Nanopart. Res. 1 397

[97]   Armendariz V, Herrera I, Peralta-Videa J R, Jose-Yacaman M, Troiani H, Santiago P and Gardea-Torresdey J L 2004 J. Nanopart. Res. 6 377

[98]   Sathishkumar M, Sneha K and Yun Y S 2009 Int. J. Mater. Sci. 4 11

[99]   Sathishkumar J and Narendhirakannan R T 2011 Dig. J. Nanomater. Biostru. 6 961

[100]   Sathishkumar M, Krishnamurthy S and Yun Y S 2010 Biores. Technol. 101 7958

[101]   Dwivedi A D and Gopal K 2010 Colloids Surf. A 369 27

[102]   Rai A, Singh A, Ahmad A and Sastry M 2006 Langmuir 22 736

[103]   Veerasamy R, Xin T Z, Gunasagaran S, Xiang T F W and Yang E F C 2011 J. Saudi. Chem. Soc. 15 113

[104]   Natrajan.D, SrinivasanK.S,.Sundar,Ravindran.A Formulation of essential oil-loaded chitosan–alginate nanocapsules J Food Drug Anal.2015;23,560-568

[105]   Luo Y, Wang Q. Recent development of chitosan-based polyelectrolyte complexes with natural polysaccharides for drug delivery. Int J Biol Macromol 2014;64:353-67.

[106]   Li P, Dai YN, Zhang JP, Wang AQ, Wei Q. Chitosan-alginate nanoparticles as a novel drug delivery system for nifedipine. Int J Biomed Sci 2008;4:221-8.

[107]   Kumari R, Saini AK, Kumar A, Saini RV (2020) Apoptosis induction in lung and prostate cancer cells through silver nanoparticles synthesized from Pinus roxburghii bioactive fraction. J Bio Inorg Chem 25:23–37.

[108]   Singh D, Yadav E, Falls N, Kumar V, Singh M, Verma A (2019) Phyto-fabricated silver nanoparticles of *Phyllanthus emblica* attenuated diethyl-nitrosamine-induced hepatic cancer via knock-down oxidative stress and inflammation. Inflammopharmacol 27:1037–1054.

[109]   Yusuf A, Casey A (2020) Liposomal encapsulation of silver nanoparticles (AgNP) improved nanoparticle uptake and induced redox imbalance to activate caspase-dependent apoptosis. Apo 25:20–134.

[110]   El-Naggar NE, Hussein MH, El-Sawah AA (2017) Bio-fabrication of silver nanoparticles by phycocyanin, characterization, in vitro anticancer activity against breast cancer cell line and in vivo cytotoxicity. Scient Rep 7:1–20

[111]   Simard JC, Durocher I, Girard D (2016) Silver nanoparticles induce irremediable endoplasmic reticulum stress leading to unfolded protein response dependent apoptosis in breast cancer cells. Apo 21:1279–1290

[112]   Lee YS, Kim DW, Lee YH, Oh JH, Yoon S, Choi MS, Lee SK, Kim JW, Lee K, Song CW (2011) Silver nanoparticles induce apoptosis and G2/M arrest via PKC_f-dependent signaling in A549 lung cells. Arch Toxicol 85:1529–1540

[113]   Erdogan O, Abbak M, Demirbolat GM, Birtekocak F, Aksel M, Pasa S et al (2019) Green synthesis of silver nanoparticles via *Cynara scolymus* leaf extracts: The characterization, anticancer potential with photodynamic therapy in MCF7 cells. PLoS ONE 14:6

[114]   Vasanth K, Ilango K, Kumar MR, Agrawal A, Dubey GP (2014) Anticancer activity of *Moringa oleifera* mediated silver nanoparticles on human cervical carcinoma cells by apoptosis. Coll Surf B: Bioint 117:354–359

[115]   Valsalam S, Paul A, Arasu MV, Al-Dhabi NA, Mohammed Ghilan AK, Kaviyarasu K, Ravindran B, Chang SW, Arokiyaraj S (2018) Rapid biosynthesis and characterization of silver nanoparticles from the leaf extract of *Tropaeolum majus* L. and its enhanced in-vitro antibacterial, antifungal, antioxidant and anticancer properties. J Photochem Photobiol, B 191:65–74

[116]   Sarkar S, Kotteeswara V (2018) Green synthesis of silver nanoparticles from aqueous leaf extract of Pomegranate (*Punica granatum*) and their anticancer activity on human cervical cancer cells. Adv Nat Sci Nanosci Nanotechnol 9(2):025014

[117] Muthukrishnan S, Vellingiri B, Murugesan G (2018) Anticancer effects of silver nanoparticles encapsulated by *Gloriosa superba* (L.) leaf extracts in DLA tumor cells. Fut J Pharma Sci 4:206–214

[118] Hashemi F, Tasharrofi N, Saber MM (2020) Green synthesis of silver nanoparticles using *Teucrium polium* leaf extract and assessment of their antitumor effects against MNK45 human gastric cancer cell line. J Mol Str 1208:127889

[119] Chen YN, Hsueh YH, Hsieh CT, Tzou DY, Chang PL (2016) Antiviral activity of graphene-silver nanocomposites against non-enveloped and enveloped viruses. Int J Environ Res 13(4):430

[120] Kim M, Nguyen DY, Heo Y, Park KH, Paik HD, Kim YB (2020) Antiviral activity of *Fritillaria thunbergii* extract against Human Influenza Virus H1N1 (PR8) In Vitro, In Ovo and In Vivo. J Microbiol Biotechnol 30(2):172–177

[121] Elechiguerra JL, Burt JL, Morones JR, Bragado BC, Gao X, Lara HH, Yacaman MJ (2005) Interaction of silver nanoparticles with HIV-1. J Nanobiotech 3(6):1–10

[122] Chen N, Zheng Y, Yina J, Lia X, Zhenga C (2013) Inhibitory effects of silver nanoparticles against adenovirus type 3 in vitro. J Vir Met 193:470–477

[123] Hu RL, Li SR, Kong FJ, Hou RJ, Guan XL, Guo F (2014) Inhibition effect of silver nanoparticles on herpes simplex virus 2. Gen Mol Res 13(3):7022–7028

[124] Sujitha V, Murugan K, Paulpandi K, Panneerselvam C, Suresh U, Roni M, Nicoletti M et al (2015) Green-synthesized silver nanoparticles as a novel control tool against dengue virus (DEN-2) and its primary vector *Aedes aegypti*. Parasitol Res 114(9):3315–3325

[125] Gaal H, Fouad H, Mao G, Tian J, Jianchu M (2017) Larvicidal and pupicidal evaluation of silver nanoparticles synthesized using *Aquilaria sinensis* and *Pogostemon cablin* essential oils against dengue and zika viruses' vector *Aedes albopictus* mosquito and its histopathological analysis. Art Cell Nanomed Biotech 46(6):1171–1179

[126] Sharma V, Kaushik S, Pandit P, Dhull D, Yadav JP, Kaushik S (2019) Green synthesis of silver nanoparticles from medicinal plants and evaluation of their antiviral potential against chikungunya virus. App Microbio Biotech 103:881–891

[127] Bekele AZ, Gokulan K, Williams KM, Khare S (2016) Dose and size-AU2 c dependent antiviral effects of silver nanoparticles on feline calicivirus, a human norovirus surrogate. Foodbor Patho Dis 13(5):239–244

[128] El-Mohamady RS, Ghattas TA, Zawrah MF, Abd El-Hafeiz YGM (2018) Inhibitory effect of silver nanoparticles on bovine herpesvirus-1. Int J Vet Sci Med 6:296–300

[129] Gaikwad S, Ingle A, Gade A, Rai M, Falanga A, Incoronato N, Russo L, Galdiero S, Galdiero M (2013) Antiviral activity of myco-synthesized silver nanoparticles against herpes simplex virus and human parainfluenza virus type 3. Int J Nanomed 8:4303–4314

[130] Aji Kumaran Nair. S., Gaythri V. (2023) Scaffold materials and toxicity, In Biomedical Applications and Toxicity of Nanomaterials, Eds; Mohanan, P.V. and Kappalli, S. ISBN:

9811978336, 9789811978333, Springer Nature Singapore.
https://books.google.co.in/books?id=4ctpzwEACAAJ

[131]   V. Gayathri and B. Sabulal. (2022) Biodegradable composites for commodities
        packaging applications and toxicity, In Biodegradable Composites for Packaging
        applications.Eds; Arbind Prasad, Ashwani Kumar, Kishor Kumar Gajrani. ISBN:
        9781003227908 CRC Press, Taylor Francis Publications, Boca Raton, Florida, USA.
        https://doi.org/10.1201/9781003227908.

Nanoparticles in Healthcare: Applications in Therapy, Diagnosis and Drug Delivery   Materials Research Forum LLC
Materials Research Foundations 160 (2024) 83-112   https://doi.org/10.21741/9781644902974-4

Chapter 4

Nano Antibiotics: Prospects and Challenges

V.P. Jayachandran*

Department of Applied Sciences, University of Technology and Applied Sciences, PO Box 74, Al-Khuwair, Postal Code 133 Muscat, Sultanate of Oman

*Jayachandran.p@utas.edu.om

Abstract

Antibiotic resistance is a major global public health concern, resulting in high healthcare costs, increased morbidity, and mortality. With a limited number of new antibiotics being developed, the emergence and spread of antibacterial resistance poses a constant threat to public health. After 1980s the antibiotic pipeline began to dry up and fewer new drugs were introduced, and microbes seem to be on the winning side in tiding over the barricades built by drugs. Adoption of an effective strategy to intervene the burning issue of MDR is a necessity of the hour. Nanotechnology gives great hope to the medical community in their fight against pathogens. As an essential byproduct nanoantibiotics (nAbts) are being developed by conjugating nanoparticles with antibiotic molecules. Various types of nanoparticles are used as drug delivery systems. This chapter explores the prospects of nAbts, the composition and types of nAbts, their mode of action, challenges faced by nAbts in terms of safety, toxicity, manufacturing complexity, regulatory hurdles etc. as well as development of resistance among bacterial pathogens against nAbts.

Keywords

Multidrug Resistance, Nano Antibiotics, Drug Delivery, Biocompatible, Liposome

Contents

1.   Introduction

The emergence of antimicrobial resistance (AMR) is a burning issue of global significance. The phenomenon of AMR can be defined as the ability of a microbe to resist the effects of an antimicrobial agent that previously had the ability to inhibit or kill it. Resistant microbes tend to survive the detrimental effects of antimicrobial agents like antivirals, antibiotics as well as antimalarials, consequently the conventional therapies become futile and infections persist and may spread to other individuals, especially in hospital settings leading to nosocomial infections. Antimicrobial resistance (AMR) is primarily a result of the inappropriate /misuse of antimicrobial medications, which can cause microorganisms to mutate or acquire genes that confer resistance. AMR is a consequence, particularly of the misuse, of antimicrobial based medicines and mostly emerges while a microbe undergoes a mutation/s or gains a resistant gene [1].

The treatment of a broad range of infections, such as extensively drug-resistant (XDR) as well as multi-drug-resistant (MDR) tuberculosis, methicillin-resistant *Staphylococcus aureus* (MRSA), anti-retroviral-resistant HIV, vancomycin-resistant *Enterococcus* (VRE) and MDR Gram-negative bacteria (including *Klebsiella, Pseudomonas, Escherichia coli, Enterobacter*, and *Acinetobacter* species, among others) is a major challenge that needs special attention as ineffective approaches lead to cataclysms in hospital settings as well as in the community. The phenomenon of MDR spreading among prominent bacterial pathogens poses a considerable toll on healthcare systems and poses a challenge to the security of nations and economies, as MDR microbes can spread through travel and trade. According to the WHO, the primary leads to AMR are i) absence of well-designed and coordinated strategies, ii) a deficient surveillance system, iii) inadequate allocation of resources iv) substandard drug quality, v) inappropriate/misuse of antimicrobials for human as well as for livestock and vi) insufficient infection prevention and control measures [2]. One of the leading factors accounting for the widespread occurrence of MDR in bacterial pathogens is thought to be bacterial recombination. Interspecific and intraspecific transfer of drug resistance genes by genetic recombination among bacteria gainsays disease management to a greater extent. In recent years, there has been a drastic modification in the drug metabolism of many bacteria. Infections mediated by resistant pathogens frequently fail to answer to the therapeutic strategies, resulting in prolonged illness and greater risk of death. The accelerated emergence of antibiotic resistance is mainly contributed by the high incidence of infectious conditions, errant entree to high- end antibiotics along with unfitting lab facilities, fall out in the health care system and deficiency in proper protocols for the disease management as well as infection control facilities. The issue is exacerbated by the shortened duration of therapy due to cost-related reasons as many times the poor cannot afford to complete the course of treatment [3].

The 21st century is marked by a significant worldwide public health issue, which is the emergence of antibiotic resistance, resulting in high healthcare costs, increased morbidity, and mortality. With a limited number of new antibiotics being developed, the dissemination of antibacterial resistance poses a constant menace to the health of the public [4,5].

Various reports underscore the impact of AMR on patient mortality, incidence, duration of hospitalization and healthcare expenses for particular pathogens as well as drug combos used in selected locations.

In a 2019 study, a statistical model predicted approximately 4.95 million deaths to be linked with bacterial AMR. This study also revealed that the prevalence of resistance was highest in western sub-Saharan Africa, with an estimated 27.3 deaths/100,000 individuals, and lowest in Australasia, with 6.5 deaths/100,000 individuals. Also, this study highlighted lower respiratory tract (LRT) infections as the cause of more than 1·5 million deaths linked to antibiotic resistance, and LRT as the most troublesome infectious syndrome. In addition, six prominent bacterial pathogens allied to AMR related mortality included *E.coli*, followed by *S. aureus, K. pneumoniae, S. pneumoniae, A. baumannii*, and *P. aeruginosa*. It is noteworthy to mention that, in 2019, a single pathogen–drug combo viz., methicillin and methicillin resistant *S aureus*, (MRSA) led to over100 000 patient deaths underscoring AMR as the root cause [6].

During the period spanning from the late 1960s to the early 1980s, the pharmaceutical industry introduced various antibiotics to tackle the problem of resistance. Though, since then, the development of new antibiotics has significantly slowed down, leading to a reduction in the number of newly introduced drugs. The resurgence of bacterial infections has emerged as a serious threat, despite several decades of using antibiotics for treating patients [7].

Antibiotics were first used commercially with the production of penicillin in the late 1940s, which was celebrated as a significant achievement until the development of new and promising antibiotics in the 1970s and 1980s. However, after that period, the number of newly developed antibiotics declined, and microbes seem to be on the winning side by tiding over the barricades built by antibiotics. In reality, the low return on investment in the development of antibiotics has contributed to the current crisis in the battle against prevalent MDR pathogens. Consequently, several decades after the first administration of antibiotics to patients, bacterial infections have once again become a major threat [8].

All the aforementioned aspects point to the fact that there is necessity to introduce an effective approach to intervene the burning issue of MDR microbes.

The turn of 19th century has witnessed great revolutions in the arena of science and technology and nanotechnology is one among those fields contributed to gigantic facelifts in various fields including physics, chemistry, medicine, engineering, food, agriculture etc. Nanotechnology deals with the perceiving as well as controlling of matter at the nanoscale which typically ranges from 1 to 100 nanometers. Through atomic-level manipulation of nanoparticles' (NPs) size and shape, their distinctive characteristics can be utilized for innovative and novel applications in various domains. The word *nano* has its root in Greek, means dwarf. A nanometer is one thousand millionths of a meter (10^{-9} m). The concept of using nanoparticles can be traced back to the 4th century when the Romans made the Lycurgus cup [9] placed in the British museum that depicts King Lycurgus from the sixth book of Homer's Iliad. This is a dichroic glass, made by proportionately mixing silver and gold colloidal NPs distributed in the glass material. The cup undergoes changes in colour, displaying red-purple as light transmits through it and green in reflected light. In true sense, the studies of Michael Faraday on the distinctive optical and electronic features of colloidal suspensions of "Ruby" gold in 1857 has paved way to the exploration of this field [10].

Nanotechnology has been comprehensively explored in the medical field, generating approved nutritional supplements, chemotherapeutics, imaging agents, anesthetics, and antimicrobials. The very two aspects, a dried pipeline of antibiotic discovery and the search of scientists for curbing drug resistance among pathogens ship shaped a group of novel antimicrobial drugs called nanoantibiotics (nAbts) with the proficient utilization of nanotechnology. In true sense this is an alternate therapeutic strategy established by utilizing a combinatorial method that combines molecules of antibiotics and nanomaterials.

2.   NAbts: Nanomaterial based infection control agents

The famous microbiologist Selman Waksman is credited with coining the term "antibiotic" in 1942 and this term describes any chemical. This term describes any chemical produced by a microbe that has static (inhibitory) or cidal (killing) effect on another microbial species. With the emergence of nanotechnology and drug delivery systems the term nanoantibiotics shortly abbreviated nAbts has come into being and it is defined as tiny particles that are physicochemically conjugated with antibiotics/synthetically made pure molecules of antibiotics that measures ≤100 nm at least in single dimension [11,12]. Otherwise, these are nanomaterials, which possess antimicrobial effect individually or uplift the safety as well as efficacy of the administration of antibiotics [13]. Generally speaking, nAbt refers to a nanoparticle (NP) that is specifically designed to transport an antibiotic payload to a designated infection site. These are emerging candidates to defend antibiotic resistant bacteria where conventional antibiotics are proved to be ineffective. The NP component of nAbts can be made from a variety of materials. These novel antimicrobials encapsulate a variety of nanostructured pathogen combatants in the form of nanocarriers loaded with antibiotic molecules, metal/metal oxide NPs, graphene/ graphene oxide, carbon nanotube, organic–inorganic composite NPs, dendrimers, unimolecular micelles, self-assembled micelles, supramolecular nanostructures, polymer molecular brushes or bottlebrush polymers, solid lipid NPs, nanoliposomes, nanocapsules, biodegradable polymeric NPs and are being employed as vehicles for drug delivery [14]. Otherwise, NPs are fastened to antibiotics to enhance their efficacy. The very synergistic approach of utilizing antibiotics in combo with nanostructures notably enhances antibiotic functionality that in turn can minimize antibiotic dosage, exposure time, as well as bacterial resistance. NPs possess added utilities in comparison to the effectiveness of a single/ multiple drugs' combination while treating prominent bacterial pathogens possessing with MDR armamarium. MDR microbes possess extraordinary versatility to tide over the effect of different classes of antibiotics. In reality, synergistic release of antibiotics from NPs or by loading multiple antibiotics onto NPs, capacitate rapid and efficient destruction of bacterial pathogens [12]. In this novel approach, the antimicrobial effect is concurrently achieved by synergistic effect of NPs and antibiotics. Nanotechnology presents a promising solution for the difficulties associated with antibiotics. Various forms of nanoparticles have been employed as vehicles for drug delivery.

3.   NAbts (nAbts): Composition and types

The structure of a nanoantibiotic is tailored to optimize its effectiveness in combating antibiotic-resistant infections by enhancing antibiotic delivery, reducing toxicity to healthy cells, and promoting targeted antimicrobial action. The antibiotic component of a nanoantibiotic can be any type of antibiotic, such as penicillin, cephalosporin, or tetracycline. The nanoparticle can be designed to have a specific shape, such as spherical, rod-shaped, or dendritic, which can affect its properties and behavior in biological systems. The antibiotic is typically attached or encapsulated within the nanoparticle structure, which can protect it from degradation and enhance its delivery to the targeted site of infection. Characterization of the resultant nAbts can be carried out by a series of analytical studies using Fourier Transform Infrared Spectroscopy (FTIR), UV–Vis Spectroscopy, Energy Dispersive X-Ray Analyzer (EDX), Scanning Electron Microscopy (SEM), Dynamic Light Scattering (DLS) Particle Size Analyzer, X-ray Diffraction (XRD) Analysis etc.

NAbts do not have a specific or fixed structure, as they can be composed of different materials and have various shapes as well as sizes depending on the intended use. The structure of NPs varies depending on the specific type of nanomaterial that is used. Metallic NPs, for instance, possess distinctive physicochemical characteristics, such as a small size and a larger surface area, that allow them to generate reactive oxygen species that can efficiently eliminate bacteria. In contrast, carbon nanotubes have a tubular shape that allows them to penetrate bacterial cells and release their antibacterial contents. Dendrimers, which have a tree-like structure, can be engineered to carry multiple copies of antibiotic molecules. Meanwhile, liposomes are spherical vesicles composed of phospholipids that can encapsulate antibiotic drugs within their structures.

3.1   NAbts based on metallic and metal oxide nanoparticles

Heavy metals including silver, gold, copper etc., are being used as antimicrobial agents. NAbts based on metallic NPs are a type of antibacterial agent that uses NPs made of metals. The inorganic metal NPs viz., Cu, Ag, Au, Pt, Pd, Zn and metal oxide like CuO, ZnO, TiO$_2$, Al$_2$O$_3$ etc. are utilized as antibacterial or anti-infective agents [15]. These NPs have unique physicochemical properties that make them effective at targeting and killing bacteria, while minimizing damage to surrounding healthy cells. Some of the advantages of metallic NPs as nAbts include their small size, high surface area, and capacity to lead to the production of reactive oxygen species that can damage bacterial cells. Also spherical shape, ease in synthesis, plasmonic features, surface functionalization make them attractive models in the generation of nano-based designing of drug, in the process of delivery of drug and mechanistic studies.

Various methods are adopted in the preparation of metal NPs and these include microwave-assisted synthesis, microemulsion method, biochemical method, irradiation reduction, photo-induced or photo-catalytic reduction, electrochemical reduction, ultrasonic-assisted reduction, aqueous/non aqueous chemical reduction, template method etc. [16]. Preparation of metal NP- antibiotic conjugate, the generally followed methods fall under the categories i) category- 1: synthesis of metal NPs and their posterior mix up with antibiotic solutions' ii) category- 2: synthesis of metal NPs in the presence of the combining agents iii) category-3: synthesis of metal NPs and functionalization followed by the posterior conjugation step; iv) category 4: synthesis of metal NPs using the conjugating agents as reducing agents.

The surface functionalization of metals NPs and metaloxide NPs is an efficient approach to create bonds with antibiotic/ molecules as well as cells leading to the localized accumulation of NPs in the target. Surface functionalization is done with an array of molecules like antibiotics, polymers, ligands etc. The chemical reactions adopted include covalent, electrostatic or physical adsorption. The NPs functionalized with antibodies or peptides can recognize specific microbial proteins. In general, metal NPs were modified using amines (–NH2), nitriles(−C≡ N), carboxylic acids(-COOH), thiols(-SH), disulfides(-CSH), phosphines(-PH 3) etc. and metal oxide NPs were functionalized primarily using silanes/phosphonates. In fact, metals, metalloids, epoxides, metal alkoxides can form a film on the NP surface as an oxide film. Subsequently, the metal NPs are mixed with respective conjugating antibiotics at particular proportion. The synthesis of the metal NPs mediated by antimicrobials as reducing agents is an explorative strategy and it has the added advantage of utilizing lesser number of chemicals, but it demands longer reaction. Functionalization of AgNPs and AuNPs with ampicillin was showed in which ampicillin acted simultaneously as reducing, conjugating, stabilizing and reducing agents [15,17]. In a recent study, a single-step strategy of synthezing AuNP using kanamycin (an aminoglycoside antibiotic) as reducing and capping agents to form kanamycin-conjugated Au NPs. The scientific community has taken an interest in AuNPs due to their inherent features like biosafety ease of functionalization, facile synthesis, multiple targets of bactericidal effects [18][19].

The virtue of the antibacterial mechanisms of antibiotics and NPs are synergistically utilized to bind to the bacterial cells, /enter the cells/ to release ROS that damage the bacterial cell membrane and disrupt its normal functions or interfere with bacterial DNA replication and synthesis of protein resulting in the death of bacteria. One of the most widely studied metallic NPs as nAbts is the AgNPs. The broad spectral activity of AgNPs to gram-positive and gram-negative bacteria, anti-inflammatory as well as wound-healing properties of AgNPs make them attractive candidates for antibiotic conjugates to be used especially in wound dressings.

Metal-oxide NPs possess stable net negative surface charge, resistance to swelling, facile entrapment of hydrophilic and hydrophobic antibiotics, etc. Based on the purpose, vacancies of oxygen as well as variable defect sites can be made in metal oxide NPs, in which catalytic active and electrochemical sites are present. One of the advantageous thing in this approach is the allotment of multiple oxidation states and polyvalency as well as crystallographic coordination in combo with broad functionalization of molecules of antibiotics. Paramagnetic species like iron oxide NPs prefer significant entrapment efficiency of antibiotic molecules because of the van der Waals interactions, hydrogen bonding and dipole-dipole attraction. Consequently, high surface energy as well as net negative multivalent surface charge permits passive diffusion leading to cell membrane damage and oxidative stress [12,20].

3.2   NAbts based on carbon nanotubes

The number of published works on carbon-based nAbts is comparatively lesser than the metal or metal oxide NPs. The enhanced adsorption capacity makes them attractive candidates as drug delivery system. A prominent example is graphene which is typically 200–500 nm in size and is extremely thin. When carbon atoms arranged in a hexagonal lattice are rolled into cylindrical structures, they form carbon nanotubes (CNTs). The large surface area-to-volume ratio present in CNTs provides multiple sites on surface where drug molecules get bounded. The extremely thin nature of grapheme enables simple and rapid diffusion of antibiotic molecules across its surface and interact with its functional groups. In addition, the high electrical conductivity and large pi-electron cloud of grapheme enables it to form strong pi-pi stacking interactions with drug molecules. Their unique tubular structure makes them able to enter the bacterial cells and release the antibacterial cargo. This makes them a promising alternative to conventional antibiotics, which often have limited penetration into bacterial cells. Various studies have provided evidence of the antimicrobial properties of carbon nanotubes, showcasing their effectiveness against both gram-positive and gram-negative strains. Both nanographene oxide and CNTs conjugated with antibiotics like tetracycline/vancomycin/ linezolid are found to be effective against prominent bacterial pathogens.[12,21]. Additionally, carbon nanotubes are found to be effective against MDR bacteria, which are a growing problem in clinical settings.

3.3   NAbts based dendrimeric structures

The word dendrimer has its root in Greek., *dendron* means tree and *meros* means part. Dendrimers can be defined as homogeneous monodisperse, stable nanosized, and radially symmetrical 3D and highly branched structure with possibly high structural functionalities provided with exceptional properties completely distinct from linear polymers. The polymer molecule is formed by a central core that serves as an anchor for the monomeric branches. These branches extend from the core and create a globular structure, typically ranging in size from 2 to 5 nm. Each section of the structure, referred to as a generation and the number of generations is proportionate to the number of branches present.

As the number of branches increases, there is a corresponding increase in the number of exposed functional groups available for conjugation with molecules of the antibiotic. A clear correlation can be observed between generations and the arrangement of peripheral groups, which in turn impacts the presentation of loads for antimicrobial or antiinfective purposes. The attachment of targeting moieties to antibiotics is closely related to the synthetic properties of terminal functionalities, which serve as an optimal solution for conjugating multiple copies of drugs and/or ligands to the periphery of dendrimers. By conjugating multiple antibiotic molecules to a dendrimer scaffold, it is possible to convert the structure into a molecule with a high-affinity [22,23]. Fullerenes are closed-cage carbon molecules and found to possess broad spectral antibacterial and its effect was found to be enhanced of drug conjugation. Dendrimers are found to address the solubility, toxicity, and stability issues that prevent many drugs from being effectively utilized, thereby enhancing their potential for clinical applications.

Various studies highlighted the antibacterial activity of dendrimers against both gram-positive as well as gram-negative bacterial pathogens. To mention, while conjugating an antibiotic to a dendrimer, two methods are adopted viz., non-covalent or covalent. Non-covalent attachment involves creating a stable guest-host complex between the dendrimer and antibiotic, which can withstand the gastrointestinal system and bloodstream. This complex can have an antimicrobial effect on its own or release the antibiotic upon entering the bacteria [24]. Also *in vitro* studies demonstrated the improved activity of combination of sulfamethoxazole and amino-terminated EDA-core Poly(amidoamine) (PAMAM) dendrimers against *E. coli*. In an in vitro study involving, *Chlamydia trachomatis* the intracellular pathogen related to sexually transmitted infections and eye infections, the hydroxyl-terminated generation 4 PAMAM dendrimer, after being conjugated with Azithromycin, effectively treat inhibited intracellular *C. trachomatis* by releasing the antibiotic upon cellular uptake. In yet another study, compared to free erythromycin, the nanodendrimer conjugated with erythromycin exhibited notably greater antibacterial efficacy against various bacteria [12,24,25].

3.4   Lipid based nAbts

Lipid-based NPs are utilized as vehicles for delivery of antibiotics can be categorized into liposomes, self-nanoemulsifying drug delivery system (SNEDDS), niosomes, solid lipid NPs (SLNs), and nanostructured lipid carriers (NLCs). In these types of nAbts liposomes are mainly utilized for delivering compounds with antibacterial properties. Liposome is a spherical structure made up of phospholipids and it possesses a hydrophilic head and a hydrophobic tail and is seen as a bilayer vesicle structure with size ranging between 25-1000 nm. This special structure allows liposomes to encapsulate both hydrophilic and hydrophobic compounds. Based on their lipophilicity antibiotic molecules are dispersed in the hydrophilic/hydrophobic compartments of liposomes, making them an ideal carrier for antibacterial agents. The biocompatibility as well as biodegradability of lipids make them attractive models. Lipids act as a protective barrier for drugs from GI damage and aids in controlled drug release. Additionally, they enhance the absorption of drugs by facilitating transcellular/paracellular/ lymphatic transport, and ultimately enhancing the bioavailability of antibiotic/s [26].

Liposome and Niosome

For more than four decades liposomes have been utilized as oral delivery systems for antibiotics. The structural resemblance to cell membrane makes liposomes attractive drug delivery systems with excellent biocompatibility. Their hall marks include convenient and controlled drug loading and release, high entrapment efficiency, agreeable safety profile, convenient surface modification. Their disadvantages include structural damage and premature drug release in the GIT due to the interference of bile salts, digestive enzymes as well as the gastric acid. Many times, their larger size may interfere with the their penetration through the GI barriers. Surface modification was found to be an effective way out from the aforementioned glitches. For e.g., vancomycin conjugated folic acid and poly ethylene glycol (PEG) surface modified liposome showed enhanced stability. Linolenic acid, Tween 80 and dioleoylphosphatidylethanolamine (DOPE) modified liposomes encapsulatedg amoxicillin resulted in better stability, increased resident time and appealing tolerance to acidic pH [26,27]. As mentioned earlier, phospholipids in liposomes tend to undergo oxidative degradation in GIT. Also, their manufacture is an expensive affair. Thence scientists tried to substitute phospholipids with various nonionic surfactants/amphiphilic molecules that are renewable viz. creatinine, creatinine derivatives, cetyl alcohol, stearyl alcohol.

Niosomes refer to the drug delivery agents comprising of non-ionic alkyl/polyglycerol etheric surfactant as well as cholesterol, which form microscopic lamellar structures in an aqueous phase. Niosomes are noted for desirable features like biodegradability, biocompatible low cost, low toxicity as well as stable physical/chemical properties. Niosomal cefixime, niosomal daptomycin, niosomal levofloxacin as well as niosomal levofloxacin are proved to be effective on oral administration [27].

Self-nanoemulsifying drug delivery system (SNEDDS) is basically a stable nanoemulsion consisting of water, oil, Tween-80/Span-20(surfactants), 1, 2-octanediol and propylene glycol as cosurfactants and fat-soluble drugs. The particle size of this complex typically ranges between 20 and 200 nm. Nanoemulsions exhibit sensitivity and metastability, whereas SNEDDS spontaneously generate oil-in-water nanoemulsions when orally administered, triggered by mechanical forces in the gastrointestinal tract (GIT). SNEDDS facilitates the transport of drugs through the intestinal lymphatic system and safeguards them against degradation within the digestive tract. For e.g., Cefpodoxime proxetil (CP) is an oral prodrug of cefpodoxime, which belongs to the BCS IV class of β-lactam antibiotics and exhibits low bioavailability. To enhance its oral delivery, CP has been encapsulated in different formulations of self-nanoemulsifying drug delivery systems (CP-SNEDDS). These formulations consist of surfactants/co-surfactants such as Tween-80, propylene glycol, and tocopherol polyethylene glycol succinate (TPGS). Cefpodoxime proxetil (CP) is a prodrug of cefpodoxime, a BCS IV class β-lactam antibiotic with low oral bioavailability. CP encapsulated with various SNEDDS (CP-SNEDDS) formulations with oil, surfactants and co-surfactants, showed faster dissolution, higher permeability and needed only half of the MIC required by the conventional formulations in vitro. Also, the oral bioavailability CP-SNEDDS was 4 times higher than CP [12,27].

Solid Matrix Mediated Lipid Nanoparticle Formulations (LNFs)

Solid lipid nanoparticles (SLNs) as well as nanostructured lipid carriers (NLCs) are LNFs primarily utilized for delivering antibiotics. The very distinctive shell-core structure enabled LNFs as appropriate delivery systems for antibiotics with varying composition, lipophilicity. Also, LNFs enhances the solubility, permeability as well as the bioavailability of encapsulated drugs. SLNs are colloidal carrier systems that combine the advantages of polymeric NPs with liposomes. SLNs are utilized for water-soluble drug delivery and are spherical NPs composed of a solid lipid core that holds the antibiotics. The lipid components of SLNs matrix are in solid state at body/room temperatures; they mostly include medium- or long-chain lipids like anisodylglycerol, stearic acid, tripalmitin, triglycerides, super-purified waxes, and complex glyceride emulsions. Their size range lies between 10-1000 nm. These LNFs are designed for sustenance of drug release thereby curtailing the chances of repeated administration of drugs. SLNs are principally efficacious in tiding over the first-pass metabolism (The first pass effect can be defined as a phenomenon in which an antibiotic/ drug is metabolized at a specific site in the body of human that causes a reduction in the concentration of the active drug upon reaching its location of effect) and leading to enhanced intestinal lymphatic transport via paracellular and transcellular pathways of enterocytes, as well as the endocytosis of phagocytes. Along with the solid lipids, surfactants like poly vinyl alcohol (PVA) /Tween-80 are also utilized in SLNs [12,27].

One of the advantages of liposome structure-based nAbts is their high stability and low toxicity. Liposomes have been shown to have low toxicity to mammalian cells, making them a promising candidate for use in medical applications. Moreover, liposomes offer the opportunity to incorporate diverse drugs and molecules, thereby augmenting their antibacterial efficacy. Additionally, liposomes can be functionalized with a variety of drugs and other molecules to enhance their antibacterial activity. Various works highlighted the antibacterial activity of liposome structure-based nAbts against disease causing bacteria.

3.5   Polymeric based nAbts

Polymeric NPs utilized in antibiotic delivery refer to NPs comprising of polymers with biocompatibility and biodegradability. Common polymers used for polymeric NPs include poly(lactic-co-glycolic acid) (PLGA), polyethylene glycol (PEG), chitosan etc. Due to its strong affinity to cell membranes, chitosan serves as a potent targeting agent for the precise delivery of antibiotics to bacterial cells. Additionally, it effectively traps antibiotic molecules during the delivery process.

Due to the protonation of its primary amine groups, chitosan carries a primarily +ve charge, allowing for covalent and electrostatic bonding and creating nanosystems characterized by a dense concentration of surface cations. Cationic polymers are identified as significant antibacterial agents as bacteria possess a low propensity to develop resistance against them. Due to its strong affinity to cell membranes, chitosan can effectively target bacteria and deliver antibiotics while trapping the drug molecules within its structure. The polycationic nature of chitosan allows for strong electrostatic attraction to the anionic components of the cell wall of bacteria. When combined with chitosan nanocarriers, antibiotics can be released slowly and continuously, resulting in sustained drug delivery and sudden death of the cell due to cell membrane disruption. Chitosan based nAbts are generally spherical and their size ranges between 30–130 nm. Polymeric chitosan crosslinked folic acid-based carbon quantum dots (CQDs) supplemented with gentamicin showed improved antibacterial efficacy with other properties [12,28]. Ionic gelation mediatedly synthesized chitosan NPs in combination with gentamicin resulted in gentamicin-loaded chitosan NPs (Gen-Cs) of 100 nm. The MIC and MBC of Gen-Cs nAbt were half as compared to free gentamicin against *Brucella* [29].

Many times, polymeric NPs possess antimicrobial nature by themselves. For e.g., NPs of PLGA—poly (D, L-lactide-co-glycolide) a lactic acid-glycolic acid copolymer, are biodegradable and biocompatible and are opted as drug delivery systems. Modifying the surface of PLGA NPs can result in responsive behavior to specific stimuli or targeting capabilities, influencing their biodistribution in cells, and uptake pathways in bacteria. One can understand that PLGA-based antibiotics are mainly affected by the enhanced permeability and retention (EPR) effects, which enable the targeting of multiple structural/functional components of the cell of bacteria and the infected wound site [30][31]. PLGA NPs conjugated azithromycin –rifampin combo exhibited enhanced antibacterial efficacy to *C. trachomatis* [32].

3.6   Mesoporous nAbts

Mesoporous NPs are amorphous solids. They have varying void volumes contributed by inner pore diameters. But in case of an ordered mesoporous structure, the pore diameters as well as void volumes are uniform. Various studies explored the efficacy of mesoporous silica for effective delivery of antibiotics. Mesoporous silica has garnered significant interest since its discovery in the late 1970s due to its distinct characteristics, including highly ordered pore structures, high specific surface areas, and ability to synthesize in spherical, rod, disc shapes as well as in the form of powder. In contrast to traditional porous silica, mesoporous silica has remarkably ordered pores, resulting from the application of nanotemplating method while they are synthesized. These porous NPs are considered to possess superior drug sequestering capacity in their voids, resulting in high drug loading capacities [33,34,35]. Antibiotics can be conjugated by incorporating functional organics into their inner walls as well as pore spaces. By incorporating long-chain alkyl chains onto mesoporous silica, the functionalization of macrolide antibiotics led to a reduction in degradation rate and an increase in hydrophobicity of the resulting conjugate, resulting in efficient drug delivery [36,37]. Mesoporous NPs possess better potential for differential drug release compared to others. Also, a greater number of inner cavities results in electron-deficient/electron-rich areas, based on which negative or positive charged antibiotics molecules can be entrapped. Even the concentrations of the drugs can be maintained by size and shape control of the voids. The mesopores are predominantly governed by electrostatic interactions between the antibiotic molecules and the surface of the particle. Also, H- bonds are formed between the surface of silica (Si-O bonds) and molecules of antibiotics. Mesoporous silica NPs with gelatin/colistin coating hindered premature release of colistin from the mesopores, efficiently disrupting *S. aureus* biofilm. This finding can have a great impact in the treatment of osteomyelitis (a bone infection) principally aggravated by *S. aureus* [38].

4.   Mode of action of nAbts

In general, antibiotics are categorized based on their effect on the pathogen's cellular component/function they affect, whether they are bactericidal or merely bacteriostatic drugs. While considering bactericidal or those antibiotics induce the killing of bacteria mostly interfere with DNA/RNA synthesis, synthesis of proteins or cell wall. It can be understood that generally bacterial death is an essential outcome of the antibiotic mediated disruption of proton motive force across the cell membrane [39]. Naturally, the cell wall of bacteria is peptidoglycan (a covalently cross-linked polymer matrix formed by peptide-linked β-(1–4)-N-acetyl hexosamine) layers. The mechanical strength contributed by cell wall determines a bacterium's ability to withstand environmental challenges including osmotic pressure. Hence antibiotics that interfere with the cell wall synthesis can detrimentally affect the cell integrity [40]. Certain antibiotics affect the composition of cytoplasmic membrane through integration of mistranslated membrane proteins resulting in increased cellular permeability thereby permeating enhanced access of the antibiotic molecule into the cell via the distorted cell membrane [41].

Similarly, metal NPs have been shown to have antimicrobial properties. The biophysical interactions between bacteria and NPs including uptake by bacterial cells as well as aggregation of NPs result in cytoplasmic membrane damage and toxicity. The mode of action against bacterial pathogens can be attributed to several mechanisms, including i) Disruption of cell membrane: Metal nanoparticles can interact with bacterial cell membranes, leading to their disruption and eventual cell death. ii) Oxidative stress: Metal nanoparticles can generate reactive oxygen species (ROS) that can cause oxidative stress in bacterial cells, leading to compromised integrity of the cytoplasmic membrane, cell damage and death. iii) DNA damage: Metal nanoparticles can interact with bacterial DNA, leading to DNA damage and eventual cell death. iv) Protein denaturation: Metal NPs can interact with bacterial proteins, leading to their denaturation and loss of function. v) Enzyme inhibition: Metal NPs may interfere with bacterial enzymes and this may lead to disruption of metabolic pathways consequently resulting in bacterial cell death.

Unlike mammalian cells that possess the ability to engulf (endocytosis) NPs on contact, generally bacteria (even though exceptions do exist) do not uptake nanomaterials. Even though various reports suggested the nAbts mediated disruption of bacterial cell membrane, still there exists a haziness whether the NPs or the toxic chemical moiety associated with the nAbts that enables the fatal outcome. In the revolutionizing approach of tailoring the benign nanostructures to enable the busting of cell membrane of bacteria while sparing mammalian cells. This can be achieved through a nanoengineering strategy in which judiciously structured nAbts bearing biocompatible chemical moieties possessing bactericidal effect on physical contact. However, these nanostructures remain safe for mammalian cells when engulfed [14]. On a closer look it can also be perceived that NPs conjugated with antibiotics lead to biomechanical damage mediated by dissolution and release of metal-ions resulting in physical disruption of cellular structures in turn blocking the bacterial electron transport chains. Ultimate suppression of growth and cell death take place due to the signal transduction that hinders ATPase activity leading to the ribosome subunits to bind with tRNA modifying mechanisms like protein regulation/metabolism of carbohydrate, fat and energy.

Reports suggested that NPs are found to intervene bacterial resistance in pathogens with considerable number of mutant genes. Studies underscore the successful cumulative effect of silver NPs along with antibiotics like cefuroxime, fosfomycin, cefoxime, azithromycin and chloramphenicol against *Escherichia coli*. Similarly, zinc oxide NPs (ZnONPs) along with the aforementioned antibiotics showed improved antimicrobial efficacy against *E. coli* and *S. aureus* than when the antibiotic was used alone. On the contrary, silver NPs in combination with oxacillin and neomycin displayed low efficiency on *S. aureus* whereas individually antibiotic showed better effect. Also, most antibiotics displayed an effect of anatagonism along with ZnO NPs against *Salmonella* spp [42].

It has been demonstrated that when carbon based engineered nanoparticles (ENPs) as well as inorganic ENPs come in direct contact with bacteria, the cell membrane becomes damaged leading to bacterial inactivation. During the dissolution of silver NPs, Ag^+ ions are set out leading to cytotoxic effect [43]. Either the close proximity to AgNPs or its attachment to the bacterial cell membrane facilitate the interaction of microbes to the ionic Ag^+ species. It was noted that AgNPs accumulation in the cell membrane of *E. coli* resulted in pit formation in its cell wall. Transmission electron microscopy analysis showed that, ZnO NPs could damage the membrane and gain entry into *E. coli cells* [44].

Overall, the effect of metal NPs on bacterial pathogens is an intriguing complex mechanism comprising of multiple processes, which may vary depending on the type of metal, size, and shape of the NPs and the type of bacteria being targeted. Fig 2 shows the possible mode of action of nAbts on bacteria.

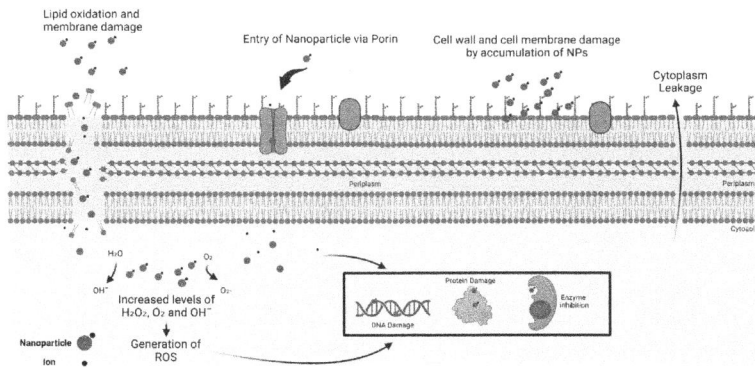

Fig 2. Diagrammatic representation of the mechanistic aspects of nAbts against bacteria (created with Biorender.com- April,20,2023)

5.   Functional properties of nanoantibiotics

NAbts possess unique functional characteristics compared to traditional antibiotics. Morphological features like variations in size, shapes as well as surface chemistry can greatly influence biodistribution, delivery of therapeutics as well as uptake by bacterial cells. To mention, as an example, the mechanism of action of lipid based nAbts involves several steps. First, the liposomes must be functionalized by targeting moieties such as antibodies or peptides that recognize specific bacterial proteins. Once the liposomes are bound to the bacterial cells, they can enter the cells and release their antibacterial cargo. The antibacterial activity of liposome structure-based nAbts is thought to be due to their ability to disrupt bacterial membranes, interfere with bacterial DNA replication and protein synthesis, and induce oxidative stress in bacterial cells.

Therefore, optimizing these aspects capacitate improved accumulation of nAbts in the targeted tissue. Some of the functional characteristics of nAbts include:

5.1   Reduced toxicity

NAbts can be designed to target particular cell types, curtailing off-target effects and decreasing toxic effects. One of the key reasons that nAbts are less toxic to mammalian cells is that the nanoparticles used in these antibiotics can help to target the antibiotic specifically to bacterial cells, while minimizing exposure to healthy cells. For example, AgNPs used in nAbts have been shown to bind to the cell membranes of bacteria, disrupting their structure and killing them. This targeting mechanism allows the antibiotic to be more effective against bacteria while reducing its impact on healthy cells. Also reduction in antibiotic dosage plays pivotal role in decreasing the degree of toxicity rendered by drugs [45]. It is well understood that comparatively lower concentration of nano-enabled antibiotics extends better efficacy compared to their parent antibiotics owing to the differential drug release kinetics [12].

5.2   Enhanced antimicrobial activity

NAbts can penetrate microbial cell walls more effectively than traditional antibiotics, allowing them to exert a more potent antimicrobial effect. NAbts have a larger surface area than traditional antibiotics, which permits to interact with bacteria more effectively and the large surface area of NPs enhances the number of interactions between the antibiotic and bacteria, improving the antibiotic's bactericidal effect. The small size of NPs allows them to penetrate through bacterial biofilms, which are protective layers of extracellular material secreted by bacteria that contribute greatly to drug resistance as well as pathogenesis as these biofilms are notoriously challenging to penetrate, making traditional antibiotics less effective against them. In addition, the synergistic effect extended by the antibiotic-NPs combo hikes the antibiotic's effect on the target microbe. For example, AgNPs-amoxycillin combo enhanced the antibiotic's effect on *P. aeruginosa* and in case of vancomycin a threefold increment in activity was noted against *Sterptococcus mutans* which was previously resistant to this antibiotic. Also, in the presence of Ag NPs clinical isolates possessing antibiotic resistance to one or more β-lactam class antibiotics showed significant reduction in the MIC (minimum inhibitory concentration) underscoring the fact that NPs addition with antibiotics not only curtails the MICs, but also make the resistant bacteria susceptible to particular antibiotics. In reality, this underlines the effect of antibiotics conjugated with AgNPs in hampering resistance among prominent pathogens as well as the dosage reduction of antibiotic to be administered [45]. From general observation it can be understood that combination therapy would open doors to the strategy of using NPs as adjuncts to presently available antibiotics which in turn can curtail MDR associated with prominent pathogens.

5.3   Extended circulation time

The size, morphology as well as solubility of the micelles significantly impact the circulation. The small size of nAbts permits longer retention in circulation, increasing their effectiveness. Circulation half-life can be elaborated as the time period taken for half the concentration of a drug/ compounds to be cleared from the blood circulation. This is a measure of how long the antibiotic remains in our circulation prior to being metabolized, excreted, or eliminated from the body. Modifying the antibiotics by stabilizing in a core made up of hydrophobic polymeric/lipid matrix encapsulated in a hydrophilic polymer lengthens the duration of nAbts in the blood circulation. It is noted that the small micelles possess extended blood circulation and they do not tend to the immune system to be considerably activated. In case the size is less than 20 nm, our renal system remediates them swiftly out of circulation prior to the achievement of the supposed action. In general, the hydrodynamic diameter is recommended between 60-150 nm so as to carry out successful delivery and release of the drug. Surface charge of the nAbts is a crucial aspect in deducing the cell adsorption and interactions. Certain *in vitro* studies highlighted the enhanced circulation time and uptake of charged micelles by cells. Conjugation of antibiotic to a nano polymer reduced metabolic clearance by enhancing the chemical stability of the ring structure and reduced renal clearance by size increment thereby extending the circulation half-life [46]. The circulation half-life is an important pharmacokinetic parameter that impacts the potency and toxicity of a drug. A longer circulation half-life generally means that the drug persists actively in the body for a longer period, which can increase its intended use and efficacy. However, it also may boost the chances of toxic effects, in case the drug accumulates in the body or is not eliminated properly.

5.4   Improved solubility and bioavailability

NAbts can be structured to enhance the solubility and bioavailability of antibiotics that are not easily soluble, making them more effective. Various modes like intravenous, oral, transdermal as well as transpulmonary routes are adopted for administration of antibiotics. Due to convenience, safety as well as superb patient compliance, oral route is the most opted one for long term medications. Various physical and chemical hurdles of the gastrointestinal tract (GIT) like drug metabolic enzymes, P-glycoprotein and mucus layer, epithelial tight junction etc. influence the efficiency of oral administration [12,47]. Many antibiotics like ciprofloxacin, rifampicin, vancomycin and cefopodoxime have low solubility and low permeability. The low solubility and permeability impede their GIT absorption thereby restricting bioavailability [26] [48]. The significant surface area of nanoparticles (NPs) not only improves the solubility of antibiotics but also increases the contact area within the gastrointestinal (GI) tract. The large specific surface area of NPs enhances solubility of antibiotics as well as the GI contact area, simultaneously the nanoscale carrier encapsulated antibiotics are also protected from degradation by endo- or exopeptidase in GIT or cytochrome P450 enzymes in liver or small intestine which could otherwise result in untimely destruction of the drug molecules. The Lipid-based NPs utilized in antibiotic delivery includes niosomes, liposomes, solid lipid nanoparticles (SLNs), self-nanoemulsifying drug delivery system (SNEDDS), and nanostructured lipid carriers (NLCs). Encapsulation by nanocarriers acts as a protective shell for the antibiotic. The traits of biocompatibility and biodegradability borne by lipids enhance drug absorption via paracellular/transcellular/lymphatic transport leading to increased bioavailability. To cite, liposomes made up of bile salt-loaded cefotaxime efficiently hindered bile acid mediated damage toward the payload consequently improving the stability of the antibiotic. Drug solubility is enhanced by the nanoformulations' large surface area [26] [49]. Polymeric micelles can enhance a10- to 5000-fold increase in the solubility of water insoluble antibiotics.

5.5   Controlled release

NAbts can be structurally modified to release their active ingredients in an orderly mode, providing sustained antimicrobial activity over time. In fact, the precise delivery of drug molecules at a prolonged rate of release to the infection site is boosted up by the form and its nano level size. As mentioned in one of the sections above, compared to the molecular form of the antibiotics lower concentrations of nano-enabled multiple antibiotics exhibited enhanced efficacy due to differential release kinetics [12].

5.6 Targeted delivery: NAbts can be targeted to specific cells or tissues, increasing their specificity and reducing the risk of resistance. For example, as in the case of carboxyl-modified mesoporous silica NPs bearing polymyxin B and vancomycin displayed enhanced synergistically high antibacterial efficacy as well as biocompatibility. Usage of nanocarriers for antibiotic delivery leads to enhanced targeted delivery of comparatively higher doses of drug molecules to the cell of bacteria causing enhanced efficacy as well as therapeutic potential [50]. Also, while considering for clinical use, the drug carrier must be inert as well as possess comparatively lower toxicity while effectively interfering with the pathogen's metabolism. Porous inorganic NPs like silica are considered to be well suited for microbiological applications [51].

Overall, antimicrobial therapy can potentially benefit from the ability of nAbts to enhance effectiveness and safety. Also, institution of nAbts could be a favorable strategy to combating the growing issue of antibiotic resistance.

The smaller size attributes to the effect of NPs. Increment in the ratio of surface area to volume exponentially rises along with size reduction consequently leading to enhanced chemical as well as biological reactivities. For e.g., as the size of the NP is reduced from 30-3 nm, the number of surface molecules expressed to be amplified from 10%- 50% [52,53]. Otherwise, the surface area of the conjugate's colloidal system rises with decrease in the size of NPs enabling functionalization of more antibiotic molecules onto the surface of nanoparticles. In fact, the particle characteristics are greatly relied upon the approach of synthesis based on which the shape, size and surface functional groups are altered. The presence of efflux pumps in AMR bacteria eliminates the antibiotic molecules prior to their entry to the target site ending up in unsuccessful drug delivery and outcome [54].

6.   Challenges of nAbts

6.1   Safety Concerns

As in the case of any other new era technology or process, one of the major challenges associated with nAbts is ensuring their safety based on the lack of toxicity on the mammalian system to ensure that they are safe for human use. The interactions of NPs with mammalian cells remains largely incomplete, and there is a risk that they may cause unintended harm to human cells or tissues. Here are some of the safety concerns associated with nAbts:

Toxicity: NAbts may possess toxic effects including inducing oxidative stress and inflammation, damaging DNA, and interfering with cellular processes of mammalian cells. In fact, NPs and nanomaterials interact with the soft surfaces of biological systems like cells, thereby playing pivotal roles in implementing their biomedical functions and causing toxicity. The toxic manifestations of NPs and engineered nanomaterials are largely dependent on their physicochemical characteristics as shape, size, chemical composition, crystal structure, physiochemical stability, surface area, surface energy and roughness of the surface. In reality, the distinctive property of nanomaterials and NPs is their high surface-to-volume ratio that can contribute to their beneficial features, paradoxically this very same property is also linked to unique mechanisms of toxicity. Cytotoxic effects mediated by NPs/nanomaterials on normal tissue and organs can be a stark limiting factor that deters their clinical use [55]. Especially metallic and carbon-based nAbts, are found to cause severe toxicity due to prolonged exposure. Silver NPs based formulation may result in irreversible pigmentation in the skin and eyes. Toxicity is concentration dependent, and it was also suggested that not just nanotubes and fullerenes extended toxicity but even the solvent contaminants present in the formulation can be contributive [56]. There is a great probability of metal NPs to gain entry into the blood circulation once inside human body. The possibility of higher reactivity between the positively charged AgNPs with the RBCs possessing negative surface charge point in the direction of cellular toxicity. Furthermore, the biocompatibility of orthopedic implants is being undermined by the cytotoxicity of AgNPs against osteoblasts and osteoclasts. [57,58,59]. The safety of nanosized antibiotics drug carriers and NPs, especially in long run can be a prevailing safety concern and must be looked upon from the perspective of therapeutic value [13] as certain studies reported that protein based NPs possess hepatotoxicity, nephrotoxicity, cardiotoxicity, hypersensitivity and aggregation whereas lipid based NPs possess cardiopulmonary complications and anaphylactoid reactions. Additionally, metal NPs are found to cause anxiogenic and depression effects, reproductive and developmental toxicity, genotixicity DNA damage and cellular inflammation [60]. Hence the toxicity, translocation to secondary target organs, non-specific protein interaction etc. of nAbts must be closely studied.

Biodistribution: On administration of a drug molecule, it gets distributed throughout the body. Otherwise biodistribution is referred to be the movement of the drug from the site of administration into the bloodstream, and its subsequent distribution to various organs, tissues, and cells in the body. Biodistribution is a significant term in pharmacokinetics that deals with mechanism pertaining to the drug absorption, distribution, its metabolism and elimination from the body. A proper perception about the biodistribution of a drug molecule is crucial for determining its efficacy as well as adverse effects, as well as for optimizing its dosing regimen the biodistribution of NPs depends on their physicochemical properties, which can impact their potential to cross the blood-brain barrier (BBB). Especially the smaller size of nAbts enable them to transit the biological barriers and build up in tissues and organs, which may lead to potential toxicity, as NPs may cross the BBB and accumulate in the brain, which could lead to neurotoxicity [61]. Reports suggest the ability of NPs into the human body through injection, ingestion as well as respiration and consequently accumulating into different tissues and organs. Also, the potential of NPs entering the brain by breaching the strong link between cells and transit the BBB. NPs bind with the cells containing CXCR6 chemokine receptor and transit the tight junction in the BBB causing potential effects [55]. These effects may be applicable in the case of nAbts too pointing to the chances of adverse effects.

Immunological Response: Particle specific features viz., shape, size or surface chemistry of NPs/nanomaterials contribute to the interactions with the immune system. The NPs- immune system is complex and not fully understood. NPs are capable of stimulating an immune response, including the release of cytokines and chemokines, which can contribute to inflammation and tissue damage. Another challenge is that the immune system may recognize the NPs as foreign and mount an immune response, which could potentially limit their effectiveness or cause adverse effects. Additionally, the long-term NPs-immune system interaction effects are yet unclear, which could raise concerns about their safety e.g., NPs are reported to either suppress or stimulate immune responses. Also, magnetic NPs are reported to overpower selected inflammation related immune responses in mice models as well as modulated LPS related inflammatory outcomes in primary human monocytes [62].

Interactions with other drugs: Many reports on the generation of adverse side effects of a drug to another type. In general, antibiotics which are involved in such adverse effects are those which themselves have some degree of toxicity to certain cephalosporins, aminoglycosides, tetracyclines, colistin etc. NPs can interact with other drugs, which can impact their efficacy and toxicity. The co-administration of NPs with other drugs needs to be thoroughly evaluated to ensure their safety and efficacy.

Environmental Impact: The free discharge of NPs into various environmental compartments can pose a risk to human health and the ecosystem. The environmental impact of NPs needs to be thoroughly evaluated, including their potential to accumulate in soil, water, and living organisms, and their impact on the ecosystem. The potential environmental impact of nAbts will rely on a number of factors, like the specific properties of the NPs used, the dose and duration of exposure, and the environmental conditions in which they are released. As far as NPs are concerned, their environmental fate is greatly influenced by a number of factors, including their size, shape, surface charge, surface chemistry, and aggregation state, their ability to absorb on to surfaces as soil, sediments and plant roots. Many times, their stability and behavior could be impacted by the presence of natural matter. Absorption to such surfaces enables them to be available to organisms thereby gaining entry into the food chain leading to biomagnification. Various environmental factors, such as temperature, pH, and salinity, also influence the fate of NPs. As a general rule, NPs reaching the environmental compartments may undergo physical, chemical and biological transformations that can affect their behavior which may have far reaching effects of the biota as well as the biosphere. NPs may aggregate or agglomerate mainly based on the pH and ionic strength and this in turn influences their mobility/ transport in aquatic or terrestrial compartments. They may be subjected to chemical transformation as oxidation/reduction, which have impact on their toxicity and reactivity. Overall, the potential environmental impacts of nAbts require careful consideration and monitoring to ensure their safe and sustainable use. The development of environmental friendly manufacturing processes, proper waste management practices, and the implementation of regulatory measures to control their release into the environment are critical steps towards mitigating the potential environmental impacts of nAbts.

Without doubt one can tell that the scientific community and regulatory bodies must take necessary steps to diligently assess the potential risks associated with nAbts. It is important to thoroughly study the safety of nAbts in both in vitro and *in vivo* models to ensure their safety before they can be used in clinical settings. The effectiveness and long-term safety issues pertaining to nanoantibiotcs have to be established by intensive laboratory, computer aided as well as clinical level research. For the purpose such evaluative research, sophisticated and high-end technologies including nanoproteomics, high-resolution single-particle scanning flow cytometry, nano secondary ion mass spectrometry imaging etc., can be adopted [12,63,64] prior to be applied in clinical settings. The disposal of nAbts and their manufacturing waste may pose a risk to the environment, particularly if not handled properly. The release of nanoparticles into the environment through wastewater or solid waste disposal can have unintended consequences. Appropriate guidelines must be adopted to impede the risks to the environment and its biota.

6.2   Manufacturing Complexity

Another challenge associated with nAbts is the complexity of their manufacturing process. Scale-up for the production of NPs requires specialized equipment and expertise, which can be expensive and time-consuming. It can also be difficult to ensure the reproducibility of nanoparticle production, which can impact the consistency of their antimicrobial activity. Achieving a uniform and consistent size and shape of the NPs as well as controlling their surface chemistry and functionalization need special attention. The methods like microemulsion, sol-gel and precipitation techniques require proper control over reaction conditions, such as temperature, pH, and concentration to achieve the desired size and shape of the NPs. In order to prepare the nAbts, the synthesized NPs need to be functionalized with antibiotics or other drug molecules, which requires additional processing steps and expertise. Due to the high surface area and reactivity of the NPs, functionalization via, surface modification or coating of the NPs with a layer of antibiotics can be a challenging process. Stringent quality control procedures to be followed for ensuring the safety and efficacy of the final product and require rigorous testing and characterization of the NPs and include size/ shape analyses, surface chemistry, and stability biological activity and toxicity. The manufacturing of liposomal drug formulation may be technically difficult to an extent. In total, nAbts manufacturing process require specialized equipment, expertise, and careful attention to quality control. Even then, the potential benefits of nAbts over antibiotics make them an attractive model confronting current research.

6.3   Regulatory Hurdles

The development and approval of nAbts may be hindered by regulatory hurdles, nAbts may require special considerations that are not currently included in regulatory guidelines for traditional antibiotics. It is the necessity of the hour to institute clear regulatory reccomendations for the development and approval of nAbts so as to ensure their efficacious and safe use in clinical settings, as there does exist conflicts in findings pertaining to the toxic/adverse effects of NPs and nanobased formulations. Often these can be contributed to the deficient standardized experimental measures [65]. It can be noted that sometimes similar experiments may lead to dissimilar outcomes and conclusions. Each NP based nanoparticle formulation must be tested based on their portal of entry [66]. This issue needs to be addressed appropriately and needs the active involvement and cooperation of local, national or international regulatory bodies. The PP emphasizes that scientific uncertainty should not hinder action when there is a possibility of significant adverse effects are present.

To encourage the production and safe clinical use of nAbts, it is suggested to exchange the data on the toxicity of nanomaterials between various sectors, which in turn can considerably curtail the time lag for products to enter the market. The precautionary principle (PP) which has been a subject of extensive debate in international politics was initially incorporated into EU environmental regulation through Article 174 (formerly Article 130r) of the Maastricht Treaty. The PP emphasizes that scientific uncertainty should not hinder action when there is a possibility of significant adverse effects are present. Kessler has elaborately discussed the present status of research and regulatory policies related to nanomaterials [67] [68]. One of the most critical concerns regarding threats is whether demonstrating the safety of nanoformulations for humans alone is adequate or if it is necessary to also establish their environmental fate. The reality is that an emerging number of NP based formulations are being dealt by the scientific community with an expanding number of materials of whose properties are unclear to a greater extent, because of which the currently available assessment protocols and statutes could yield appropriate results. The unique properties of nAbts require special testing and evaluation strategies that add up to the challenge of implementing in manufacturing processes. The manufacturing of nAbts must comply with regulatory requirements, including Good Manufacturing Practices (GMP) and Quality Control (QC) standards. NAbts must be approved by regulatory agencies such as the Food and Drug Administration (FDA) in the USA or the European Medicines Agency (EMA). Such regulatory agencies will evaluate the safety and efficacy of the medication before granting approval. This makes it challenging for researchers and manufacturers to navigate the regulatory process and comply with the relevant regulatory requirements.

6.4   Scalability and Cost

The sizeable production of nAbts can be challenging due to the elaborateness of the manufacturing process and the cost of ingredients and equipment. The scalability and cost of manufacturing are critical factors in the commercialization of nAbts. The cost and expenditure pertaining to the developmental aspects and use of nAbts are a significant concern. Some of the factors that add up to the expensive disposition of nAbts are contributed by the R&D expenses as the development of nAbts including the development of novel mode of synthesis, optimization of NPs' properties as well as testing of their safety and efficacy need substantial investment. Manufacturing cost of nAbts can be considerably more than traditional antibiotics due to the necessity for specialized equipment, expertise as well as additional quality control requisites.

7.   Bacterial resistance mechanisms towards nAbts

Although nAbts have shown promising results in the laboratory, bacterial resistance to these agents is still a concern. Recent years have witnessed evolution of defense strategies in bacteria to cope up with metal NPs. Reports on clinical and non clinical isolates having elevated resistance to Ag+ ions and Ag-NPs are available. This may be due to the recurring exposure to sub-lethal concentrations of NPs. Also, the widespread use of Ag/Cu NPs may potentiate the co-selection of metal-AMR traits. Environmental isolates of *Bacillus* spp as well as opportunistic pathogens like *P. aeruginosa, E. coli* and *S. aureus* showed such a tendency, especially those strains with antibiotics resistance. All these point to the direction that repeated exposure to sublethal/subinhibitory doses of NPs contribute to development of NP resistance [69].

Various studies suggested that bacteria utilize a multitude of defense mechanisms to tackle the toxic effects mediated by NPs. Genetically encoded defense mechanisms as well as collective response are utilized by bacteria to resist nAbts and their toxicity. These strategies include i) decreased NP uptake/adsorption, ii) enhanced efflux iii) upregulation of antioxidant mechanism iv) formation of biofilm and microbial aggregates v) production of substance that immobilize NPs vi) Co-selection of Antibiotic Resistance Genes (ARGs) on exposure to metal NPs.

Decreased NP uptake/adsorption

Cation-selective porins mostly present in Gram-negative cell wall normally allow the smaller NPs/NP-released ions transit via the outer membrane thereby yielding to toxic impacts of NPs. On the other hand, if expression of porin is down regulated the access routes of NPs will be restricted by rendering the microbe insusceptible to NPs. Loss of porins combined with and overexpression of efflux systems can be an effective resistance strategy. Downregulation of genes encoding porins is an effective mode. Gene mutations that result in OmpC or OmpF porin deficiency resulted in increased NP resistance in bacteria like *E. coli.* This can be considered as a genetically encoded defense mechanisms:

Enhanced efflux:

Efflux pumps are specialized membrane proteins found in the bacterial cell membranes that are responsible for pumping a variety of molecules including antibiotics out of the cell. When bacteria are exposed to drug molecules, they may activate their efflux pumps in order to eliminate the toxic molecules from the cell. This in turn restricts the entry of NPs as well as accumulation of NPs in the cytosol. Utilization of various efflux systems to extrude metal ions out of the cell is present in bacteria. Upregulation of genes encoding efflux pumps is a prominent strategy Transcriptomic studies in *P. aeruginosa* PAO1 showed the overexpression of genes encoding resistance-nodulation-cell division (RND) pumps and other metal efflux systems on exposure to moderate amount of Cd quantum dots and metal NPs. *E. coli* showed copper/silver (cusA) and copper (copA) efflux systems encoding gene expressions, while induced with Ag-NPs. RND efflux systems are found to drive out drug molecules and build up insuspectibility to nAbts [69].

Upregulation of antioxidant mechanism

Antimicrobial agents can generate reactive oxygen species (ROS) in bacteria leading to oxidative stress thereby damaging the bacteria. In order to tide over the detrimental effects of ROS bacteria have evolved various antioxidant mechanisms that render them with the ability to neutralize ROS. One of the modes to upregulate their antioxidant mechanisms is by triggering their regulatory pathways in response to antimicrobial stress. Certain bacteria increasingly produce antioxidant enzymes like catalase, superoxide dismutase (SOD) peroxidase and reductases. These enzymes can help neutralize ROS and protect the bacterial cell from oxidative damage. The production of enzymatic antioxidants are often regulated by systems called regulons and these are clusters of genes that are regulated together to control specific responses. For e.g., the OxyR regulon seen in certain Gram+ve and Gram-ve bacteria. An influx of hydrogen peroxide (H_2O_2) in the bacterial cell will induce a response from the OxyR transcription factor resulting in an oxidized form. On oxidation the transcription factor will positively regulate the genes associated with the OxyR regulon. This leads to recruitment of RNA polymerase. The OhrR regulon is yet another antioxidant regulon that identify and eliminates organic peroxides formed at the time of oxidative stress. *E. coli* and *P. aeruginosa* PAO1 exposed to Ag-NPs as well as Cd-QDs showed ROS scavenger genes' over expression. Additionally, bacterial resistance to Ag-NP was contributed by gene mutations in nucleotide synthesis and oxidative stress defenses [69][70][71].

Formation of biofilm and microbial aggregates

In biofilm formation, microbes irreversibly adhere to the surface and grow. The extracellular polymeric substances (EPS) formed by the bacteria aid in this attachment as well as matrix formation resulting in microbial aggregates. Subsequently alteration in its phenotype may take place. In addition to their own biofilms, bacteria are also capable of forming aggregates with other bacteria, archae, protists or fungi as a consortia. Such organisms are seen to be embedded inside a self-secreted EPS made up of polysaccharides, peptides or DNA. Both biofilms as well as microbial aggregates enable bacteria to tide over the environmental stress as well as NP toxicity. The EPS in the form of a matrix turns into a barricade that obstructs the NP transit and even entraps NPs at the peripheral layer, curtailing exposure / modifying the NPs and reducing the reactivity as well as the antibacterial effect. In addition, this reduced NP penetration may enable bacterial cells in the inner layers of the matrix to detect sub-lethal concentrations. Consequently, these bacteria may generate an adaptive response that may result in enhanced NP resistance and increased growth of biofilm. This phenomenon can be termed as hormesis, in which the cell experiences an adaptive beneficial effect when exposed to a lower dosage of a drug which would otherwise be detrimental at higher doses. For e.g., sub-lethal doses of Ag-NPs and ZnO-NPs enhanced *P. aeruginosa* and *P. putida* growth/biofilm formation, resulting in quorum sensing, LPS biosynthesis genes as well as signal molecule secetion [69,72].

Production of substance that immobilize NPs

Bacteria can produce substances that can immobilize NPs. One example of such a substance is exopolysaccharides. These are complex polysaccharides that are secreted by bacteria and form a matrix-like structure around bacterial cells. This matrix can trap NPs and immobilize them. EPS may also agglomerate and deactivate the NPs. In a study using selected strains of *E. coli* and *P. aeruginosa*, it was found that the flagellin matrix overexpressed by them resulted in agglomeration of 20 nm AgNPs impeding the direct interaction between NPs-bacteria. *E. coli* capable of producing ECS are found to amend the zeta potential and size of the NPs thereby agglomerating them. In certain pathophysiological conditions NPs were reported to be covered with the biomolecule coronas that in turn interfered with the NPs physical contact with the bacterial pathogens thereby hindering their bactericidal effect. In bacteria like *P. aeruginosa*, ECS like flagellin leads to the agglomeration of NPs, whereas the pigments like pyocyanin, pyoverdin and pyochelin, inactivates the ions released by the NPs [69,73].

Co-selection of Antibiotic Resistance Genes (ARGs) on exposure to metal NPs

In reality, heavy metal exposure contributes to antibiotic tolerance in bacteria which is justified by the frequent presence of antibiotic-resistant bacteria in sites polluted with metals. Field studies in farms/ agriculture lands as well as in wastewater treatment plants, underscored the very aspect of prolonged exposure to metals as a co-selecting factor for ARGs and metal-resistant genes. Chances of metallic NPs to facilitate the dissemination of ARGs among bacterial species via co-selection and horizontal gene transference (HGT) are more. Therefore, the increasing usage of nAbts can be a matter of concern as prolonged bacterial exposure to sublethal levels of NPs could be a catalyzing factor inducing selective pressure that in turn exaggerate the dissemination of antibiotic resistance by bacterial conjugation and transformation in the various environmental compartments [69].

Conclusion

The use and overuse of antibiotics and other chemotherapeutics have contributed to development to drug resistance among prominent bacterial pathogens. Industrialization as well as pollution of the various environmental compartments have effects like adding oil to fire on the global issue of the emergence of MDR. The main challenge is that the pharmaceutical companies are currently stuck in a continuous cycle of trying to discover new antibiotics. Even though a limited number of drug manufacturers invest in the discovery of new antibiotics, they often do not receive a sufficient return on their investment. Hence at the financial standpoint, developing novel antibiotics appears highly unattractive because of the paucity of return on investment. However, the emergence of the new millennium brought promise to the field of antimicrobial therapy through the utilization of nanoparticles and their conjugated forms, resulting in the development of nAbts. A substantial positive attribute of nAbts is the dexterity to pass over the traditional antibiotic discovery pathways. The development of nAbts holds great promise in addressing the pressing issue of antimicrobial resistance. The antimicrobial agent/s-nanomaterials combo approach has resulted in enhanced efficacy and targeted delivery, making nAbts a potentially valuable therapeutic strategy. However, there are still substantial impediments that need resolutions prior to the wide use of nAbts in clinical settings. Overall, the future of nAbts as a viable alternative to traditional antibiotics looks promising, but further research and development are necessary to fully realize their potential. NAbts need to undergo extensive testing and evaluation to ensure their safety and efficacy in treating infections. The manufacturing process needs to be closely monitored, and the medication should be specifically targeted to the site of infection. Regulatory approval is necessary before the medication can be commercially available to the public.

References

[1] Information on (http://www.who.int/mediacentre/factsheets

[2] Information on (http://www.who.int/world-health-day/2011).

[3] Kamini, W. (2006). Emerging Problem of Antimicrobial Resistance in Developing Countries: Interwining Socioeconomic Issues. Regional Health Forum WHO South-East Asia Region, 7, [8-12].

[4] Nikaido, H. (2009). Multidrug Resistance in Bacteria. Annual Review of Biochemistry, 78(1), [119-146]. doi: 10.1146/annurev.biochem.78.082907.145923

[5] Laxminarayan, R., Matsoso, P., Pant, S., Brower, C., Røttingen, J.-A., Klugman, K., et al. (2016). Access to Effective Antimicrobials: A Worldwide Challenge. The Lancet, 387(10014), [168-175]. doi: 10.1016/S0140-6736(15)00474-2

[6] Christopher, J. L. M., Kevin S I., Fablina, S et al., Global burden of bacterial antimicrobial resistance in 2019: a systematic analysis. The Lancet (2022) 399(10325) P[629-655], https://doi.org/10.1016/S0140-6736(21)02724-0

[7] Ventola, C. L. (2015). The antibiotic resistance crisis: Part 1: Causes and threats. P&T, 40(4), [277-283]. PMID: 25859123; PMCID: PMC4378521.

[8] Spellberg, B., & Gilbert, D. N. (2014). The future of antibiotics and resistance: A tribute to a career of leadership by John Bartlett. Clinical Infectious Diseases, 59(Suppl 2), S71-S75. https://doi.org/10.1093/cid/ciu392

[9] Freestone, I., Nigel, M., Margaret, Sax., Catherine, Higgitt.,(2007) The Lycurgus Cup-A Roman Nanotechnology.2007. Gold Bulletin 40(4):[270-277] DOI:10.1007/BF03215599

[10] Information on https://www.azonano.com/

[11] Soares, S., Sousa, J., Pais, A., & Vitorino, C. (2018). Nanomedicine: Principles, Properties, and Regulatory Issues. Frontiers in Chemistry, 6. doi: 10.3389/fchem.2018.0036

[12] Mamun, M. M., Sorinolu, A. J., Munir, M., & Vejerano, E. P. (2021). Nanoantibiotics: Functions and Properties at the Nanoscale to Combat Antibiotic Resistance. Frontiers in Chemistry, 9, 687660. doi: 10.3389/fchem.2021.687660

[13]   Huh, A. J., & Kwon, Y. J. (2011). "Nanoantibiotics": A new paradigm for treating infectious diseases using nanomaterials in the antibiotics resistant era. Journal of Controlled Release, 156(2), [128-145]. doi: 10.1016/j.jconrel.2011.07.002

[14]   Jiang, Y., Zheng, W., Tran, K., et al. (2022). Hydrophilic nanoparticles that kill bacteria while sparing mammalian cells reveal the antibiotic role of nanostructures. Nature Communications, 13, 197. https://doi.org/10.1038/s41467-021-27193-9

[15]   Ana Isabel Ribeiro, Alice Maria Dias, and Andrea Zille. (2022). Synergistic Effects Between Metal Nanoparticles and Commercial Antimicrobial Agents: A Review. ACS Appl. Nano Mater., 5, [3030-3064]. https://doi.org/10.1021/acsanm.1c03891

[16]   Kotrange, H., Najda, A., Bains, A., Gruszecki, R., Chawla, P., Tosif M. M.(2021). Metal and Metal Oxide Nanoparticle as a Novel Antibiotic Carrier for the Direct Delivery of Antibiotics. Int J Mol Sci. 4;22(17):9596. doi: 10.3390/ijms22179596.

[17]   Neouze, M.-A., & Schubert, U. (2008). Surface Modification and Functionalization of Metal and Metal Oxide Nanoparticles by Organic Ligands. Monatshefte für Chemie - Chemical Monthly, 139(3), [183-195]. https://doi.org/10.1007/s00706-007-0775-2

[18]   Payne, J. N., Waghwani, H. K., Connor, M. G., et al. (2016). Novel Synthesis of Kanamycin Conjugated Gold Nanoparticles with Potent Antibacterial Activity. Front Microbiol., 7, 607. doi: 10.3389/fmicb.2016.00607.

[19]   Patil, T., Khot, V., & Pandey-Tiwari, A. (2022). Single-step antibiotic-mediated synthesis of kanamycin-conjugated gold nanoparticles for broad-spectrum antibacterial applications. Lett Appl Microbiol., 75(4), [913-923]. doi: 10.1111/lam.13764. PMID: 35689349.

[20]   Armijo, L. M., Wawrzyniec, S. J., Kopciuch, M., Brandt, Y. I., Rivera, A. C., Withers, N. J., et al. (2020). Antibacterial Activity of Iron Oxide, Iron Nitride, and Tobramycin Conjugated Nanoparticles against Pseudomonas Aeruginosa Biofilms. Journal of Nanobiotechnology, 18(1), 35. doi: 10.1186/s12951-020-0588-6

[21]   Carver, J. A., Simpson, A. L., Rathi, R. P., Normil, N., Lee, A. G., Force, M. D., et al. (2020). Functionalized Single-Walled Carbon Nanotubes and Nanographene Oxide to Overcome Antibiotic Resistance in Tetracycline-Resistant Escherichia Coli. ACS Applied Nano Materials, 3(4), [3910-3921]. doi: 10.1021/acsanm.0c00677

[22]   van Dongen, M. A., Silpe, J. E., Dougherty, C. A., Kanduluru, A. K., Choi, S. K., Orr, B. G., Low, P. S., & Banaszak Holl, M. M. (2014). Avidity mechanism of dendrimer-folic acid conjugates. Molecular Pharmaceutics, 11, [1696-1706]. doi: 10.1021/mp5000967

[23]   Falanga, A., Del Genio, V., & Galdiero, S. (2021). Peptides and dendrimers: How to combat viral and bacterial infections. Pharmaceutics, 13(1), 101. doi: 10.3390/pharmaceutics13010101

[24]   Chis, A. A., Dobrea, C., Morgovan, C., & Bălşeanu, T. A. (2020). Applications and limitations of dendrimers in biomedicine. Molecules, 25(17), 3982. doi: 10.3390/molecules25173982

[25]   Kannan, R. M., Nance, E., Kannan, S., & Tomalia, D. A. (2014). Emerging concepts in dendrimer-based nanomedicine: From design principles to clinical applications. Journal of Internal Medicine, 276, [579-617]. doi: 10.1111/joim.12280.

[26]   Wu, Z. L., Zhao, J., & Xu, R. (2020). Recent Advances in Oral Nano-Antibiotics for Bacterial Infection Therapy. International Journal of Nanomedicine, 15, 9587-9610. https://doi.org/10.2147/IJN.S279652

[27]   Lopes-de-Campos, D., Pinto, R. M., Lima, S. A. C., et al. (2019). Delivering amoxicillin at the infection site - a rational design through lipid nanoparticles. International Journal of Nanomedicine, 14, [2781-2795]. https://doi.org/10.2147/IJN.S193135

[28]   Kazeminava, F., Javanbakht, S., Nouri, M., Gholamzadeh, P., Rajabnia, R., & Mohammadi-Samani, S. (2022). Gentamicin-loaded chitosan/folic acid-based carbon quantum dots nanocomposite hydrogel films as potential antimicrobial wound dressing. Journal of Biological Engineering, 16(1), 36. doi: 10.1186/s13036-022-00318-4.

[29]   Razei, A., Cheraghali, A. M., Saadati, M., Khosravi, A. D., & Lotfi, M. (2019). Gentamicin-loaded chitosan nanoparticles improve its therapeutic effects on Brucella-infected J774A.1 murine cells. Galen Medical Journal, 8, e1296. doi: 10.31661/gmj.v8i0.1296.

[30]   Richards, S.-J., Isufi, K., Wilkins, L. E., Lipecki, J., Fullam, E., Gibson, M. I., ... & Fairen-Jimenez, D. (2018). Multivalent antimicrobial polymer nanoparticles target mycobacteria and gram-negative bacteria by distinct mechanisms. Biomacromolecules, 19(1), [256-264]. doi: 10.1021/acs.biomac.7b01561.

[31]   Liu, J., Gefen, O., Ronin, I., Bar-Meir, M., & Balaban, N. Q. (2020). Effect of tolerance on the evolution of antibiotic resistance under drug combinations. Science, 367(6474), [200-204]. doi: 10.1126/science.aay3041

[32]   Toti, U. S., Guru, B. R., Hali, M., McPharlin, C. M. Susan. M., Wykes, S. M., Panyam, J., & Lehrer, R. I. (2011). Targeted delivery of antibiotics to intracellular chlamydial infections using PLGA nanoparticles. Biomaterials, 32(27), [6606-6613]. doi 10.1016/j.biomaterials.2011.05.038

[33]   Sigma-Aldrich. (n.d.). Mesoporous silica. Retrieved from https://www.sigmaaldrich.com/OM/en/technical-documents/technical-article/environmental-testing-and-industrial-hygiene/waste-water-and-process-water-testing/mesoporous-silica

[34]   Croissant, J. G., Fatieiev, Y., Almalik, A., & Khashab, N. M. (2018). Mesoporous silica and organosilica nanoparticles: Physical chemistry, biosafety, delivery

strategies, and biomedical applications. Advanced Healthcare Materials, 7(4), 1700831. doi: 10.1002/adhm.201700831

[35] Selvarajan, V., Obuobi, S., & Ee, P. L. R. (2020). Silica nanoparticles-a versatile tool for the treatment of bacterial infections. Frontiers in Chemistry, 8, 602. doi: 10.3389/fchem.2020.00602

[36] Doadrio, J. C., Sousa, E. M. B., Izquierdo-Barba, A. L., Perez-Pariente, J., & Vallet-Regí, M. (2006). Functionalization of mesoporous materials with long alkyl chains as a strategy for controlling drug delivery pattern. Journal of Materials Chemistry, 16(5), [462-466]. doi: 10.1039/b510101h

[37] Dinos, G. P., & George, P. (2017). The macrolide antibiotic renaissance. British Journal of Pharmacology, 174(18), [2967-2983]. doi: 10.1111/bph.13936

[38] Aguilera-Correa, M., Gisbert-Garzarán, A., Mediero, M. J., Fernández-Aceñero, D., de-Pablo-Velasco, D., Lozano, D., Esteban, J., & Vallet-Regí, M. (2022). Antibiotic delivery from bone-targeted mesoporous silica nanoparticles for the treatment of osteomyelitis caused by methicillin-resistant Staphylococcus aureus. Acta Biomaterialia, 154, [608-625]. doi: 10.1016/j.actbio.2021.11.017

[39] Kohanski, M. A., Dwyer, D. J., & Collins, J. J. (2010). How antibiotics kill bacteria: From targets to networks. Nature Reviews Microbiology, 8(6), [423-435]. https://doi.org/10.1038/nrmicro2333

[40] Bugg, T. D., & Walsh, C. T. (1992). Intracellular steps of bacterial cell wall peptidoglycan biosynthesis: Enzymology, antibiotics, and antibiotic resistance. Natural Product Reports, 9, [199-215]. https://doi.org/10.1039/NP9920900199

[41] Davis, B. D., Chen, L. L., & Tai, P. C. (1986). Misread protein creates membrane channels: An essential step in the bactericidal action of aminoglycosides. Proceedings of the National Academy of Sciences of the United States of America, 83, [6164-6168]. https://doi.org/10.1073/pnas.83.16.6164

[42] Abo-Shama, U. H., El-Gendy, H., Mousa, W. S., Hamouda, R. A., Yousuf, W. E., Hetta, H. F., & Abdeen, E. E. (2020). Synergistic and Antagonistic Effects of Metal Nanoparticles in Combination with Antibiotics Against Some Reference Strains of Pathogenic Microorganisms. Infection and Drug Resistance, 13, [351-362]. doi: 10.2147/IDR.S234425

[43] Xiu, Z. M., Zhang, Q. B., Puppala, H. L., Colvin, V. L., & Alvarez, P. J. J. (2012). Negligible particle-specific antibacterial activity of silver nanoparticles. Nano Letters, 12(8), [4271-4275]. https://doi.org/10.1021/nl301934w

[44] Sondi, I., & Salopek-Sondi, B. (2004). Silver nanoparticles as antimicrobial agent: A case study on E. coli as a model for gram-negative bacteria. Journal of Colloid and Interface Science, 275(1), [177-182]. https://doi.org/10.1016/j.jcis.2004.02.012

[45]   Franci, G., Falanga, A., Galdiero, S., Palomba, L., Rai, M., Morelli, G., & Galdiero, M. (2015). Silver Nanoparticles as Potential Antibacterial Agents. Molecules, 20(5), [8856–8874]. https://doi.org/10.3390/molecules20058856

[46]   Guzmán Rodríguez A, Sablón Carrazana M, Rodríguez Tanty C, Malessy MJA, Fuentes G, Cruz LJ. Smart Polymeric Micelles for Anticancer Hydrophobic Drugs. Cancers (Basel). 2022 Dec 20;15(1):4. doi: 10.3390/cancers15010004.

[47]   Moss, D. M., Curley, P., Kinvig, H., Hoskins, C., & Owen, A. (2018). The biological challenges and pharmacological opportunities of orally administered nanomedicine delivery. Expert Review of Gastroenterology & Hepatology, 12(3), [223–236]. https://doi.org/10.1080/17474124.2018.1399794

[48]   Olivera ME, Manzo RH, Junginger HE, et al. Biowaiver monographs for immediate release solid oral dosage forms: ciprofloxacin hydrochloride. J Pharm Sci. 2011;100(1):[22–33]. doi: 10.1002/jps.22259

[49]   Patra, J. K., Das, G., Fraceto, L. F., Campos, E. V. R., Del Pilar Rodriguez-Torres, M., Acosta-Torres, L. S., Diaz-Torres, L. A., Grillo, R., Swamy, M. K., Sharma, S., Habtemariam, S., & Shin, H. (2018). Nano based drug delivery systems: recent developments and future prospects. Journal of Nanobiotechnology, 16(1). https://doi.org/10.1186/s12951-018-0392-8

[50]   Gounani, Z., Asadollahi, M. A., Pedersen, J. N., Lyngsø, J., Skov Pedersen, J., Arpanaei, A., & Meyer, R. L. (2019). Mesoporous silica nanoparticles carrying multiple antibiotics provide enhanced synergistic effect and improved biocompatibility. *Colloids and Surfaces B: Biointerfaces*, *175*, [498–508]. https://doi.org/10.1016/j.colsurfb.2018.12.035

[51]   Vallet-Regí, M., Schüth, F., Lozano, D., Colilla, M., & Manzano, M. (2022). Engineering mesoporous silica nanoparticles for drug delivery: where are we after two decades? Chemical Society Reviews, 51(13), [5365-5451]. https://doi.org/10.1039/d1cs00659b

[52]   Johnston, H. J., Hutchison, G. R., Christensen, F. M., Peters, S., Hankin, S. M., & Stone, V. (2010). A review of the in vivo and in vitro toxicity of silver and gold particulates: Particle attributes and biological mechanisms responsible for the observed toxicity. Critical Reviews in Toxicology, 40(4), [328–346]. https://doi.org/10.3109/10408440903453074

[53]   Oberdörster, G., Oberdörster, E., & Oberdörster, J. (2005). Nanotoxicology: An Emerging Discipline Evolving from Studies of Ultrafine Particles. Environmental Health Perspectives, 113(7), [823–839]. https://doi.org/10.1289/ehp.7339

[54]   Li, X., Plésiat, P., & Nikaido, H. (2015). The Challenge of Efflux-Mediated Antibiotic Resistance in Gram-Negative Bacteria. Clinical Microbiology Reviews, 28(2), [337–418]. https://doi.org/10.1128/cmr.00117-14

[55]   Ajdary, M., Moosavi, M. A., Rahmati, M., Falahati, M., Mahboubi, M., Mandegary, A., Jangjoo, S., Mohammadinejad, R., & Varma, R. S. (2018). Health Concerns of Various Nanoparticles: A Review of Their in Vitro and in Vivo Toxicity. Nanomaterials (Basel), 8(9), 634. https://doi.org/10.3390/nano8090634

[56]   Edson, J. A., Kwon, Y.J. (2016). Design, challenge, and promise of stimuli-responsive nanoantibiotics. Nano Converg, 3(1):26. doi: 10.1186/s40580-016-0085-7.

[57]   Albers, C. E., Hofstetter, W., Siebenrock, K. A., Landmann, R., Klenke, F.M.(2013) In vitro cytotoxicity of silver NPs on osteoblasts and osteoclasts at antibacterial concentrations. Nanotoxicology. 7(1):30-6. doi: 10.3109/17435390.2011.626538.

[58]   Chen, L. Q., Fang, L., Ling, J., Ding, C. Z., Kang, B., & Huang, C. Z. (2015). Nanotoxicity of silver nanoparticles to red blood cells: size dependent adsorption, uptake, and hemolytic activity. Chem Res Toxicol, 28(3), [501–509]. https://doi.org/10.1021/tx500411z

[59]   Vimbela GV, Ngo SM, Fraze C, Yang L, Stout DA. Antibacterial properties and toxicity from metallic nanomaterials. Int J Nanomedicine. 2017 May 24;12:3941-3965. doi: 10.2147/IJN.S134526. Erratum in: Int J Nanomedicine. 2018 Oct 16;13:6497.

[60]   Sharma, S., Parveen, R., & Chatterji, B. P. (2021). Toxicology of Nanoparticles in Drug Delivery. Current Pathobiology Reports, 9(4), [133-144]. https://doi.org/10.1007/s40139-021-00227-z

[61]   Gualtierotti, R., Guarnaccia, L., Beretta, M., Navone, S. E., Campanella, R., Riboni, L., Rampini, P., & Marfia, G. (2017). Modulation of Neuroinflammation in the Central Nervous System: Role of Chemokines and Sphingolipids. Advances in Therapy, 34(2), [396–420]. https://doi.org/10.1007/s12325-016-0474-7

[62]   Wnorowska, U., Fiedoruk, K., Piktel, E., et al. (2020). Nanoantibiotics containing membrane-active human cathelicidin LL-37 or synthetic ceragenins attached to the surface of magnetic nanoparticles as novel and innovative therapeutic tools: current status and potential future applications. Journal of Nanobiotechnology, 18, 3. https://doi.org/10.1186/s12951-019-0566-z

[63]   Goddard, G., Martin, C. J., Naivar, M. A., Goodwin, P. M., et al.(2006) Single particle high resolution spectral analysis flow cytometry Cytometry Part A 69(8):842-51 DOI:10.1002/cyto.a.20320

[64]   Tiambeng, T. N., Roberts, D. S., Zhu, Y., Chen, B., Wu, Z., Mitchell, S. D., ... & Kelleher, N. L. (2020). Nanoproteomics enables proteoform-resolved analysis of low-abundance proteins in human serum. Nature Communications, 11(1), 3903. https://doi.org/10.1038/s41467-020-17643-1

[65]   Tao, C. (2018). Antimicrobial Activity and Toxicity of Gold Nanoparticles: Research Progress, Challenges and Prospects. Lett. Appl. Microbiol. 67 (6), [537–543]. doi:10.1111/lam.13082

[66]   De Jong, W. H., & Borm, P. J. (2008). Drug delivery and nanoparticles: applications and hazards. International Journal of Nanomedicine, 3(2), [133-49]. https://doi.org/10.2147/ijn.s596

[67]   Kessler, R. (2011). Engineered nanoparticles in consumer products: understanding a new ingredient. Environmental Health Perspectives, 119, A120–A125. https://doi.org/10.1289/ehp.1103687

[68]   Elliott, K.C. (2011). Nanomaterials and the precautionary principle. Environmental Health Perspectives, 119(6), A240. https://doi.org/10.1289/ehp.1103687

[69]   Amaro, F., Morón, Á., Díaz, S., Martín-González, A., & Gutiérrez, J.C. (2021). Metallic Nanoparticles—Friends or Foes in the Battle against Antibiotic-Resistant Bacteria? Microorganisms, 9(2), 364. https://doi.org/10.3390/microorganisms9020364

[70]   Sydney Rose Addorisio, Shteynberg, R., Dasilva, M., Mixon, J., Mucciarone, K., Vu, L., Arsenault, K., Briand, V., Parker, S., Smith, S., Vise, C., Pina, C. and Laranjo, L. (2022) "Oxidative Stress Response in Bacteria: A Review", *Fine Focus*, 8(1), pp. 36–46. doi: 10.33043/FF.8.1

[71]   Addorisio, S.R., Shteynberg, R., Dasilva, M., Mixon, J., Mucciarone, K., Vu, L., Arsenault, K., Briand, V., Parker, S., Smith, S., Vise, C., Pina, C., & Laranjo, L. (2022). Oxidative Stress Response in Bacteria: A Review. Fine Focus, 8(1), [36-46]. https://doi.org/10.33043/FF.8.1

[72]   Ouyang, K.; Mortimer, M.; Holden, P.A.; Cai, P.; Wu, Y.; Gao, C.; Huang, Q. Towards a better understanding of *Pseudomonas putida* biofilm formation in the presence of ZnO nanoparticles (NPs): Role of NP concentration. Environ. Int. 2020, 137, 105485. https://doi.org/10.1016/j.envint.2020.105485

[73]   Niño-Martínez, N., Salas Orozco, M.F., Martínez-Castañón, G.A., Torres Méndez, F., & Ruiz, F. (2019). Molecular Mechanisms of Bacterial Resistance to Metal and Metal Oxide Nanoparticles. International Journal of Molecular Sciences, 20(11), 2808. https://doi.org/10.3390/ijms20112808

Materials Research Foundations 160 (2024) 113-144   https://doi.org/10.21741/9781644902974-5

Chapter 5

Applications of Nanoparticles in Bioimaging

Riya Thomas, Meera Varghese, Manoj Balachandran*

Department of Physics and Electronics, CHRIST (Deemed to be University), Bengaluru, Karnataka 560029, India

*manoj.b@christuniversity.in

Abstract

Mounting stipulation for early identification and diagnosis of illnesses has constantly compelled the need for the development of non-invasive imaging techniques. These imaging methods work by exposing bodily tissues to a variety of energies, including magnetic fields, sound waves, radioactive chemicals, and high-energy radiations. Changes in the energy pattern that occur from these interactions are then used to create an image or picture. In order to provide more precise anatomical and functional information, contrast agents are utilized in imaging modalities to differentiate between normal tissue and pathological lesions. Compared to traditional contrast agents, nanoparticles (NPs) have attracted a lot of interest in the field of bioimaging due to their unique physicochemical traits and low toxicity profiles. Here in this chapter, we discuss the structure-related properties, benefits, and significant advancements of nanoparticle-based contrast agents used in the most popular biomedical imaging modalities.

Keywords

Biomedical Imaging, Nanoparticles, Contrast Agents, Tumor Detection

Contents

1.   Introduction

The impact of nanotechnology has embraced diverse facets of medical field such as drug delivery, targeted therapy, biosensors, tissue engineering, bioimaging etc. Owing to their small size, and high surface to volume ratio, NPs possess unique physicochemical properties and their usage in bioimaging applications demands high stability, good dispersibility, high retention time, better image contrast, long circulation time in the bloodstream, high biocompatibility and low cytotoxicity [1]–[4].

Introduction of novel nanoprobes have significantly improved the sensitivity and specificity of bioimaging technologies including magnetic resonance imaging (MRI), ultrasound (US) imaging, positron emission tomography (PET), computed tomography (CT), single photon emission CT, photoacoustic imaging (PAI), and near-infrared optical imaging. Suitable functionalization, and modifications can highly improve the imaging qualities, biosafety and target-based applications ensuring safety and effectiveness of the diagnosis approaches. Certain NPs are suitable for multi model imaging and offers less invasive and accurate detection and diagnosis. Imaging guided therapies also grabbed immense research attention. In this article, recent progress in NPs based bioimaging and the unique properties of NPs that allow their successful bioimaging applications are discussed in detail. The discussions develop with the significant contributions of metallic NPs, polymeric NPs, carbonaceous NPs and lipid based NPs in the field of bioimaging.

2.   NPs in bioimaging

Most of the biomolecules are in nanosize which favors the interaction of the NPs to the biological system. Extensive research attempts were taken to enhance the desired properties of NPs to revolutionize the bioimaging techniques. There are a wide variety of NPs examined for their potential applications in vivo and vitro imaging [5].

2.1   Metallic NPs

With their appreciable X-ray attenuation ability and great density Metallic NPs are widely utilized as contrast agent in CT. However, considering the toxic nature of Metal NPs, certain modification strategies like the introduction of functional groups or capping with organic molecules will be performed to ensure biosafety and biocompatibility which is also favorable for target based applications [6].

2.1.1   Rare-Earth Metal Based NPs

Rare-earth based nanomaterials exhibit excellent optical, electrical and magnetic properties which enable their bioimaging applications with a high resolution. The superior magnetic moment and X-ray absorption coefficients of these materials allow enhanced electron relaxation time which improves the spatial resolution of MRI and CT imaging. The down-conversion and up conversion properties of Lanthanides favor their usage in luminescence imaging. Imaging possibilities in the near-infrared region widely observed with rare earth metal NPs has great importance as infrared wavelength satisfy safety concerns. Moreover, rare earth metal NPs are highly biocompatible and photostable with long-term fluorescence lifetime and narrow emission bandwidth. Hence, their effective application as excellent bioimaging agents is highly assured [7].

Zhu et al. realised lifetime multiplexed imaging of a living mice using Lanthanide NPs (combination of Er^{3+}, Ce^{3+} and Yb^{3+}). The fluorescence intensity and lifetime of Er^{3+} was highly increased by Ce^{3+} and Yb^{3+}. At an excitation wavelength of 1532 nm, similar fluorescence intensity and distinguishable lifetime were observed with low signal to noise ratio and high quantitative accuracy [8]. In a recent work, $GdVO_4$ NPs co-doped with Eu^{3+} and Bi^{3+} were used as contrast agent for MRI and luminescence imaging. The energy transfer between VO_4^{3-} to the Eu^{3+} and Bi^{3+} dopants increased the stability and intensity of emission which facilitated the quality of imaging. The composite material was used to image the liver tumor of a mice with Zero toxic effects [9]. Matos et al. prepared Gd^{3+}-Functionalized Lithium Niobate NPs as contrast agents in MRI. By the excitation in NIR region and resultant harmonic generation the contrast of imaging is highly improved. Moreover, a concentration dependent contrast enhancement was observed with the material [10].

Araichimani et al. designed rare-earth ions (europium/gadolinium) integrated silica NPs through the microwave combustion of rice husk. The paramagnetic properties of gadolinium NPs offered potential capability for T1-weighted MRI which is reproduced for different concentrations of the sample in water medium. The intensity of imaging was improved with the increasing concentration of gadolinium integrated silica NPs sample. Whereas, europium integrated silica NPs offered fluorescence imaging properties facilitated by a red luminescence [11]. Jia et al. reported that cerium oxide on up conversion NPs as efficient theragnostic agents having potential application in CT, MRI and up conversion luminescence imaging [12]. Shan et al. designed Polypyrrole-based double rare earth hybrid NPs with high biocompatibility, solubility, good stability, less toxicity and capable of multimodal imaging. The paramagnetic properties and X-ray attenuation capability of the material supports MRI and CT imaging in vitro and vivo [13]. Ren et al. tested the potential application of Er-based NPs in targeted imaging of skull and also for the imaging-guided surgery of orthotopic glioma [14].

Song et al. reported the successful application of Ag_2Se quantum dot-sensitized lanthanide-doped nanocrystals as contrast enhancing agent for the NIR-IIb bioimaging of brain Injury. The emission of these nanomaterials was observed above 1500 nm which showed a 100-fold enhancement by the addition of quantum dots. The penetration depth of the material was reported as 11mm and it is proven for an excellent cerebrovascular imaging of the whole brain [15]. Similarly, Wang et al. examined the potential of dye sensitized rare earth NPs as contrast agents in NIR-IIb imaging of mouse cerebral vessels. The dye sensitization provided 40-folded enhancement to the performance of the NPs. The material exhibited a reduced perfusion rate and longer clipping time in the short-time ischemic hindlimbs. Whereas, at the long-time ischemic hindlimbs, changes in the vascular anatomy were delineated, which helped the display of regeneration vessels. Moreover, the material could be clearly image femoral artery and the formation of thrombus in atherosclerosis [16].

Dong et. al. utilized Ytterbium NPs as contrast agents in CT and spectral photon-counting CT. These NPs showed superior attenuation properties and contrast compared to Au NPs [17]. $NdVO_4$/Au nanocrystals prepared by Chang et al. offered outstanding performance in photothermal and photoacoustic dual-modal imaging owing to their light-to-heat conversion efficiency and thermal expansion properties [18]. Yu et al. realized multi model imaging (NIR-II fluorescence/PA/MRI) of tumor sites with the aid of CaF_2:Y, Gd, Nd NPs synthesized via a hydrothermal route. The degradation of performance and material loss were highly hindered by the simplified synthesis route. Y^{3+} and Gd^{3+} doping saved Nd^{3+} from concentration quenching threshold and improved the NIR-II luminescence and paramagnetic properties [19].

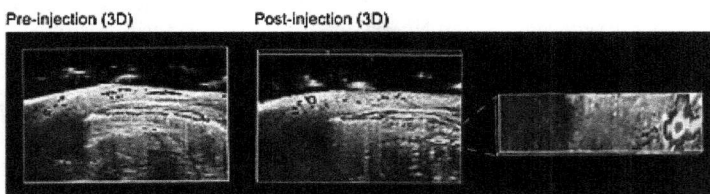

Figure 1.   Pre-injection and post-injection photoacoustic images of 10 mg/mL of CaF_2:Y,Gd,Nd NPs into a mouse [19].

Shi et al. synthesized human serum albumin encapsulated GdF_3 NPs and utilized them for the enhancement of T1 MRI contrast. Clear visualization and quantification of tumor was realized through the MRI images which offered superior T1 enhancement after 24 hours of injecting the obtained NPs [20].

2.1.2 Noble metallic NPs

AgBiS$_2$ NPs synthesized by Cheng et al. showed potential capability for enhanced In vitro and in vivo CT imaging by improving the contrast of the tumor cells. The sample could offer better Hounsfield units slope compared to clinical CT contrast agents namely iobitridol and iopromide [21]. In a recent work, the authors reported ferrocene capped Au NPs that offered an enhanced CT imaging of cancer cells [22]. Dumani et al. used glycol-chitosan-coated gold NPs for US-guided PAI to image sentinel lymph node. These NPs improved the contrast of the images, and the metastatic and non-metastatic lymph nodes were well distinguishable in the images. Through this immunofunctional imaging with the aid of biocompatible glycol-chitosan-coated gold NPs, the immune cells could be isolated by which their spatio-temporal distribution within the sentinel lymph node was analyzed

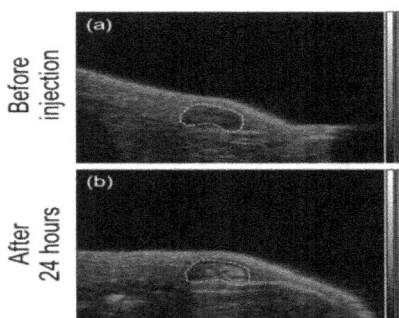

[23].

Figure 2. US-guided photoacoustic images of a non-metastatic sentinel lymph node (yellow contour) (a) pre-injection and (b) 24 hours post-injection of glycol-chitosan-coated gold NPs [23].

Mariquez et al. functionalized commercial Au NPs with mannose and radiolabeled it with ^{99m}Tc. These NPs were then utilized for the single-photon emission CT imaging of lymph node in a rat model [24]. Li et al. reported the cytoplasmic labeling of living cells and CT imaging of tumor in a mice model using Silver@quercetin NPs prepared by the by redox activity of quercetin and silver ions. These NPs are highly biocompatibe with good photostability. The aggregation-induced emission luminogens of the obtained particles was used for the fluorescence imaging and CT imaging of tumor in a mice [25].

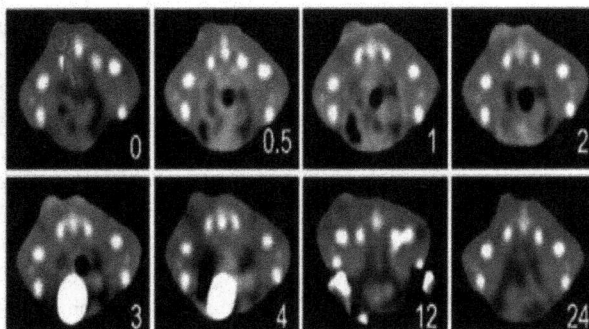

Figure 3.   *CT images of tumor in a mouse model (red arrow) [25].*

Bouché et al. synthesized poly[di(carboxylatophenoxy)phosphazene]-Au NPs that could help deep-tissue imaging via CT and photoacoustics. The obtained nanoprobe was capable of imaging the overproduced ROS through the comparison of the contrasts of CT and photoacoustics signals. This is possible because the poly[di(carboxylatophenoxy)phosphazene] polymer has a tendency to degrade if it is exposed to ROS which in turn reduce the photoacoustics signal [26]. Lyu et al. synthesized cysteamine -coated FePd nanodots that can be used as concentration-dependent contrast agent in tri-modal CT, magnetic resonance and PAI. The experiments were conducted on the tumour site of a BALB/c mice model and the obtained nanodots with high biocompatibility and stability provided high contrast images [27]. Li et al. designed TiZrRu metal-organic nanostructure for CT/magnetic resonance imaging-Guided X-ray induced dynamic therapy. 2,2'-Bipyridine-5,5'-dicarboxylic acid was used as co-organic ligands and Gd(III) was chelated with it. The material showed T1-weighted MRI property which can be useful for a guided X-ray induced dynamic therapy for the selective destruction of cancer cells [28].

2.1.3 Magnetic NPs

Magnetic metal NPs can be easily directed around the body with a magnetic and can be effectively used to improve the quality of bioimaging [29].

Carregal-Romero et al. synthesized Manganese ferrite NPs with potential capability to use as contrast agents in MRI by which positive or dual-mode usage is possible by proper tuning of the

NPs [30]. Gull et al. synthesized Manganese-doped cesium iodide NPs and examined its potential application in multi-model bioimaging. The obtained highly fluorescent, biocompatible and water-soluble NPs exhibited excellent computed X-ray tomography and MRI(MRI) imaging properties [31]. Hobson et al. explained the potential of positively charged raspberry iron oxide NPs for organ-targeted magnetic resonance images. These superparamagnetic and colloidally stable NPs could produce high contrast T2 weighted magnetic resonance images of the liver and spleen. The raspberry iron oxide NPs enhanced the r2 relaxivity and reduced the r1 relaxivity to act as a strong negative contrast agent. The biodistribution studies showed that, these NPs have the potential to be used for whole body MRI with shorter scan times. However, for longer scan times, they are more appropriate for imaging the organs of the reticuloendothelial system [32].

2.2   Polymeric nanomaterials

Polymeric NPs for bioimaging is a fast-developing research field. Flexible functionalization capability of polymeric NPs allows their easy alternations according to the research and application interests. Polymeric NPs possess many beneficial characteristics such as they are biocompatible, biodegradable, renewable and easily tunable. Both natural and synthetic polymer NPs have been extensively utilized for various bioimaging techniques.

2.2.1   Natural polymer based NPs

Varieties of natural polymers are present in nature, and they can be extracted by the implementation of appropriate methods. Various natural polymer based NPs have been developed for their potential applications in diverse biomedical fields [33],[34].

Patra et al. used stable, nontoxic, biodegradable and autofluorescent albumin NPs having emission at 555 and 665 nm for in vitro bioimaging of MOLT4 cells [35]. Wang et al. designed a human serum albumin structure-based fluorescent probe, DNPM for bioimaging in living cells. DNPM exhibited the capability to bind to both Sudlow site I and site II in human serum albumin, with fluorescence enhancement influenced by electromagnetic coupling [36]. Radford et al. designed [^{55}Co]Co-cm10 and [^{55}Co]Co-rf42 albumin-binding folate derivatives, which are suitable for PET imaging. The biodistribution studies and imaging for the samples were conducted on groups of four mice at 4 hours and 24 hours post-injection. Biodistribution studies conducted on KB tumor-bearing mice showed that both samples can offer better tumor uptake. The PET / CT images revealed the clear delineation of folate receptors and highest activity uptake was observed with kidneys and tumours [37].

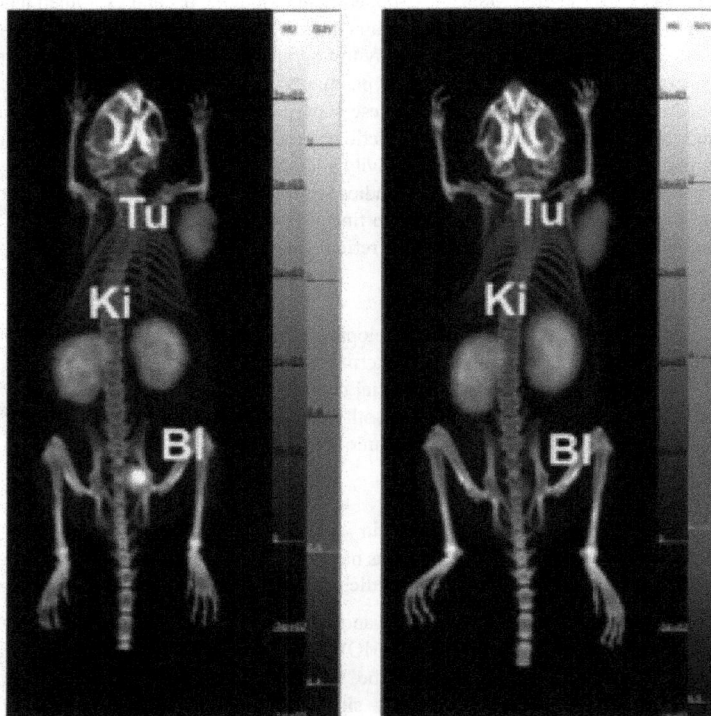

Figure 4.   The PET CT images (shown as maximal intensity projections) of KB tumor-bearing mice injected with (left) [55Co]Co-cm10 and (right) [55Co]Co-rf42 after 4 h of injection. Tumor (Tu), kidney (Ki) and bladder(Bl) are marked [37].

Gao et al. designed an albumin-consolidated aggregation-induced emission probes to overcome the fluorescence limitations ans targeting issues of such nanoprobes. Using this noval nanoprobes, high resolution NIR-II images of cerebrovascular and brain tumors in mouse model were obtained with a high signal-to-background ratio [38]. Zhang et al. prepared albuminpaclitaxel nanoglue in which sinoporphyrin sodium photosensitizer was loaded by a simple mixing method. The as prepared material exhibited enhanced fluorescence imaging potential compared to sinoporphyrin sodium alone which is attributed to the reduced quenching of the photosensitizer after dispersed in the albumin [39].

Li et al. modified mesoporous organosilica NPs by CuS@bovine serum albumin (CuS@BSA) nanocomposites to obtain a noval material suitable for photoacoustic imaging guided chemo-photothermal therapy. CuS@BSA nanocomposites are biocompatible and efficient contrast agent for PA imaging. The experiments performed in a mouse model showed that the material has a durable PAIpotential even after 24 hours intravenous injection [40]. Yang et al. fabricated a biomimetic catalase-integrated-albumin phototheranostic nanoprobe with simultaneous encapsulation of Indocyanine green/Au nanorods. The as prepared nanoprobe was successfully utilized for deep multimodal imaging (fluorescence imaging, photoacoustic imaging, and infrared thermal imaging) of glioma with high signal-to background ratio. Also, an imaging guided therapy was realized via magnetic resonance imaging, and PET [41].

Baki et al. reported Albumin-Coated Single-Core Iron Oxide NPs for enhanced MRIand magnetic particle imaging. The steric stabilization with the bovine serum albumin offered colloidal stability to the magnetic NPs and maintained the overall imaging performance [42]. El-Sayed et al. reported that Au nanoclusters modified with Gelatin NPs induced a redshift in the emission of nanoclusters and offered stability against varying pH conditions and enzymatic degradation. The material was further successfully utilized for the fluorescence imaging of skin tissue [43].

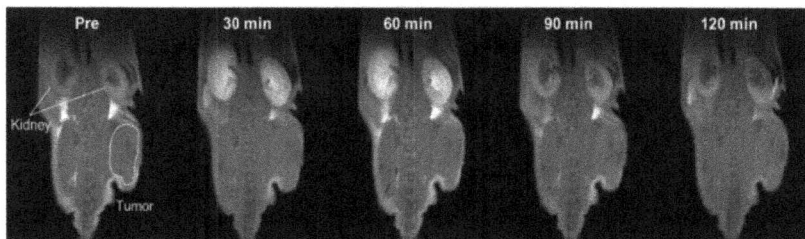

Figure 5.   Horizontal slices of the T1-weighted magnetic resonance imaging: pre and post application of Gd-coordinated gelatin nanogels [46].

Xue et al. reported a gelatin methacryloyl -based NPs conjugated with rhodamine B as fluorescent label for bioimaging of 3T3 fibroblasts and B16F10 cancer cells. Compared to pure rhodamine B, the sample offered higher biocompatibility with improved cell diffusion and enhanced brightness properties [44]. Paul et al. hydrothermally synthesized gelatin quantum dots and successfully use them as non-toxic biomarkers for the imaging of bacterial cells (Escherichia coli and Staphylococcus aureus), yeast cells (*Candida albicans, C. krusei, C. parapsilosis,* and *C. tropicalis*), mycelial fungi (*Aspergillus flavus* and *A. fumigatus cells*), and cancer cell lines (A549, HEK293 and L929) [45]. Kimura et al. fabricated Ultra-small Gd-coordinated gelatin nanogels and reported it as a non-toxic contrast agent for magnetic resonance imaging. The imaging performance of the sample was analyzed by in vivo experiments conducted in the brain of a mouse model. Rapid renal excretion and short-term accumulation of the sample was observed in the kidney and liver of the tumor-bearing mice [46].

Yang et al. designed a composite by immobilizing indocyanine green photosensitizer and doxorubicin drug into silk fibroin NPs and mineralized the particle surface by MnO_2. The in vitro and in vivo studies of the material were demonstrated on tumor cell lines and tumor-bearing mouse models respectively. The obtained NPs were capable of accumulating in tumor region and were used for MRI and fluorescence imaging-guided PDT/PTT/chemotherapy [47]. Ma et al. reported indocyanine green-based silk fibroin nanoprobes for the NIR-I/II fluorescence imaging of cervical diseases. These nanoprobes were tested on a rat model and confirmed their excellent photo-stability, biocompatibility and long circulation time [48].

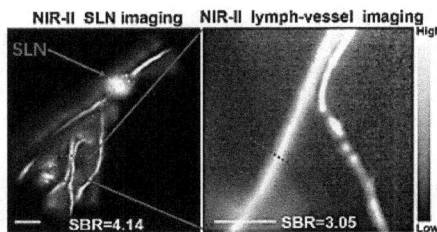

Figure 6.   NIR-II sentinel lymph node fluorescence imaging of in vivo anatomy structure of a SF@ICG NPs treated nude mice (Left) and the NIR-II lymph-vessel fluorescence microscopic images (Right) [48].

Wang et al. used NIR cypate-induced silk fibroin labeled with 99mTc for photoacoustic/single-photon emission CT imaging. The in vivo experiments were performed on a osteosarcoma bearing female BALB/c mice. These nanoagents were confirmed with their characteristic properties of biocompatibility, good in vivo cycle time, and tumor enrichment efficiency which assures their fruitful usage as a multimodal imaging probe with tumor targeting [49]. Salarian et al. developed a collagen targeting protein which can be used as a contrast agent for MRI for the sensitive and early detection of liver metastasis [50]. Zhou et al. designed a biocompatible doxorubicin hydrochloride-loaded Chitosan Nanobubbles and used it for US enhancement and assisted Doxorubicin targeted delivery [51]. Xiao et al. synthesized fluorescent starch NPs from broken-rice and investigated live-cell imaging properties of these NPs. The material offered a stable fluorescence and served as an excellent fluorescent probe for the HeLa cell imaging [52].

2.2.2   Synthetic polymer based NPs

Zhou et al. synthesized Iodine-rich semiconducting polymer NPs that can be used for CT and fluorescence dual-modal Imaging-guided photodynamic therapy. The experiments were conducted on a tumor bearing mouse. The low sensitivity of CT was bridged by combining fluorescence imaging. The role of iodine is to improve the $^1O_2$ formation and the semiconducting polymer NPs served as photosensitizer absorbing near-infrared light and a source of fluorescence signal [53].

Yang et al. synthesized semiconducting polymer NPs excited by a 1064 nm pulsed laser source that as contrast agent for PAIof deeper tissues and gliomas. Because of the high photothermal conversion efficiency, even for low concentrations of the sample, clear photoacoustic images of deeper chicken breast tissues and glioma tumor in a mouse were obtained [54].

Liu et al. tested the bioimaging potential of Polymethine-based semiconducting polymer dots having excitation and emission in the NIR-II region. Using the material the authors could successfully developed Whole-body fluorescence images of living mice. Compared to the traditional NIR-II imaging, the signal-to background ratio and spatial resolution of NIR-IIb imaging were superior [55]. Guo et al. reported successful usage of conjugated polymer NPs for NIR-II optical-resolution PA microscopy imaging of cerebral and tumor vasculatures of HepG2 tumor-bearing mouse. These conjugated polymer NPs exhibited promising properties such as high sensitivity, superior photoacoustic stability, and excellent biocompatibility to be used as efficient contrst agents for the 3D vasculature in vivo imaging. Moreover, these NPs were capable of deep penetration and produced images with high resolution, signal-to background ratio and depth resolution ratio for images in 3D volume [56].

Liu et al. designed fluorine substituted semiconducting polymer dots having red-shifted emission and high NIR-II fluorescence which allowed their application in in vivo fluorescence imaging. The material could produce deep tissue imaging of the brain-tumor vasculature with remarkable signal to background ratio [57]. Zhao et al. synthesized highly stable and biocompatible Mn-AH nanoscale coordination polymer nanodots having potential application in PAIT1-weighted MRI of subcutaneous murine colon CT26 tumor and orthotopic bladder tumor [58]. Hu et al. designed gadolinium-chelated conjugated polymer-based nanotheranostic agents and verified its potential application in multimodal imaging (photoacoustic/magnetic resonance/NIR-II Fluorescence Imaging)-guided tumor photothermal therapy in a 4T1 tumor-bearing mice [59].

2.3   Carbonaceous nanoparticles

The family of carbon-based nanoparticles contains a large number of members with unique optical, electronic, mechanical, and chemical properties that make them excellent choices for use in clinical medicine. The most well-known (and extensively studied) carbon nanoparticles involve carbon nanotubes, carbon dots, graphene and its derivatives, and graphene quantum dots. The primary application area is in the field of medical diagnostics, which encompasses bioimaging and the identification of substances or metabolites that exist in the body.

2.3.1   Carbon nanotubes

Carbon nanotubes (CNTs), including single-walled nanotubes (SWNTs) and multi-walled nanotubes (MWNTs), have peculiar physical characteristics that are exceedingly intriguing for imaging applications. CNTs have proven to be capable of overcoming biological barriers, therefore they can penetrate cells regardless of the kind and functional groups present on the surface. It also possesses some characteristics such as a large interfacial area with the cellular membrane, the ability to integrate numerous functionalization, and transportability in biological fluids which are highly beneficial for bioimaging modalities [60].

An emerging and potential bioimaging application of CNTs is in photoacoustic imaging (PAI). Being one of the darkest materials with substantial absorption in the near-infrared region, CNTs are anticipated to be a potential contrast agent for PAI [61]. Compared to other carbon materials like fullerenes and graphitic particles, SWNTs are able to produce maximum signals, hence making them applicable in PAI as an ideal contrast medium. Pramanik et al. developed an SWNTs-based photoacoustic contrast agent for the detection of malignant tumor tissues [62]. The results showed a large contrast enhancement, high signal-to-noise ratio, and sensitivity, which allowed further applications in tissue phantom imaging studies. It is believed that functionalized CNTs can be a more promising imaging agent to improve solubility, biocompatibility, and performance. Because pristine CNTs are naturally hydrophobic, and nonfunctionalized CNTs have considerable cytotoxicity owing to their insolubility or remaining metal catalysts. Xie et al. fabricated albumin/Chlorin 6 (Ce6) functionalized SWNTs (ACEC) as a PAI contrast agent for the imaging of tumor cells (figure) [63]. In their study, strong PAI signals could be observed from functionalized SWNTs, while only weak signals were obtained from the pristine SWNTs. This is because the encapsulation of photosensitizer Ce6 enhances the fluorescence characteristics of SWNTs beneficial for in-vivo imaging, while albumin improves its stability and hydrophilicity.

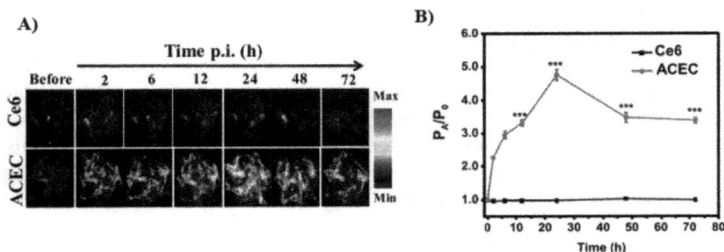

Figure 7. a) In vivo photoacoustic images of mice intravenously administrated with free Ce6 and ACEC at 100 mg Ce6 concentration at different time intervals. (b) Signal intensity of tumors treated with free Ce6 and ACEC at different time intervals (PA/PA0 ratio) [64]

Owing to SWNTs' strong resonance Raman scattering with extremely large scattering cross-section, they are also considered as excellent Raman probe useful in biological imaging. Compared to other organic Raman dyes, SWNTs possess several advantages such as (a) Sharp, intense, and narrow (FWHM ~2 nm) Raman peaks facilitating the clear distinction from autofluorescence background (b) Long-term stability with negligible quenching or bleaching (c) Multiple colour imaging by modulating the Raman shifts via changing the isotope combination, which makes them a novel Raman tag [64]. Liu et al. demonstrated colon cancer cells' five-colour multiplexed molecular imaging using SWNTs functionalized with different C13/C12 isotope compositions [65]. Taking advantage of the low tissue absorption and autofluorescence background, the developed CNT Raman tag could achieve near zero interfering background of imaging. However, the acquisition time of Raman imaging for the biological samples labeled with SWNTs is quite long. The ideal method for cutting down on imaging time is to increase the Raman signals of those tags, which can be done by using surface enhanced Raman scattering (SERS) to cover the nanotube surface with metal nanoparticles. Wang and his group synthesized DNA-functionalized metal (Ag and Au) nanoparticles coated SWNTs modified with strong SERS effect for the Raman imaging of cancer cells [66]. Due to the highly enhanced Raman signals of nanotubes by SERS, cancer cell labeling, and Raman 60imaging were achieved in a remarkably shorter imaging time compared to that of non-enhanced SWNTs-nanoprobe.

Also, CNTs have great potential for fluorescence imaging of tumor cells owing to their unique emission characteristics. Especially, 1D-semiconducting SWNT possesses a narrow bandgap of 1eV, which permits fluorescence emission in the near-infrared (NIR)-II region (900-1600 nm) while being excited in the NIR-I region (550-850 nm) [67]. Moreover, the considerable Stoke shift between the excitation and emission wavelengths will significantly reduce the autofluorescence of biological tissues during imaging, providing improved sensitivity. The intrinsic fluorescence properties of SWNTs were adopted by Robinson et al. for the live cell imaging of tumors [68]. In the proposed study, dynamic contract imaging of breast tumor cells was carried out utilizing the combined high tumor accumulation ability and fluorescence excitation/emission properties of SWNTs without the need for any additional labeling agents. Although SWNTs have shown favourable results in fluorescent imaging, their low quantum yield (QY) remains the fundamental obstacle to their further usage in in vivo imaging. Welsher et al. introduced a novel method for synthesizing highly biocompatible SWNTs with enhanced QY ideal for biological systems [69]. Furthermore, the as-developed bright nanotube fluorophore was then used as a high-magnification NIR photoluminescence contrast agent for tumor cell imaging.

2.3.2   Carbon dots

Carbon dots (CDs), the nanometer-sized fluorescent carbon particles with a unique set of optical and electronic properties, particularly, their high QY, solubility, chemical and physical stability, tunable functional properties, and biocompatibility are now creating significant opportunities in the bioimaging of live cells.

Recently researchers have shown that CDs have been widely used as fluorescent imaging probes in cancer diagnosis. Yang et al. synthesized green fluorescent CDs for the imaging of human breast cancer and colorectal cancer cells [70]. The imaging performance of CDs was found to be comparable to that of commercial synthesis fluorophores and no toxicity effects were noted. But, in contrast with unmodified CDs, cells can be labeled more effectively and precisely with functionalized CDs-based biomarkers. Song et al. created green fluorescent CDs modified with folic acid to target and detect HeLa cancer cells. The CDs were able to distinguish between normal cells and cancer cells that overexpressed the folate receptor (FR) protein because of the particular interaction between folic acid and FR molecules. However, extensive tissue permeation, minimal tissue intake, negligible photodamage, and low autofluorescence hindrance to biological material, red or near-infrared (NIR) emissive (>600 nm) CDs are more recommended as a fluorescence imaging probe [71]. To date, a variety of heteroatoms, including N, S, P, and F, have been used with various doping techniques to investigate CDs as bioimaging probes in the red/NIR emissive area. Parvin and Mandal developed N, P doped CDs with dual emissions like green and red where QY is 30% and 78%, which have a lot of potential for cancer diagnosis research [72].

CDs are also becoming an increasingly popular contrast agent in photoacoustic imaging. Xu et al. used CDs exhibiting high absorption in the near-infrared (NIR) region was explored as a contrast agent for PAI analysis of tumor growth [73]. The study revealed that the infused CDs accumulated in tumor tissue facilitated unchallenging imaging. Besides, no appreciable signs of damage to other internal organs were observed, suggesting the low systemic toxicity of as-developed CDs. To augment the PA signals of CDs in cancer cells, Sun et al. synthesized red-emissive Chlorin e6 (Ce6) modified CDs with excellent photostability and biocompatibility [74]. The prepared CDs possessed an enhanced PAI signal and maintained a strong signal for 8 h, indicating its abundant accumulation in cancer cells.

Lately, CDs have also gained growing interest in serving as T1-weighted magnetic resonance imaging (MRI) contrast agents, due to the numerous benefits, including their excellent renal clearance capacity, incredibly low cytotoxicity, and superior photoluminescence. Two approaches have primarily been used up to adapt CDs for MR imaging applications: (1) Inclusion of metal ions in CDs (2) Use of metal-free CDs. Bouzas-Ramos et al. prepared CDs co-doped with lanthanides (Gd and Yb) and nitrogen as an improved contrast agent for MRI purposes [75]. The obtained doped CDs exhibited outstanding MRI contrast properties as well as intense fluorescence emission of excellent quantum yields (66 %), without significantly increasing cytotoxicity after being exposed to cell lines for 72 hours. However, biosafety concerns about the potential leakage of free Gd^{3+} and Yb^{3+} ions in-vivo limit its more extensive usage. As a result, novel metal-free CDs are discovered to be used as secure contrast agents in T1-weighted MR imaging. Zhao and co-workers synthesized metal-free boron-doped magentofluorescent CDs as an MRI contrast agent (figure) [76]. B-CDs displayed apparent red-shift fluorescence emission with enhanced intensity, and better magnetic resonance when compared to undoped CDs. It proved that adding boron to CDs can increase their fluorescence intensity and possibly create paramagnetic centers desirable for dual imaging modalities.

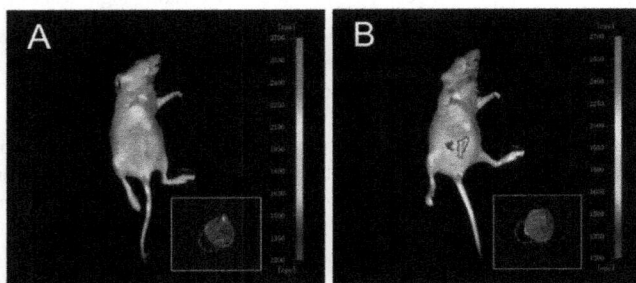

Figure 8. The in vivo fluorescence and MR (inset) imaging of nude mice (A) before and (B) after subcutaneous injection of B–CDs [76].

2.3.3 Graphene and its derivatives

Due to their unique physical and chemical characteristics, adaptable surface modification, extremely high surface area, and superior biocompatibility, graphene-based nanomaterials have attracted a lot of interest in bioimaging [77]. Among the various graphene derivatives considered, graphene oxide (GO), and reduced graphene oxide (rGO) are effusively explored in the area of cellular probing and imaging.

The idea of utilizing inherent Raman characteristics of graphene derivatives in Raman imaging has opened up an ultrasensitive platform for in-vivo tumor detection in biosystems. Though GO exhibits intrinsically strong D and G peaks, their Raman scattering signals can be greatly improved by Surface-enhanced Raman scattering (SERS) by the introduction of metal nanoparticles, which enables Raman imaging at considerably faster speeds. Liu et al. synthesized GO-Au nanocomposite-based Raman probes for the imaging of HeLa cells with a very quick integration time of roughly 0.06 s per pixel [78]. In comparison to cells incubated with pure GO, the as-obtained nanocomposite shows outstanding SERS effect, Raman signal augmentation of factor about 48, and more distinct Raman pictures. In the case of rGO, their chemical flexibility permits tunable adsorption of molecules or cells on SERS active surfaces. Chen et al. constructed an rGO-porous silica composite encapsulated with Au nanoparticles and Rhodamine for the selective SERS imaging of lung cancer cells [79]. It is found that the as-fabricated nanocomposite has an appropriate pore size to separate the Au nanoparticles in sub-nanometer gaps that can serve as "hot spots" for increasing the Raman signals up to a magnification of 5×10^6.

Graphene derivatives have also gained popularity in fluorescence imaging owing to their intrinsic photoluminescence. Especially, the sp2 and sp3 carbons in graphene-based nanomaterials can transform into electrical and optical band gaps required for the photoluminescence phenomena [80]. Besides, the fluorescence emission and intensity can be enhanced by tailoring the size, dopants, surface functional groups, and oxidation degree. As an attempt to enhance the QY of graphene derivatives for cellular imaging of breast cancer cells, Wate et al. functionalized GO with an organic fluorescent cyanine dye (Cy5) [81]. The in-vivo investigation clearly demonstrated that the GO-Cy5 nanosystem behaved as a bright and stable fluorescent marker and was also successfully uptaken by cancer cells. Huang and his group demonstrated a surface functionalization strategy of rGO by loading aromatic photosensitizer Chlorin e6 (Ce6) via hydrophobic interactions and π–π stacking of nanosystems [82]. The study highlighted that the Ce-6-rGO composite had better tumor cell accumulation efficiency than pristine rGO. However, due to the probable photobleaching of fluorescent dyes or photosensitizers, heteroatoms like N, S, or P are doped into GO/rGO system. Thus, Thomas and Balachandran synthesized a nitrogen-doped oxidized graphene derivative as an imaging agent in the visible region for the fluorescence bioimaging of melanoma cancer cells (figure) [83]. It is discovered that nitrogen atom doping could have a significant impact on photophysical parameters, leading to an increase in quantum yield (16%), a lengthened fluorescence lifespan (8.51 ns), a higher level of photostability (92%), and remarkable cytotoxicity that is highly anticipated in a biomarker.

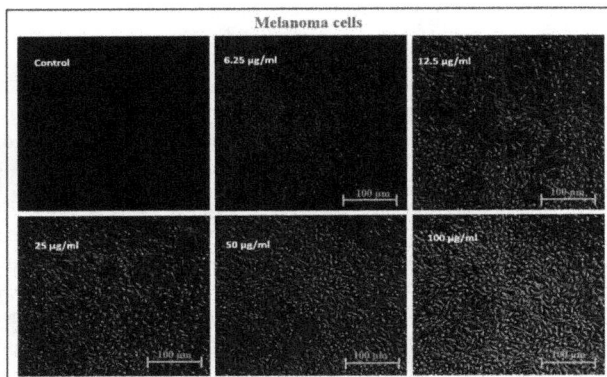

Figure 9. Fluorescent microscopic images of Melanoma cells incubated with nitrogen-doped graphene derivative under excitation filters of 450–480 nm [83].

Among the graphene derivatives, rGO plays a crucial function as PA contrast agents because of their bigger sp2 domains which facilitate the absorption of NIR light more effectively than GO [84]. However, due to the reduction process, the hydrophilicity of GOs is greatly reduced resulting in the poor water solubility of rGOs. To overcome this challenge, Lee et al. synthesized a partially reduced graphene oxide-Au nanorod composite for near-infrared induced PA imaging of tumor cells [85]. The sample exhibited excellent stability in an aqueous solution and showed PA signal was 2.4 times more intense than that of raw Au nanorods. The absorption cross-section of rGO was further enhanced by loading indocyanine green (ICG) in a study by Chen et al. [86]. In-vivo experiments demonstrated that the nanocomposite possesses low toxicity, excellent PA signals, superior tumor-targeting ability, and extended blood circulation time (up to 6h) without any accumulation in other major organs excluding the tumor regions.

The most common MRI contrast agents, such as the paramagnetic metal ions (Gd^{3+}, Mn^{3+}), are often hazardous because they coordinate non-selectively with biomolecules. GOs with several oxygen functionalities and cavities can be easily coupled with these metal ions by binding or isolating the ions between graphene sheets, which ultimately reduces their toxicity. Yang et al. prepared dendrimer-grafted gadolinium-functionalized nanographene oxide (Gd-GO) as a contrast agent for MR imaging of cancer cells [87]. Gd-GO could clearly identify and determine the location of tumor and its extent of growth. Because iron oxide nanoparticles (IONP) may be produced directly on graphene or connected with graphene using its encapsulated ligands, the superparamagnetic graphene/IONP hybrids have received a lot of attention as MRI contrast agents [88]. Yang and his group developed an rGO/IONP composite-based nanoprobe for the MR imaging of breast tumors [89]. It was discovered that rGO/IONP might gradually accumulate at the tumor, due to the relatively long blood circulation time, which favoured the enhanced permeability and retention impact of malignant tumors.

2.3.4 Graphene quantum dots

The zero-dimensional structure, graphene quantum dots (GQDs) possess intriguing properties derived from two-dimensional (2D) graphene and typically exhibit extraordinary properties of the QDs, such as edge effects, non-zero band gap, and quantum confinement effects, which give them great potential in the optical applications. Interestingly, fluorescent GQDs have overtook traditional organic and semiconductor QD-based fluorophores due to their superior photostability, widened fluorescence, compact dimension, biological suitability, affordable price, simplicity of preparation, and water solubility, and have become an all-encompassing fluorophore that provides unheard-of opportunities in bioimaging for accurate diagnosis [90].

The superior photoluminescence properties of GQDs make them an encouraging fluorescent probe with great potential in tumor imaging. Wang et al. synthesized GQDs of astonishing optical properties (QY-30%, lifetime-4.6ns, and no obvious photobleaching) for the cellular imaging of HeLa cells [91]. The GQDs provided a high-contrast fluorescence image and exhibited non-cytotoxicity even at high concentrations and for an extended period of time, making them appropriate for long-term studies in cancer cells. Gao et al. introduced polyethyleneimine (PEI)-coated multicolor emissive GQDs for tumor cell imaging [92]. The produced GQDs displayed red, yellow, and blue emissions depending on the molecular weight of the PEI, enabling multicolor imaging of tumor cells. In contrast, heteroatom doping of GQDs has also been employed to alter their electronic energy, resulting in a broad emission. Wang et al. prepared N, B-doped GQDs exhibiting an excitation wavelength-dependent PL emission in the NIR-II window (950 to 1100 nm) [93]. The work demonstrated the possibility of having in-vivo NIR-II imaging with a metal-free quantum dot most favorable for imaging-guided cancer therapy.

The functionalization and modification of MRI contrast agents using GQDs have attained prominence in the field of cancer diagnosis due to their improved water permeability, less photobleaching, and lower deep-tissue penetration depth capabilities [94]. The wang group developed Gd_2O_3 nanoparticles-modified GQDs showing significantly improved longitudinal relaxivity in comparison with commercial Magnevist [95]. This is because the larger surface-to-volume ratio of the nanocomposites may have enhanced the synergistic effect of Gd ions and GQDs, thus speeding up the r1 relaxation process. Wang et al. reported metal-free boron-doped GQDs as a safe T1 contrast agent for MR imaging of tumor cell lines [96]. The work showed that boron doping is more effective than metal doping at improving relaxivity. This might be because doped boron atoms with smaller diameters produce more paramagnetic centers in nanostructured carbon materials.

GQDs have also gathered attention as an imaging probe in CT and PAI, significantly expanding their bioimaging applications in tumor detection. Badrigilan et al. synthesized GQDs-coated Bimuth nanoparticle composite exhibiting excellent physiological dispersibility, biocompatibility, and steady absorption profile in the NIR region for high-performance CT imaging of cancerous cells [97]. No discernible difference in the X-ray attenuation magnitude was observed even at higher voltages, demonstrating the optimal function of GQDs-Bi composite with identical contrast for any CT imaging protocols. Xuan et al. prepared an N-doped GQDs-based PA imaging nanoprobe having great application prospects in the early detection of tumors [98]. It has been seen that HeLa cells incubated with N-GQD showed enhanced PA response compared to normal cells, confirming their excellent capability of tumor targeting and PA imaging.

2.4   Lipid-based nanoparticles

For a very long time, lipid-based nanoparticles have been used as in-vivo nanocarriers in biomedical applications because of their capability to entrap pharmacological molecules, enhancing bioavailability and overall activity. Lipids are amphiphilic molecules that self-assemble in an aqueous milieu and contain both hydrophilic and hydrophobic components [99]. Especially liposomes, lipid micelles, and solid lipid nanoparticles have gained tremendous attraction in the past years and have been used as contrast agents for bioimaging. Generally, lipids serve as a protective layer to enclose contrast-enhancing materials like quantum dots, fluorescent dyes, and iron oxide nanoparticles. Lipid nanoparticles' fundamental characteristics are the basis for their broad therapeutic usefulness, which in turn makes them suitable biomolecules for hybrid formation.

2.4.1   Liposomes

Liposomes- small phospholipid bubbles- with a bilayered membrane structure, have drawn a lot of attention as a diagnostic agent. Given their ability to retain the photoactivity of contrast agents (QDs), increase endocytosis and damage to cancer cells, extend the time that drugs remain in tumors, and be beneficial for examining molecular pathways and cellular examinations, liposome-based hybrid nanocarriers are one among the widely used commercially available nanocarriers for imaging applications [100].

Liposome-based fluoroprobes opened up exciting opportunities for the biolabelling and detection of tumor cells. Wen et al. prepared a nanocomposite of liposomes loaded with QDs for effective fluorescence imaging of melanoma cells [101]. According to the outcomes of in-vivo imaging, QDs-loaded liposomes exhibited a greater propensity to concentrate in solid tumors for exhibiting high-intense fluorescent signals in comparison to alternative carriers. Dong et al. used amphiphilic far-red squaraine dye-modified liposomes for the bioimaging of ovarian cancer cells [102]. The as-engineered liposomes demonstrated better bioavailability in circulation and decreased non-specific absorption by non-cancerous cells, demonstrating its practical applications in the diagnostic detection of tumors.

Interestingly, magneto-liposomes prepared by the encapsulation of iron-oxide nanoparticles in the liposomes core exhibited improved imaging capabilities as an MRI contrast agent. Nina and his co-workers developed an IONP-embedded liposome having unprecedented improvements in MRI properties compared to free iron-oxide nanoparticles [103]. The study demonstrated that magneto-liposomes had a strong potential for use as an MRI contrast agent even at extremely low concentrations, making it simpler to distinguish between healthy and malignant tissues. Rita et al. fabricated an Arg-Gly-Asp (RGD) motif functionalized magentoliposome with targeting abilities toward ovarian cancer cells [104]. The high tolerance of cells towards magentoliposomes with high concentrations of iron oxide concentrations suggests their intriguing possibilities for upcoming clinical investigations.

According to research, liposome-based CT contrast agents have the most promise for imaging extremely lethal, rapidly developing tumors and may even provide a more accurate assessment of the tumor's perfusion. Badea et al. suggested a liposomal-Iodinated CT Contrast agent assess the dynamic enhancement of lung nodules [105]. It was found that liposome-based CT agents have higher vascular permeability to rapidly growing tumors nanoparticle contrast agents when compared to slow-growing tumors which facilitate the distinction of tumors depending on their pace of growth. Polyethylene glycol (PEG) and tumor-specific antibody (mAb 2C5) modification of liposomes to attain enhanced CT contrast was proposed in a study by Silindir et al. [106]. The as-modified liposomes designated almost 3-8fold uptake in tumor cell lines than conventional ones, and are found to be considerably durable even after two weeks of incubation, indicating the benefit for in-vivo research to acquire longer imaging periods.

2.4.2 Lipid Micelles

Lipid molecules produce spherical structures called micelles, which contain a hydrophobic core with a hydrophilic shell. Because of their small size (from 5 to 100 nm), they showed a very effective and spontaneous accumulation in affected areas with weakened vasculature and thus, they have been successfully employed as molecular imaging probes applied in cancer theragnostic for the elucidation of biological mechanisms [107].

Over the course of the last few years, several works have compiled the progress of using lipid micelles in fluorescence imaging. Despite the encouraging results of FDA-approved fluorophores, the poor aqueous solubility, low quantum yield, and higher binding affinity to non-specific plasma proteins limit their further applications in in-vivo imaging. Therefore these fluorophores were encapsulated with lipid micellar systems as an attempt to increase the optical properties and prolonged stability in aqueous solutions [108]. Zheng et al. developed an indocyanine-phospholipid micellar (ICG-PL) nanostructure-based probe for targeted optical imaging [109]. The nanoprobe possessed several unique features such as high stable structure, absorption and emission peaks in the NIR range, non-toxic components, and internalization in targeted cells that could be used for cancer cell imaging (Figure 10). Duconge and his group proposed a phospholipid quantum dot micelles-based nanoprobe that holds great promise for optical fluorescence in-vivo imaging [110]. The encapsulation of quantum dots by phospholipids offers very adaptable surface chemistry to conjugate a variety of chemicals or biomolecules, dynamic body distribution, slow absorption by the reticuloendothelial system, and a long blood circulation half-time.

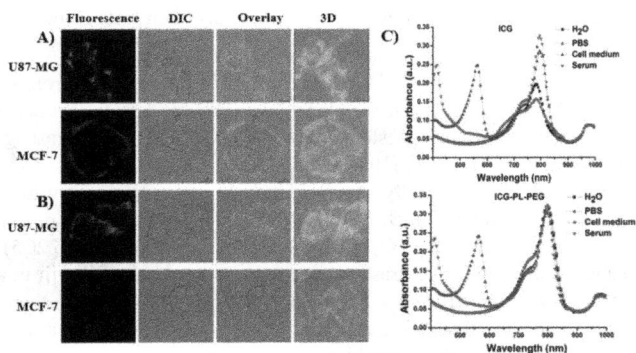

Figure 10 (A) Confocal images of human glioblastoma cancer cells (U87-MG) and human breast cancer cells (MCF-7) after incubation with targeting ICG-PL probe before being washed by PBS. (B) Confocal images of U87-MG and MCF-7 cells after incubation with targeting ICG-PL probe after being washed by PBS (C) Absorption and fluorescence spectra in different solvents [107].

The delivery of photosensitizers and photo-absorbing agents particularly for photoacoustic imaging has also been made possible by the development of lipid micelles-based nanocarriers with outstanding biocompatibility and loading capacity. Zhang's group fabricated lipid micelles incorporated semiconducting polymer dots and photosensitizer nanohybrid with excellent photo-acoustic effects for PA imaging of liver cancer cells [111]. The study reported that as-prepared micellar nanoformulation exhibited enhanced imaging contrast ability, low cytotoxicity, and high NIR absorption (670 nm) which could provide information on the morphology and anatomical structure of carcinoma cells. Even though micellar nanoparticles are known for their imaging contrast ability and better penetration of tumors, some of the PAI contrast agents are difficult to load into the micellar structure due to their physicochemical properties. Therefore, Li et al. proposed an α-lipoic acid stabilization technique for increasing the loading capacity of micelles to near-infrared dye, IR780 by decreasing the energy of molecular interaction [112]. According to the in-vivo performance, an enhanced PA signal intensity (2.1-fold higher) was detected for IR780-loaded micelles at tumors compared to that of free IR780.

Superparamagnetic iron oxide nanoparticles can have their magnetic characteristics tuned through biofunctionalization with lipid micelles, producing materials with improved T2-weighted MRI properties. Larsen et al. report a novel synthetic method for the facile and precise synthesis of micelles-encapsulated SPIO nanoparticle (mmSPIO) aggregates to achieve both T2 relaxation enhancement and size control [113]. The work also demonstrated the efficacy of mmSPIO as an ideal T2-weighted molecular contrast agent for in-vitro MR imaging of ovarian carcinoma cell lines. In order to improve the biocompatibility and resolution of the T1-weighted MRI imaging contrast agent, Dixit et al. prepared gadolinium oxide (Gd_2O_3) nanoparticles enclosed within phospholipid micelles [114]. The findings showed that the as prepared nanohybrid is capable of doing MR imaging in a manner comparable to commercial Magnevist without any apparent cytotoxicity.

2.4.3 Solid lipid nanoparticles

Among the nanoparticles mentioned above in medical imaging, solid lipid nanoparticles (50-1000 nm) have drawn the most attention from researchers in recent years based on their ability to cross challenging physiological barriers, like the blood-brain barrier, even without surface modification [115]. High skin compatibility, greater skin penetration, high physical and chemical adaptability, biological degradation, lipophilicity, and shielding of the active ingredient from breakdown processes brought on by external factors advantageous qualities provided by lipid nanoparticles. The majority of preparation techniques for the lipid nanoparticles are simply scalable, and they exhibit good conformance with biological substances and minimal toxicity.

Utilizing fluorescent solid lipid nanoparticle-based nanoformulation for cancerous cell imaging is becoming an increasingly popular tool. Yuan et al. prepared an octadecylamine–fluorescein isothiocyanate labelled solid nanoparticles for in-vivo fluorescence imaging of lung carcinoma cells [116]. The cellular uptake of lipid nanoparticles showed concentration and time dependence, and was correlated with the melting point of the lipid material, the length of the lipid's carbon chain, and the particle size. The tumor targeting ability and biodistribution of indocyanine dye encapsulated RGD peptide conjugated solid lipid nanoparticles (IR-780-RGD SLN) were investigated in glioblastoma cancer cells through NIR fluorescence imaging in a study by Kuang and his co-workers [117]. The fluorescence signals produced by IR-780-RGD SLN in the tumor region were significantly higher than those produced by free IR-780, indicating that the cRGD ligand could effectively enhance the lipid nanoparticles accumulating in the tumour region in vivo.

Solid lipid nanoparticles have also been studied for integrating with contrast agents such as superparamagnetic iron oxide for MRI applications with the improvement of bio-safety. Sun et al. synthesized solid lipid nanoparticles loaded with gadolinium diethylenetriaminepentaacetic acid (Gd-DTPA) and octadecylamine (ODA) fluorescein isothiocyanate (FITC) for differentiating tumors from normal colorectal walls via MRI findings [118]. In contrast to the traditional method, this methodology permits the distinction of colorectal cancers from the normal intestinal walls through the various absorption capabilities of the nanohybrid contrast agent. Ghiani et al. developed a paramagnetic solid lipid nanoparticles (pSLNs) from crystalline triglyceride that showcased a large Gd (III) payload and strong accessibility of the paramagnetic metal ions due to their surface orientation, which was driven by the solid lipid core [119]. Additionally, the work highlighted the active targeting properties of pSLNs towards cancerous tissues like ovarian carcinoma cells.

In another interesting study, Cai et al. prepared iodixanol-loaded lipid nanocapsule as a contrast agent for the CT imaging of lung cancer cells [120]. The nanostructure attained high stability, effective targeting of tumors, and demonstrated enhanced contrast CT imaging of tiny orthotopic lung tumor facilitating rapid, delicate, and accurate diagnosis.

Conclusion

Owing to the extensive applicability, potency, and rather cheap price-to-performance ratio, nanoparticles are one of the most interesting and dynamic nanocarriers in scientific studies, particularly for bioimaging applications. In view of the fact that nanoparticles seem to possess remarkable abilities to transcend the current limits of biomedical research. In nanostructure, the surface chemistry can be easily tuned and thus their specific activity can be tailor-made according to the imaging requirements. However, the usefulness of nanoparticle-based probes in the bioimaging-guided diagnosis of tumors needs to be further explored. Meanwhile, the mechanism by which nanoparticles with cancerous cells are still not fully understood, and hence necessitates advanced research in the future. It is anticipated that new therapeutic approaches that combine nanoparticles with therapeutic agents such as anticancer medications, photodynamic therapy agents, and photothermal therapy agents would increase the effectiveness of cancer treatment. Although the results of these preclinical investigations indicate that carbon nanoparticles have a great deal of potential for diagnosis and therapy and that the utilized nanoparticles do not have substantial harmful effects, extreme caution must be exercised before nanoparticles are implemented into ordinary clinical practice.

References

[1] R. Rani, K. Sethi, and G. Singh, "Nanomaterials and Their Applications in Bioimaging," Nanotechnol. Life Sci., pp. 429-450, 2019. https://doi.org/10.1007/978-3-030-16379-2_15

[2] R. Thomas, J. Unnikrishnan, A. V. Nair, E. C. Daniel, and M. Balachandran, "Antibacterial performance of GO-Ag nanocomposite prepared via ecologically safe protocols," Appl. Nanosci., vol. 10, no. 11, pp. 4207-4219, 2020. https://doi.org/10.1007/s13204-020-01539-z

[3] A. V. Ramya and M. Balachandran, "Valorization of agro-industrial fruit peel waste to fluorescent nanocarbon sensor: Ultrasensitive detection of potentially hazardous tropane alkaloid," Front. Environ. Sci. Eng., vol. 16, no. 3, p. 27, 2021. https://doi.org/10.1007/s11783-021-1461-z

[4] A. N. Mohan and S. Panicker, "Facile synthesis of graphene-tin oxide nanocomposite derived from agricultural waste for enhanced antibacterial activity against Pseudomonas aeruginosa," Sci. Rep., vol. 9, no. 1, pp. 1-12, 2019. https://doi.org/10.1038/s41598-019-40916-9

[5] S. S. Mughal, "DIAGNOSIS AND TREATMENT OF DISEASES BY USING METALLIC NANOPARTICLES-A REVIEW," Authorea Prepr., 2022. https://doi.org/10.22541/au.166401168.84305772/v1

[6] N. Aslan, B. Ceylan, M. M. Koç, and F. Findik, "Metallic nanoparticles as X-Ray computed tomography (CT) contrast agents: A review," J. Mol. Struct., vol. 1219, p. 128599, 2020. https://doi.org/10.1016/j.molstruc.2020.128599

[7] Z. Wei et al., "Rare-earth based materials: an effective toolbox for brain imaging, therapy, monitoring and neuromodulation," Light Sci. Appl., vol. 11, no. 1, p. 175, 2022. https://doi.org/10.1038/s41377-022-00922-5

[8] X. Zhu et al., "High-fidelity NIR-II multiplexed lifetime bioimaging with bright double interfaced lanthanide nanoparticles," Angew. Chemie, vol. 133, no. 44, pp. 23737-23743, 2021. https://doi.org/10.1002/ange.202108124

[9] G. Zhu et al., "GdVO4: Eu3+, Bi3+ nanoparticles as a contrast agent for MRI and luminescence bioimaging," ACS omega, vol. 4, no. 14, pp. 15806-15814, 2019. https://doi.org/10.1021/acsomega.9b00444

[10] R. De Matos et al., "Gd3+-Functionalized Lithium Niobate Nanoparticles for Dual Multiphoton and Magnetic Resonance Bioimaging," ACS Appl. Nano Mater., vol. 5, no. 2, pp. 2912-2922, 2022. https://doi.org/10.1021/acsanm.2c00127

[11] P. Araichimani et al., "Rare-earth ions integrated silica nanoparticles derived from rice husk via microwave-assisted combustion method for bioimaging applications," Ceram. Int., vol. 46, no. 11, pp. 18366-18372, 2020. https://doi.org/10.1016/j.ceramint.2020.04.125

[12] T. Jia et al., "Mesoporous cerium oxide-coated upconversion nanoparticles for tumor-responsive chemo-photodynamic therapy and bioimaging," Chem. Sci., vol. 10, no. 37, pp. 8618-8633, 2019. https://doi.org/10.1039/C9SC01615E

[13] X. Shan et al., "Polypyrrole-based double rare earth hybrid nanoparticles for multimodal imaging and photothermal therapy," J. Mater. Chem. B, vol. 8, no. 3, pp. 426-437, 2020. https://doi.org/10.1039/C9TB02254F

[14] F. Ren et al., "NIR-II Fluorescence imaging for cerebrovascular diseases," View, vol. 2, no. 6, p. 20200128, 2021. https://doi.org/10.1002/VIW.20200128

[15] D. Song, M. Zhu, S. Chi, L. Xia, Z. Li, and Z. Liu, "Sensitizing the luminescence of lanthanide-doped nanoparticles over 1500 nm for high-contrast and deep imaging of brain injury," Anal. Chem., vol. 93, no. 22, pp. 7949-7957, 2021. https://doi.org/10.1021/acs.analchem.1c00731

[16] Q. Wang, T. Liang, J. Wu, Z. Li, and Z. Liu, "Dye-sensitized rare earth-doped nanoparticles with boosted NIR-IIb emission for dynamic imaging of vascular network-related disorders," ACS Appl. Mater. Interfaces, vol. 13, no. 25, pp. 29303-29312, 2021. https://doi.org/10.1021/acsami.1c04612

[17] Y. C. Dong et al., "Ytterbium Nanoparticle Contrast Agents for Conventional and Spectral Photon-Counting CT and Their Applications for Hydrogel Imaging," ACS Appl. Mater. Interfaces, vol. 14, no. 34, pp. 39274-39284, 2022. https://doi.org/10.1021/acsami.2c12354

[18] M. Chang et al., "Enhanced photoconversion performance of NdVO4/Au nanocrystals for photothermal/photoacoustic imaging guided and near infrared light-triggered anticancer phototherapy," Acta Biomater., vol. 99, pp. 295-306, 2019. https://doi.org/10.1016/j.actbio.2019.08.026

[19] Z. Yu et al., "Achieving effective multimodal imaging with rare-earth ion-doped CaF2 nanoparticles," Pharmaceutics, vol. 14, no. 4, p. 840, 2022. https://doi.org/10.3390/pharmaceutics14040840

[20] X. Shi, K. Gao, G. Zhang, W. Zhang, X. Yang, and R. Gao, "Signal Amplification Pretargeted PET/Fluorescence Imaging Based on Human Serum Albumin-Encapsulated GdF3 Nanoparticles for Diagnosis of Ovarian Cancer," ACS Biomater. Sci. Eng., vol. 8, no. 11, pp. 4956-4964, 2022. https://doi.org/10.1021/acsbiomaterials.2c00374

[21] J. Cheng et al., "AgBiS2 nanoparticles with synergistic photodynamic and bioimaging properties for enhanced malignant tumor phototherapy," Mater. Sci. Eng. C, vol. 107, p. 110324, 2020. https://doi.org/10.1016/j.msec.2019.110324

[22] Q. Cheng et al., "Supramolecular Tropism Driven Aggregation of Nanoparticles In Situ for Tumor-Specific Bioimaging and Photothermal Therapy," Small, vol. 17, no. 43, p. 2101332, 2021. https://doi.org/10.1002/smll.202101332

[23] D. S. Dumani, I.-C. Sun, and S. Y. Emelianov, "Ultrasound-guided immunofunctional photoacoustic imaging for diagnosis of lymph node metastases," Nanoscale, vol. 11, no. 24, pp. 11649-11659, 2019. https://doi.org/10.1039/C9NR02920F

[24] O. J. Estudiante-Mariquez, A. Rodríguez-Galván, D. Ramírez-Hernández, F. F. Contreras-Torres, and L. A. Medina, "Technetium-radiolabeled mannose-functionalized gold nanoparticles as nanoprobes for sentinel lymph node detection," Molecules, vol. 25, no. 8, p. 1982, 2020. https://doi.org/10.3390/molecules25081982

[25] Y. Li, D. Xiao, S. Li, Z. Chen, S. Liu, and J. Li, "Silver@ quercetin nanoparticles with aggregation-induced emission for bioimaging in vitro and in vivo," Int. J. Mol. Sci., vol. 23, no. 13, p. 7413, 2022. https://doi.org/10.3390/ijms23137413

[26] M. Bouché et al., "Activatable hybrid polyphosphazene-AuNP nanoprobe for ROS detection by bimodal PA/CT imaging," ACS Appl. Mater. Interfaces, vol. 11, no. 32, pp. 28648-28656, 2019. https://doi.org/10.1021/acsami.9b08386

[27] M. Lyu, D. Zhu, Y. Duo, Y. Li, and H. Quan, "Bimetallic nanodots for tri-modal CT/MRI/PA imaging and hypoxia-resistant thermoradiotherapy in the NIR-II biological windows," Biomaterials, vol. 233, p. 119656, 2020. https://doi.org/10.1016/j.biomaterials.2019.119656

[28] D. Li et al., "Metal-organic nanostructure based on TixOy/Ruthenium reaction Units: For CT/MR Imaging-Guided X-ray induced dynamic therapy," Chem. Eng. J., vol. 417, p. 129262, 2021. https://doi.org/10.1016/j.cej.2021.129262

[29] S. D. Anderson, V. V Gwenin, and C. D. Gwenin, "Magnetic functionalized nanoparticles for biomedical, drug delivery and imaging applications," Nanoscale Res. Lett., vol. 14, no. 1, pp. 1-16, 2019. https://doi.org/10.1186/s11671-019-3019-6

[30] S. Carregal-Romero et al., "Ultrasmall manganese ferrites for in vivo catalase mimicking activity and multimodal bioimaging," Small, vol. 18, no. 16, p. 2106570, 2022. https://doi.org/10.1002/smll.202106570

[31] S. Gull, F. Ahmad, W. Wu, and W. Li, "Manganese-doped cesium iodide nanoparticles for multi-model bioimaging," Mater. Lett., vol. 256, p. 126630, 2019. https://doi.org/10.1016/j.matlet.2019.126630

[32] N. J. Hobson et al., "Clustering superparamagnetic iron oxide nanoparticles produces organ-targeted high-contrast magnetic resonance images," Nanomedicine, vol. 14, no. 9, pp. 1135-1152, 2019. https://doi.org/10.2217/nnm-2018-0370

[33] V. Pandey, T. Haider, P. Agrawal, S. Soni, and V. Soni, "Advances in Natural Polymeric Nanoparticles for the Drug Delivery," in Advances in Drug Delivery Methods, IntechOpen, 2022. https://doi.org/10.5772/intechopen.107513

[34] I. Parveen, M. I. Mahmud, and R. A. Khan, "Biodegradable Natural Polymers for Biomedical Applications," Guigoz. Sci. Rev., vol. 5, no. 3, pp. 67-80, 2019. https://doi.org/10.32861/sr.53.67.80

[35] B. Patra, A. K. Mishra, and R. S. Verma, "Label-free serum albumin nanoparticles for bioimaging and Trojan horse-like drug delivery," J. Sci. Adv. Mater. Devices, vol. 7, no. 1, p. 100406, 2022. https://doi.org/10.1016/j.jsamd.2021.100406

[36] Q. Wang, J. Fan, Y. Zhou, and S. Xu, "Development of a human serum albumin structure-based fluorescent probe for bioimaging in living cells," Spectrochim. Acta Part A Mol. Biomol. Spectrosc., vol. 269, p. 120769, 2022. https://doi.org/10.1016/j.saa.2021.120769

[37] L. L. Radford et al., "New 55Co-labeled albumin-binding folate derivatives as potential PET agents for folate receptor imaging," Pharmaceuticals, vol. 12, no. 4, p. 166, 2019. https://doi.org/10.3390/ph12040166

[38] D. Gao et al., "Albumin-consolidated AIEgens for boosting glioma and cerebrovascular NIR-II fluorescence imaging," ACS Appl. Mater. Interfaces, vol. 15, no. 1, pp. 3-13, 2022. https://doi.org/10.1021/acsami.1c22700

[39] Y. Zhang, Y. Wan, Y. Chen, N. T. Blum, J. Lin, and P. Huang, "Ultrasound-enhanced chemo-photodynamic combination therapy by using albumin 'nanoglue'-based nanotheranostics," ACS Nano, vol. 14, no. 5, pp. 5560-5569, 2020. https://doi.org/10.1021/acsnano.9b09827

[40] D. Li, T. Zhang, C. Min, H. Huang, D. Tan, and W. Gu, "Biodegradable theranostic nanoplatforms of albumin-biomineralized nanocomposites modified hollow mesoporous organosilica for photoacoustic imaging guided tumor synergistic therapy," Chem. Eng. J., vol. 388, p. 124253, 2020. https://doi.org/10.1016/j.cej.2020.124253

[41] Z. Yang et al., "Albumin-based nanotheranostic probe with hypoxia alleviating potentiates synchronous multimodal imaging and phototherapy for glioma," ACS Nano, vol. 14, no. 5, pp. 6191-6212, 2020. https://doi.org/10.1021/acsnano.0c02249

[42] A. Baki, A. Remmo, N. Löwa, F. Wiekhorst, and R. Bleul, "Albumin-coated single-core iron oxide nanoparticles for enhanced molecular magnetic imaging (Mri/mpi)," Int. J. Mol. Sci., vol. 22, no. 12, p. 6235, 2021. https://doi.org/10.3390/ijms22126235

[43] N. El-Sayed, V. Trouillet, A. Clasen, G. Jung, K. Hollemeyer, and M. Schneider, "NIR-Emitting Gold Nanoclusters-Modified Gelatin Nanoparticles as a Bioimaging Agent in Tissue," Adv. Healthc. Mater., vol. 8, no. 24, p. 1900993, 2019. https://doi.org/10.1002/adhm.201900993

[44] Y. Xue et al., "Rhodamine Conjugated Gelatin Methacryloyl Nanoparticles for Stable Cell Imaging," ACS Appl. Bio Mater., vol. 3, no. 10, pp. 6908-6918, 2020. https://doi.org/10.1021/acsabm.0c00802

[45] S. Paul et al., "Hydrothermal synthesis of gelatin quantum dots for high-performance biological imaging applications," J. Photochem. Photobiol. B Biol., vol. 212, p. 112014, 2020. https://doi.org/10.1016/j.jphotobiol.2020.112014

[46] A. Kimura et al., "Ultra-small size gelatin nanogel as a blood brain barrier impermeable contrast agent for magnetic resonance imaging," Acta Biomater., vol. 125, pp. 290-299, 2021. https://doi.org/10.1016/j.actbio.2021.02.016

[47] R. Yang et al., "Biomineralization-inspired crystallization of manganese oxide on silk fibroin nanoparticles for in vivo MR/fluorescence imaging-assisted tri-modal therapy of cancer," Theranostics, vol. 9, no. 21, p. 6314, 2019. https://doi.org/10.7150/thno.36252

[48] R. Ma et al., "Clinical indocyanine green-based silk fibroin theranostic nanoprobes for in vivo NIR-I/II fluorescence imaging of cervical diseases," Nanomedicine Nanotechnology, Biol. Med., vol. 47, p. 102615, 2023. https://doi.org/10.1016/j.nano.2022.102615

[49] Y. Wang et al., "In vivo photoacoustic/single-photon emission computed tomography imaging for dynamic monitoring of aggregation-enhanced photothermal nanoagents," Anal. Chem., vol. 91, no. 3, pp. 2128-2134, 2019. https://doi.org/10.1021/acs.analchem.8b04585

[50] M. Salarian et al., "Precision detection of liver metastasis by collagen-targeted protein MRI contrast agent," Biomaterials, vol. 224, p. 119478, 2019. https://doi.org/10.1016/j.biomaterials.2019.119478

[51] X. Zhou, L. Guo, D. Shi, S. Duan, and J. Li, "Biocompatible chitosan nanobubbles for ultrasound-mediated targeted delivery of doxorubicin," Nanoscale Res. Lett., vol. 14, pp. 1-9, 2019. https://doi.org/10.1186/s11671-019-2853-x

[52] H. Xiao et al., "Preparation of fluorescent nanoparticles based on broken-rice starch for live-cell imaging," Int. J. Biol. Macromol., vol. 217, pp. 88-95, 2022. https://doi.org/10.1016/j.ijbiomac.2022.06.205

[53] W. Zhou et al., "Iodine-rich semiconducting polymer nanoparticles for CT/Fluorescence dual-modal imaging-guided enhanced photodynamic therapy," Small, vol. 16, no. 5, p. 1905641, 2020. https://doi.org/10.1002/smll.201905641

[54] Y. Yang et al., "A 1064 nm excitable semiconducting polymer nanoparticle for photoacoustic imaging of gliomas," Nanoscale, vol. 11, no. 16, pp. 7754-7760, 2019. https://doi.org/10.1039/C9NR00552H

[55] M. Liu, Z. Zhang, Y. Yang, and Y. Chan, "Polymethine-Based Semiconducting Polymer Dots with Narrow-Band Emission and Absorption/Emission Maxima at NIR-II for Bioimaging," Angew. Chemie Int. Ed., vol. 60, no. 2, pp. 983-989, 2021. https://doi.org/10.1002/anie.202011914

[56] B. Guo et al., "High-resolution 3D NIR-II photoacoustic imaging of cerebral and tumor vasculatures using conjugated polymer nanoparticles as contrast agent," Adv. Mater., vol. 31, no. 25, p. 1808355, 2019. https://doi.org/10.1002/adma.201808355

[57] Y. Liu et al., "Fluorination enhances NIR-II fluorescence of polymer dots for quantitative brain tumor imaging," Angew. Chemie, vol. 132, no. 47, pp. 21235-21243, 2020. https://doi.org/10.1002/ange.202007886

[58] J. Zhao et al., "H 2 O 2-sensitive nanoscale coordination polymers for photoacoustic tumors imaging via in vivo chromogenic assay.," J. Innov. Opt. Health Sci., vol. 15, no. 5, 2022. https://doi.org/10.1142/S1793545822500262

[59] X. Hu et al., "Gadolinium-chelated conjugated polymer-based nanotheranostics for photoacoustic/magnetic resonance/NIR-II fluorescence imaging-guided cancer photothermal therapy," Theranostics, vol. 9, no. 14, p. 4168, 2019. https://doi.org/10.7150/thno.34390

[60] Z. Chen et al., "The advances of carbon nanotubes in cancer diagnostics and therapeutics," J. Nanomater., vol. 2017, 2017. https://doi.org/10.1155/2017/3418932

[61] A. De La Zerda et al., "Carbon nanotubes as photoacoustic molecular imaging agents in living mice," Nat. Nanotechnol., vol. 3, no. 9, pp. 557-562, 2008. https://doi.org/10.1038/nnano.2008.231

[62] M. Pramanik, M. Swierczewska, D. Green, B. Sitharaman, and L. V Wang, "Single-walled carbon nanotubes as a multimodal-thermoacoustic and photoacoustic-contrast agent," J. Biomed. Opt., vol. 14, no. 3, p. 34018, 2009. https://doi.org/10.1117/1.3147407

[63] L. Xie et al., "Functional long circulating single walled carbon nanotubes for fluorescent/photoacoustic imaging-guided enhanced phototherapy," Biomaterials, vol. 103, pp. 219-228, 2016. https://doi.org/10.1016/j.biomaterials.2016.06.058

[64] H. Gong, R. Peng, and Z. Liu, "Carbon nanotubes for biomedical imaging: the recent advances," Adv. Drug Deliv. Rev., vol. 65, no. 15, pp. 1951-1963, 2013. https://doi.org/10.1016/j.addr.2013.10.002

[65] Z. Liu et al., "Multiplexed five-color molecular imaging of cancer cells and tumor tissues with carbon nanotube Raman tags in the near-infrared," Nano Res., vol. 3, pp. 222-233, 2010. https://doi.org/10.1007/s12274-010-1025-1

[66] X. Wang, C. Wang, L. Cheng, S.-T. Lee, and Z. Liu, "Noble metal coated single-walled carbon nanotubes for applications in surface enhanced Raman scattering imaging and photothermal therapy," J. Am. Chem. Soc., vol. 134, no. 17, pp. 7414-7422, 2012. https://doi.org/10.1021/ja300140c

[67] A. Jorio, M. A. Pimenta, A. G. Souza Filho, R. Saito, G. Dresselhaus, and M. S. Dresselhaus, "Characterizing carbon nanotube samples with resonance Raman scattering," New J. Phys., vol. 5, no. 1, p. 139, 2003. https://doi.org/10.1088/1367-2630/5/1/139

[68] J. T. Robinson, G. Hong, Y. Liang, B. Zhang, O. K. Yaghi, and H. Dai, "In vivo fluorescence imaging in the second near-infrared window with long circulating carbon nanotubes capable of ultrahigh tumor uptake," J. Am. Chem. Soc., vol. 134, no. 25, pp. 10664-10669, 2012. https://doi.org/10.1021/ja303737a

[69] K. Welsher et al., "A route to brightly fluorescent carbon nanotubes for near-infrared imaging in mice," Nat. Nanotechnol., vol. 4, no. 11, pp. 773-780, 2009. https://doi.org/10.1038/nnano.2009.294

[70] S.-T. Yang et al., "Carbon dots as nontoxic and high-performance fluorescence imaging agents," J. Phys. Chem. C, vol. 113, no. 42, pp. 18110-18114, 2009. https://doi.org/10.1021/jp9085969

[71] Y. Song, W. Shi, W. Chen, X. Li, and H. Ma, "Fluorescent carbon nanodots conjugated with folic acid for distinguishing folate-receptor-positive cancer cells from normal cells," J. Mater. Chem., vol. 22, no. 25, pp. 12568-12573, 2012. https://doi.org/10.1039/c2jm31582c

[72] N. Parvin and T. K. Mandal, "Dually emissive P, N-co-doped carbon dots for fluorescent and photoacoustic tissue imaging in living mice," Microchim. Acta, vol. 184, pp. 1117-1125, 2017. https://doi.org/10.1007/s00604-017-2108-4

[73] G. Xu et al., "In vivo tumor photoacoustic imaging and photothermal therapy based on supra-(carbon nanodots)," Adv. Healthc. Mater., vol. 8, no. 2, p. 1800995, 2019. https://doi.org/10.1002/adhm.201800995

[74] S. Sun et al., "Ce6-modified carbon dots for multimodal-imaging-guided and single-NIR-laser-triggered photothermal/photodynamic synergistic cancer therapy by reduced irradiation power," ACS Appl. Mater. Interfaces, vol. 11, no. 6, pp. 5791-5803, 2019. https://doi.org/10.1021/acsami.8b19042

[75] D. Bouzas-Ramos, J. Cigales Canga, J. C. Mayo, R. M. Sainz, J. Ruiz Encinar, and J. M. Costa-Fernandez, "Carbon quantum dots codoped with nitrogen and lanthanides for multimodal imaging," Adv. Funct. Mater., vol. 29, no. 38, p. 1903884, 2019. https://doi.org/10.1002/adfm.201903884

[76] X. Zhao et al., "A magnetofluorescent boron-doped carbon dots as a metal-free bimodal probe," Talanta, vol. 200, pp. 9-14, 2019. https://doi.org/10.1016/j.talanta.2019.03.022

[77] J. Zhang et al., "Phosphorescent Carbon Dots for Highly Efficient Oxygen Photosensitization and as Photo-oxidative Nanozymes," ACS Appl. Mater. Interfaces, vol. 10, pp. 40808-40814, 2018, doi: 10.1021/acsami.8b15318. https://doi.org/10.1021/acsami.8b15318

[78] Z. Liu, Z. Guo, H. Zhong, X. Qin, M. Wan, and B. Yang, "Graphene oxide based surface-enhanced Raman scattering probes for cancer cell imaging," Phys. Chem. Chem. Phys., vol. 15, no. 8, pp. 2961-2966, 2013. https://doi.org/10.1039/c2cp43715e

[79] Y. Chen, T. Liu, P. Chen, P. Chang, and S. Chen, "A high-sensitivity and low-power theranostic nanosystem for cell SERS imaging and selectively photothermal therapy using anti-EGFR-conjugated reduced graphene oxide/mesoporous silica/AuNPs nanosheets," Small, vol. 12, no. 11, pp. 1458-1468, 2016. https://doi.org/10.1002/smll.201502917

[80] Y. Esmaeili, E. Bidram, A. Zarrabi, A. Amini, and C. Cheng, "Graphene oxide and its derivatives as promising In-vitro bio-imaging platforms," Sci. Rep., vol. 10, no. 1, pp. 1-13, 2020. https://doi.org/10.1038/s41598-020-75090-w

[81] P. S. Wate et al., "Cellular imaging using biocompatible dendrimer-functionalized graphene oxide-based fluorescent probe anchored with magnetic nanoparticles," Nanotechnology, vol. 23, no. 41, p. 415101, 2012. https://doi.org/10.1088/0957-4484/23/41/415101

[82] P. Huang et al., "Surface functionalization of chemically reduced graphene oxide for targeted photodynamic therapy," J. Biomed. Nanotechnol., vol. 11, no. 1, pp. 117-125, 2015. https://doi.org/10.1166/jbn.2015.2055

[83] R. Thomas and M. Balachandran, "Doable production of highly fluorescent, heteroatom-doped graphene material from fuel coke for cellular bioimaging: An eco-sustainable cradle-to-gate approach," J. Clean. Prod., vol. 383, p. 135541, 2023. https://doi.org/10.1016/j.jclepro.2022.135541

[84] G. Lalwani, X. Cai, L. Nie, L. V. Wang, and B. Sitharaman, "Graphene-based contrast agents for photoacoustic and thermoacoustic tomography," Photoacoustics, vol. 1, no. 3-4, pp. 62-67, 2013, doi: 10.1016/j.pacs.2013.10.001. https://doi.org/10.1016/j.pacs.2013.10.001

[85] S. Lee and S. Y. Kim, "Gold nanorod/reduced graphene oxide composite nanocarriers for near-infrared-induced cancer therapy and photoacoustic imaging," ACS Appl. Nano Mater., vol. 4, no. 11, pp. 11849-11860, 2021. https://doi.org/10.1021/acsanm.1c02419

[86] J. Chen et al., "Indocyanine green loaded reduced graphene oxide for in vivo photoacoustic/fluorescence dual-modality tumor imaging," Nanoscale Res. Lett., vol. 11, pp. 1-11, 2016. https://doi.org/10.1186/s11671-016-1288-x

[87] H. W. Yang et al., "Gadolinium-functionalized nanographene oxide for combined drug and microRNA delivery and magnetic resonance imaging," Biomaterials, vol. 35, no. 24, pp. 6534-6542, 2014, doi: 10.1016/j.biomaterials.2014.04.057. https://doi.org/10.1016/j.biomaterials.2014.04.057

[88] K. Yang, L. Feng, X. Shi, and Z. Liu, "Nano-graphene in biomedicine: theranostic applications," Chem. Soc. Rev., vol. 42, no. 2, pp. 530-547, 2013. https://doi.org/10.1039/C2CS35342C

[89] K. Yang et al., "Multimodal imaging guided photothermal therapy using functionalized graphene nanosheets anchored with magnetic nanoparticles," Adv. Mater., vol. 24, no. 14, pp. 1868-1872, 2012. https://doi.org/10.1002/adma.201104964

[90] M. R. Younis, G. He, J. Lin, and P. Huang, "Recent advances on graphene quantum dots for bioimaging applications," Front. Chem., vol. 8, p. 424, 2020. https://doi.org/10.3389/fchem.2020.00424

[91] L. Wang et al., "Ultrastable amine, sulfo cofunctionalized graphene quantum dots with high two-photon fluorescence for cellular imaging," ACS Sustain. Chem. Eng., vol. 6, no. 4, pp. 4711-4716, 2018. https://doi.org/10.1021/acssuschemeng.7b03797

[92] T. Gao et al., "Red, yellow, and blue luminescence by graphene quantum dots: syntheses, mechanism, and cellular imaging," ACS Appl. Mater. Interfaces, vol. 9, no. 29, pp. 24846-24856, 2017. https://doi.org/10.1021/acsami.7b05569

[93] H. Wang et al., "Nitrogen and boron dual-doped graphene quantum dots for near-infrared second window imaging and photothermal therapy," Appl. Mater. today, vol. 14, pp. 108-117, 2019. https://doi.org/10.1016/j.apmt.2018.11.011

[94] M. C. Biswas, M. T. Islam, P. K. Nandy, and M. M. Hossain, "Graphene quantum dots (GQDs) for bioimaging and drug delivery applications: a review," ACS Mater. Lett., vol. 3, no. 6, pp. 889-911, 2021. https://doi.org/10.1021/acsmaterialslett.0c00550

[95] F. H. Wang, K. Bae, Z. W. Huang, and J. M. Xue, "Two-photon graphene quantum dot modified Gd 2 O 3 nanocomposites as a dual-mode MRI contrast agent and cell labelling agent," Nanoscale, vol. 10, no. 12, pp. 5642-5649, 2018. https://doi.org/10.1039/C7NR08068A

[96] H. Wang et al., "Paramagnetic properties of metal-free boron-doped graphene quantum dots and their application for safe magnetic resonance imaging," Adv. Mater., vol. 29, no. 11, 2017. https://doi.org/10.1002/adma.201605416

[97] S. Badrigilan, B. Shaabani, N. G. Aghaji, and A. Mesbahi, "Graphene quantum dots-coated bismuth nanoparticles for improved CT imaging and photothermal performance," Int. J. Nanosci., vol. 19, no. 01, p. 1850043, 2020. https://doi.org/10.1142/S0219581X18500436

[98] Y. Xuan et al., "Targeting N-doped graphene quantum dot with high photothermal conversion efficiency for dual-mode imaging and therapy in vitro," Nanotechnology, vol. 29, no. 35, p. 355101, 2018. https://doi.org/10.1088/1361-6528/aacad0

[99] S. K. Nune, P. Gunda, P. K. Thallapally, Y.-Y. Lin, M. Laird Forrest, and C. J. Berkland, "Nanoparticles for biomedical imaging," Expert Opin. Drug Deliv., vol. 6, no. 11, pp. 1175-1194, 2009. https://doi.org/10.1517/17425240903229031

[100] S. Mosleh-Shirazi et al., "Nanotechnology Advances in the Detection and Treatment of Cancer: An Overview," Nanotheranostics, vol. 6, no. 4, pp. 400-423, 2022, doi: 10.7150/ntno.74613. https://doi.org/10.7150/ntno.74613

[101] C.-J. Wen, C. T. Sung, I. A. Aljuffali, Y.-J. Huang, and J.-Y. Fang, "Nanocomposite liposomes containing quantum dots and anticancer drugs for bioimaging and therapeutic delivery: a comparison of cationic, PEGylated and deformable liposomes," Nanotechnology, vol. 24, no. 32, p. 325101, 2013. https://doi.org/10.1088/0957-4484/24/32/325101

[102] S. Dong, J. D. W. Teo, L. Y. Chan, C.-L. K. Lee, and K. Sou, "Far-red fluorescent liposomes for folate receptor-targeted bioimaging," ACS Appl. Nano Mater., vol. 1, no. 3, pp. 1009-1013, 2018. https://doi.org/10.1021/acsanm.8b00084

[103] N. Kostevšek et al., "Magneto-liposomes as MRI contrast agents: A systematic study of different liposomal formulations," Nanomaterials, vol. 10, no. 5, p. 889, 2020. https://doi.org/10.3390/nano10050889

[104] R. S. Garcia Ribeiro et al., "Targeting tumor cells and neovascularization using RGD-functionalized magnetoliposomes," Int. J. Nanomedicine, pp. 5911-5924, 2019. https://doi.org/10.2147/IJN.S214041

[105] C. T. Badea et al., "Computed tomography imaging of primary lung cancer in mice using a liposomal-iodinated contrast agent," PLoS One, vol. 7, no. 4, p. e34496, 2012. https://doi.org/10.1371/journal.pone.0034496

[106] M. Silindir et al., "Nanosized multifunctional liposomes for tumor diagnosis and molecular imaging by SPECT/CT," J. Liposome Res., vol. 23, no. 1, pp. 20-27, 2013. https://doi.org/10.3109/08982104.2012.722107

[107] S. Valetti, S. Mura, B. Stella, and P. Couvreur, "Rational design for multifunctional non-liposomal lipid-based nanocarriers for cancer management: theory to practice," J. Nanobiotechnology, vol. 11, no. 1, pp. 1-17, 2013. https://doi.org/10.1186/1477-3155-11-S1-S6

[108] A.-K. Kirchherr, A. Briel, and K. Mäder, "Stabilization of indocyanine green by encapsulation within micellar systems," Mol. Pharm., vol. 6, no. 2, pp. 480-491, 2009. https://doi.org/10.1021/mp8001649

[109] X. Zheng, D. Xing, F. Zhou, B. Wu, and W. R. Chen, "Indocyanine green-containing nanostructure as near infrared dual-functional targeting probes for optical imaging and photothermal therapy," Mol. Pharm., vol. 8, no. 2, pp. 447-456, 2011. https://doi.org/10.1021/mp100301t

[110] F. Ducongé et al., "Fluorine-18-labeled phospholipid quantum dot micelles for in vivo multimodal imaging from whole body to cellular scales," Bioconjug. Chem., vol. 19, no. 9, pp. 1921-1926, 2008. https://doi.org/10.1021/bc800179j

[111] D. Zhang et al., "Lipid micelles packaged with semiconducting polymer dots as simultaneous MRI/photoacoustic imaging and photodynamic/photothermal dual-modal therapeutic agents for liver cancer," J. Mater. Chem. B, vol. 4, no. 4, pp. 589-599, 2016. https://doi.org/10.1039/C5TB01827G

[112] W. Li et al., "α-Lipoic acid stabilized DTX/IR780 micelles for photoacoustic/fluorescence imaging guided photothermal therapy/chemotherapy of breast cancer," Biomater. Sci., vol. 6, no. 5, pp. 1201-1216, 2018. https://doi.org/10.1039/C8BM00096D

[113] B. A. Larsen, M. A. Haag, N. J. Serkova, K. R. Shroyer, and C. R. Stoldt, "Controlled aggregation of superparamagnetic iron oxide nanoparticles for the development of molecular magnetic resonance imaging probes," Nanotechnology, vol. 19, no. 26, p. 265102, 2008. https://doi.org/10.1088/0957-4484/19/26/265102

[114] S. Dixit et al., "Phospholipid micelle encapsulated gadolinium oxide nanoparticles for imaging and gene delivery," RSC Adv., vol. 3, no. 8, pp. 2727-2735, 2013. https://doi.org/10.1039/c2ra22293k

[115] E. Musielak, A. Feliczak-Guzik, and I. Nowak, "Synthesis and potential applications of lipid nanoparticles in medicine," Materials (Basel)., vol. 15, no. 2, p. 682, 2022. https://doi.org/10.3390/ma15020682

[116] H. Yuan, J. Miao, Y.-Z. Du, J. You, F.-Q. Hu, and S. Zeng, "Cellular uptake of solid lipid nanoparticles and cytotoxicity of encapsulated paclitaxel in A549 cancer cells," Int. J. Pharm., vol. 348, no. 1-2, pp. 137-145, 2008. https://doi.org/10.1016/j.ijpharm.2007.07.012

[117] Y. Kuang et al., "Hydrophobic IR-780 dye encapsulated in cRGD-conjugated solid lipid nanoparticles for NIR imaging-guided photothermal therapy," ACS Appl. Mater. Interfaces, vol. 9, no. 14, pp. 12217-12226, 2017. https://doi.org/10.1021/acsami.6b16705

[118] J. Sun et al., "Gadolinium-loaded solid lipid nanoparticles as a tumor-absorbable contrast agent for early diagnosis of colorectal tumors using magnetic resonance colonography," J. Biomed. Nanotechnol., vol. 12, no. 9, pp. 1709-1723, 2016. https://doi.org/10.1166/jbn.2016.2285

[119] S. Ghiani et al., "In vivo tumor targeting and biodistribution evaluation of paramagnetic solid lipid nanoparticles for magnetic resonance imaging," Nanomedicine Nanotechnology, Biol. Med., vol. 13, no. 2, pp. 693-700, 2017. https://doi.org/10.1016/j.nano.2016.09.012

[120] P. Cai et al., "Inherently PET/CT dual modality imaging lipid nanocapsules for early detection of orthotopic lung tumors," ACS Appl. Bio Mater., vol. 3, no. 1, pp. 611-621, 2019. https://doi.org/10.1021/acsabm.9b00993

Nanoparticles in Healthcare: Applications in Therapy, Diagnosis and Drug Delivery   Materials Research Forum LLC
Materials Research Foundations 160 (2024) 145-162   https://doi.org/10.21741/9781644902974-6

Chapter 6

Diagnostic Nanotechnologies

Dhanya K Chandrasekharan[1*], Dani Mathew M[2], Elizabeth P Thomas[1]

[1] St. Mary's College, Thrissur, Kerala, India

[2] Alphonsa College, Pala, Kerala, India

*dhanuchandra@gmail.com

Abstract

The application of nanotechnology in clinical diagnostic purposes can be defined as nanodiagnostics. Since most of the biological molecules as well as cell organelles fall within the nano scale, nanotechnology can assure to meet the rigorous demands being put forth by the clinical laboratory in terms of sensitivity, earlier and faster detection of diseases and cost-effectiveness. Nanotechnology Based Diagnostics or Nanodiagnostics is also termed as "molecular imaging" and it is an evolutionary leap in diagnostic imaging. Nanodiagnostics assays are utilized in immunoassays, DNA diagnostics, imaging, etc for the detection of tumors and infectious diseases.

Keywords

Nanodiagnostics, Nanoparticles, Theranostics, Nanotechnology, Cancer

Contents

1.   Introduction

The search of methods or techniques for detecting minute quantities of biomolecules dates back to the mid-1970s and advanced research in the field of nanotechnology for such applications took place since the last decade. The use of nanotechnology for clinical diagnostic purposes can be defined as nanodiagnostics [1,2]. The potential of nanodiagnostics is enormous since biological molecules and cellular organelles are in the nanometer scale [3]. Nanodiagnostics meets the rigorous demands of clinical diagnostics in terms of sensitivity, earlier detection of diseases and cost-effectiveness. Nanotechnology enables detectable interaction between single molecules of analyte and signal-generating nanoparticles [2,4]. Nanoparticle-based molecular imaging has resulted in an evolutionary leap in diagnostics and nanoparticles will surely become potential therapeutic and diagnostic tools in the near future [5].

The prime step in nanoparticle-based assay is the binding of a nanoparticle probe or label to target biomolecule and a measurable signal characteristic of the target biomolecule will be produced. Examples of such nanoparticle probe or label are Quantom Dots or QDs, metal nanoparticles, nanoshells, carbon nano tubes, cantilevers, etc [6-10]. Either nanoparticle as such will be used as the recognition signal or further improvement in signal generation can be achieved by conjugating these nanoparticles to other recognition moieties such as oligonucleotides or antibodies. Absorption of light by Surface Plasmon (combined excitation of the electrons at the surface) of nanoparticles is the reason for the unique color of nanoparticle suspensions, and the color is dependent on the size and aggregation of particles [11-14] and this feature is mainly utilized in many of the nanoparticle based diagnostic assays. Several immunoassays, immunohistochemistry techniques, DNA based diagnostics, bio separation methods for specific cell populations, cellular imaging, etc have been developed based on nanodiagnostics and it offer new frontiers for detection infectious diseases, bio-terrorism agents, neurological diseases, etc. More research is necessary to fully optimize the use of nanoparticles in clinical diagnosis and also to resolve the concerns regarding potential health and environmental risks related to the extensive use of nanoparticles [15]. Nanoparticles have also peeved the interest of medical community for their potential applications in cancer diagnosis, cancer treatment, and also as delivery vectors for pharmacologic agents [16-23].

Nanoparticles in Healthcare: Applications in Therapy, Diagnosis and Drug Delivery   Materials Research Forum LLC
Materials Research Foundations 160 (2024) 145-162   https://doi.org/10.21741/9781644902974-6

2.   Different types of nanoparticles used in diagnostics

Different types of nanoparticles used in diagnostics such as Quantom dots, Gold nanoparticles, Carbon Nano Tubes, Fluorescent nanoparticles, Nanometer sized TiO_2, Superparamagnetic nanoparticles, Silver nanoparticles, Cantilevers, etc are discussed below.

2.1   Quantom dots (QDs)

QDs are semiconductor nanocrystals and possess high photostability, single-wavelength excitation, and size-tunable emission. These are the most promising nanostructures for diagnostic applications. QDs are composed of a core semiconductor which is enclosed in a shell of another semiconductor with a larger spectral bandgap and then there is a third silica shell which improves the water solubility. QDs could be used in multiplexed diagnostics, immunohistochemical assays, immunoassays, cellular imaging, neurotransmitter detection, etc. [24,25,26]. QDs can be directed toward the target analyte by conjugating it to antibodies, oligonucleotides or aptamers or coated with streptavidin. QDs can also be used as nonspecific fluorescent labels [27]. QDs do not need laser for excitation to be fluorescent, and this makes the instrumentation needed for detection very simple and fluorometers and fluorescence microscopes are enough [1]. QDs possess other advantages also such as resistance to photobleaching, optical tunability, excitation of various QDs by a single wavelength of light (suitable for multiplexing), stability of optical properties after conjugation to biomolecules and narrow emission band [3,24,28,29,30].

Advantages of QDs over conventional organic dyes are [26]

1. Optical tunability

2. Resistance to photo bleaching

3. Excitation of various QDs by a single wavelength of light allowing multiplexing

4. Narrow emission band

5. Stability of optical properties even after conjugation to a biomolecule

6. Cause no cleavage of or disruption of interactions with proteins of DNA molecules

7. Long lifetime (>10 ns). QDs emit with lifetimes of 5–40 ns, whereas conventional organic dyes emit in 0.5–2 ns range.

QDs based assays could be measured by using fluorometry or by microscopy such as confocal, fluorescence, wide-field epifluorescence, total internal reflection, atomic force, multiphoton microscopy, etc [24,28,31,32].

There are some possible technical difficulties and safety concerns with the use of QDs. QDs are subject to low luminescence activity due to large surface area [28]. The ability of QDs to reach targets within cellular compartments or molecular complexes is doubtful [29] and this impacts the sensitivity of the assay. CdSe/ZnS QDs possess a technical problem that these cannot be used for whole blood analysis due to their inherent property of not being able to emit in the near-infrared region [2]. QD diagnostics includes the risk of toxic core semiconductor materials leakage into the host system or into the environment. Higher QD concentrations (>5 X 109 QDs/cell) is shown to affect Xenopus embryo development [27,33].

Examples of assays that use QDs:

1. Detection of prostate-specific antigen (PSA) using 107-nm QDs consisting of β diketones entrapping europium molecules and coated with streptavidin. The assay could detect 0.38 ng/L biotinylated PSA using a time resolved fluorometer and visualization of individual PSA molecules was also possible with the use of a fluorescence microscope [34].

2. Water-soluble CdSe/ZnS QDs surface modified with tetrahexyl ether derivatives of p-sulfonatocalix(4)arene, can be used for optical detection of acetylcholine [35].

3. QDs conjugated to antibodies can be used to fluorescently target the membrane protein P-glycoprotein (Pgp) in HeLa cells transfected by plasmid encoding for Pgp. This conjugation could be either done via biotin - avidin conjugation where avidin-coated QDs are used or the Pgp antibody may be electrostatically attracted to the QDs [27].

4. Peptides can be conjugated to QDs to target lung endothelial cells, breast carcinoma cells, brain endothelial cells, etc *in vitro* and *in vivo* [36]. Microinjection of QDs conjugated to suitable peptides can target cellular organelles such as nucleus or mitochondria [27]. QD–streptavidin conjugates are used for the detection of intracellular targets, such as microtubules, nuclear antigens, F-actin filaments, etc. QD–IgG (emission maximum 535 nm] and QD– streptavidin [emission maximum 630 nm) detected 2 cellular targets i.e., Her2 and nuclear antigens in SK-BR-3 cancer cells, respectively at a single excitation wavelength [25].

5. QDs can be used for the detection of cancer. Streptavidin-coated QDs detects the marker Her2 on SKBR- 3 breast cancer cells via the biotinylated secondary antibody [25,27].

6. QDs have applications in cellular imaging. Antibody-coupled QDs injected into the tail vein of mice detected prostate cancer xenografts [27,31].

7. Another potential use for QDs is angiography.

8. QDs can be used to detect homogeneous point mutation based on a QD-mediated two-color fluorescence coincidence detection scheme [37].

9. The QD labels can be used for DNA detection. This approach overcomes the problems encountered while using organic dyes for DNA labeling. QDs also allow 2-color determination of orientation of single DNA molecule by using DNA labeled with biotin and/or digoxigen at ends of linear DNA [32].

2.2   Gold nanoparticles [GNPs]

The application of gold nanoparticles in diagnostics is due to their unique combination of chemical and physical properties that allow detection of low concentrations of biological molecules. Surface Plasmon resonance, resulting from the interaction of locally adjacent gold nanoparticle bound to a target, produces changes that can be used for optical detection. The characteristic red color of gold nanoparticle changes to a bluish-purple on colloid aggregation due to plasmon resonance [3]. The plasmon frequency of gold particles can be controlled [38].

Citrate stabilized gold colloids changes their absorption band to longer wavelengths upon clustering of the colloids, lead to color change from pink-red to violet-blue [39]. Clustering may be induced either by physical methods such as the increase of ionic strength of solution or chemically by addition of molecules which connect nanoparticle together. These optical features make gold nanoparticles ideal color reporting sensor for signaling molecular recognition events and to render detection of nanomolar concentrations of analytes ranging from DNA to proteins and metal ions [9]. Elegant biosensor designs utilizing these concepts have been demonstrated such as DNA functionalized GNPs for colorimetric detection of DNA [6] and Pb (II) and carbohydrate-modified GNPs for selectively sensing lectins [40]. Protein–protein interactions can be evaluated qualitatively as well as quantitatively at nanomolar concentrations of proteins by using gold nanoparticle-based competitive colorimetric assays without any special protein modifications [41].

The principles of some such biosensors are detailed below:

1. Gold nanoparticles could be used to design calorimetric metal sensors since different aggregation states can result in distinctive color changes [42]. A colorimetric metal sensor based on DNAzyme-directed assembly of gold nanoparticles developed by Liu *et al* could be used for sensitive and selective detection and quantification of metal ions, particularly lead [7].

2. Aptamers which are SS DNA or SS RNA molecules that bind target molecules with high affinity and specificity and undergo changes in conformation upon binding to its target analyte can be used in conjunction with gold nanoparticles. Aptamer-linked gold nanoparticle aggregates [purple in colour] undergo fast disassembly into dispersed nanoparticles [red colour] upon binding of target analytes [7,43,44,45].

3. GNPs can be used for the detection of proteases. Because thiols interact strongly with gold nanoparticles, thiol causes agglomeration. When a gold colloid solution is treated with a peptide Cys-(AA)n-Cys, the color of the solution turns from pink-red to violet-blue due to aggregation of the nanoparticles. The cleavage of a Cys-(AA)n-Cys peptide in two fragments by proteases would prevent aggregation of gold nanoparticles [46].

4. Protein-coated gold nanoparticles can be used to study the conformational change of proteins on a solid-liquid interface [47] and to interpret conformational change in an oligonucleotide and phase transition of a biopolymer [48-52].

5. Gold nanoparticles (GNPs) are widely used to detect DNA with high sensitivity and selectivity [3,6,29,53,54,55]. Optical properties of 13 to 17 nm diameter Au particles have been exploited in the development of a highly selective colorimetric diagnostic method for DNA [42,56]. This diagnostic method is based on the distance-dependent optical properties of gold particles [57]. Based on the electrostatic interaction of DNA and GNPs upon addition of HCl, ssDNA sequence can be easily distinguished based on single-base

mismatch. Comparing with other nanoparticle-based DNA detection assays, this method proposed by Sun *et al* [58] has the following advantages:

(1) Complicated modifications of GNPs or DNA are not required

(2) Additional DNA probes are not required

(3) Signal amplification or temperature control is not required.

Mutations, gene expression, chromosomal translocations, single nucleotide polymorphisms (SNPs), and pathogens from clinical samples can be easily detected [59]. A colorimetric method uses gold nanoparticles for rapid and sensitive direct detection of *M. tuberculosis* in clinical samples [60].

Some other gold nanoparticle-based diagnostics are detailed below:

1. Silver-coated gold particles of 40–100 nm size possess strong light-scattering properties that allows easy detection by standard dark-field microscopy [3, Georganopoulou *et al* 2005] giving a detection limit for oligonucleotides down to ~10 fmol while conventional fluorophore-based methods could detect 50-fold higher amount only [2,3,29,55].

2. Gold nanoshells possess biocompatibility which along with the tunability of their optical properties, allows development of immunoassays for the simultaneous analysis of multiple antigens [3,9].

3. Variations in the relative thicknesses of gold nano shells allow the manipulations of optical resonance to go into the mid infrared region and to the near-infrared just [3,6,7,8,9] which helps to avoid interferences from hemoglobin during direct analysis of whole blood samples.

4. The surface-plasmon resonance enhanced optical properties of gold nanoparticles helped to develop biomedical applications on cancer diagnostics and therapeutics and a method of molecular-specific diagnostics/detection of cancer [62]. SPR scattering imaging or SPR absorption spectroscopy generated from antibody conjugated gold nanoparticles was found to be useful for the diagnosis of oral epithelial living cancer cells *in vivo* and *in vitro* [63]. Gold nanorods could be used as novel contrast agents for both molecular imaging and photothermal cancer therapy since gold nanorods can absorb and scatter strongly in NIR region. Both cancer diagnostics and selective photothermal therapy are achievable by using gold nanorods since antibody-conjugated nanorods bind specifically to the surface of the malignant-type cells with a much higher affinity due to the overexpressed EGFR on the cytoplasmic membrane of the malignant cells. As a result, the malignant cells can be clearly visualized and on exposure to continuous red laser, malignant cells can be photothermally destroyed [64].

5. Gold nanoparticles appear to be superior in optical trapping assays. Relatively large gold particles yield a six-fold enhancement in trapping efficiency and detection sensitivity when compared to similar-sized polystyrene particles due to gold's high polarizability [65].

6. Optical absorption by gold at a wavelength of 1064 nm induces dramatic heating and such heating could damage biomaterials such as enzymes and thus gold nanoparticles are useful as local molecular heaters to locally unfold protein or RNA molecules [65].

7. The nano-Au electrode showed excellent sensitivity, good selectivity and antifouling properties which can be used for the selective determination of dopamine [66].

8. Organophosphate hydrolase-gold nanoparticle (OPH-gold nanoparticle) conjugates could be used along with a fluorescent enzyme inhibitor or decoy for discriminative detection of ultralow quantities of ortho phosphate (OP) neurotoxins. The fluorescence intensity of the decoy was sensitive to the proximity of the gold nanoparticle, and thus could be correlated with concentration of paraoxon (an ortho phosphate and the active metabolite of some insecticides) present [67].

2.3   Carbon nano tubes

(CNTs) are suitable substrate for *in vitro* cell growth for tissue regeneration, they are used as vectors for gene delivery, and find applications in delivery systems for diagnostics and therapeutics [68]. Yu *et al* reported use of SWCNT (single-walled carbon nanotubes) with antibody bioconjugates for highly sensitive detection of cancer biomarker from serum. This approach provided pico gram level detection of prostate specific antigen and is an excellent promise for clinical screening of cancer [69].

2.4   Superparamagnetic nanoparticles

Due to the size-dependent properties and dimensional similarities to biomacromolecules, magnetic nano bioconjugates very useful contrast agents for *in vivo* magnetic resonance imaging [70,71], as drug delivery carriers, and as structural scaffolds for tissue engineering [72].

Superparamagnetic nanoparticles are based on a core consisting of iron oxides and are termed Superparamagnetic iron oxide nanoparticles (SPION) and these can be targeted using external magnets. SPION can be coated with biocompatible materials and can be functionalized for various applications with drugs, proteins or plasmids [73]. For example, monoclonal anti-cancer nucleosome-specific antibody can be covalently attached to the surface of SPION-loaded PEGPE micelles. This could be used to specifically recognize and bind cancer cells *in vitro* and serve as MRI contrast agent for better tumor imaging.

Magnetic resonance imaging or MRI, a non-invasive clinical imaging modality widely being used in the diagnosis and/or staging of diseases use contrast agents and Superparamagnetic iron oxide particles and soluble paramagnetic metal chelates used as contrast agents for MRI [74].

Superparamagnetic iron oxide nanoparticles could be used to mark cells *in vivo,* since following intravenous injection these particles are rapidly taken up by phagocytic cells and imaging this phagocytic activity can be used for tumor staging. Magnetic particles can also be used to monitor cell migration and cell trafficking [71].

Magnetic nanoparticles have the potential to be used as a novel tool for medical imaging [75]. Since primary colorectal cancer and metastatic tumors express unique surface-bound guanylyl cyclase C (GCC), which binds the bacterial heat-stable peptide enterotoxin ST, metastatic tumor could be targeted using ST-bound nanoparticles. The incorporation of iron or iron oxide into such structures will be useful for magnetic resonance imaging [76]. Phosphorothioate-modified DNA probes combined with superparamagnetic iron oxide nanoparticles also find applications in *in vivo* MRI [77].

Magnetic nanoparticles are used for hyperthermic treatment of malignant cells and cell separation [78].

A biocompatible, dextran coated superparamagnetic iron oxide particle conjugated with a peptide sequence from the HIV-tat protein is used for magnetic labeling of different target cells. Since the labeled cells are highly magnetic they could be detected by NMR imaging [70].

Au-coated ferromagnetic nanoparticle probes tagged with HIV antibody could be used to track viral particles [79].

The **bio-barcode assay (BCA)** uses 2 sets of oligonucleotides for DNA detection: one set bound to magnetic microparticles and other set bound to gold nanoparticles. The method could detect target DNA at concentrations as low as zepto molar levels [80, 81]. The bio-barcode assay for the detection of PSA (prostate-specific antigen) [80] uses a magnetic microparticle conjugated with PSA monoclonal antibody and gold nanoparticle to which PSA polyclonal antibody and barcode oligonucleotides are attached [82]. This assay is sensitive in the attomolar range [80]. BCA is also used to detect Alzheimer disease [61,83,84].

2.5   Triangular silver nanoparticles

Triangular silver nanoparticles of ~100 nm X 50 nm possess remarkable optical properties and their localized surface plasmon resonance (LSPR) spectrum is dependent on size, shape, and local external dielectric environment. This sensitivity of LSPR to the nano environment helps in the development of nanoscale affinity biosensors for the detection of variety of biomolecules with extremely robust, simple and low-cost instrumentation and it can detect picomolar concentrations [85].

Functionalized nanoparticles covalently linked to antibodies, peptides, proteins, and nucleic acids can be used for rapid detection of viruses with high sensitivity [86]. In Nanoparticle based assay enabled the detection of about 5000 virus particles per ml in adenovirus diagnostics. Assay using fluorescent europium (III)-chelate-doped nanoparticle is sensitive and convenient in adenovirus screening [87]. Functionalized NPs conjugated to monoclonal antibodies can be used to rapidly and specifically detect Respiratory Syncytial Virus in clinical samples [88]. Nanoparticle-based biobarcode assay for early and sensitive detection of HIV-1 capsid (p24) antigen is available [89]. The HBV, HCV, and HBV/HCV gene chips with gold/silver NP staining amplification method effectively detect these viruses in clinical samples [54].

2.6   Cantilevers

Cantilevers are small beams similar to those used in atomic force microscopy and function by use of nanomechanical deflections. This bio sensor array can be used for high-throughput detection of proteins, RNA, DNA, peptides and whole cells for diagnosis. Cantilevers bond to specific reagents can be used to detect and measure the presence of particular antigens or complementary DNA sequences. Micro and nano cantilevers facilitate simultaneous analysis of multiple samples. The micro machined silicon cantilevers are used to monitor DNA hybridization [1,2]. The cantilever scans the sample and if the target DNA sequence is present, then its hybridization with the cantilever single-stranded DNA occurs. This will cause a mechanical stress on the beam and cause it to deflect and this nano deflection proportional to the amount of DNA hybridized can be measured as an optical signal [2,29,90]. Cantilevers can also be used for the detection of PSA [2], microorganisms such as *Salmonella enterica* [1], biomarkers of myocardial damage such as cardiac troponins, for breath analysis to detect acetone and dimethylamine, etc [1].

Fluorescent nanoparticles [latex nanobeads] conjugated with EcoR1 restriction enzyme can recognize and cleave at specific recognition site on lambda bacteriophage DNA. If a chelating agent such as EDTA is used, then binding and recognition of specific sequences by the bio conjugated nanoparticles is possible without the enzymatic cleavage and the binding could be directly visualized by multicolor fluorescence microscopy. This technique opens new possibilities in optical gene mapping and in the study of DNA-protein interactions through real time observation of DNA-protein binding and enzymatic dynamics during DNA replication and transcription [91].

TiO$_2$ nanoparticles allow conjugation to nucleic acids and this enables their retention in specific subcellular compartments and visualization since TiO$_2$ can be easily functionalized by fluorescent molecules. **19F perfluorocarbon nanoparticles** can be used for highly specific labelling of cells since fluorine can be directly detected on MRI. Ahrens *et al* targeted dendritic cells with cationic perfluoropolyether molecules and tracked the migration of the cells [92]. Co nanoparticles and NiCo nanoparticles can be used to enhance the signal during magnetic resonance imaging [93]

In addition to the above, other metal and semiconductor colloidal nanoparticles are also being studied for potential applications in materials synthesis, [49,94,95,96] in multiplexed bioassays and in ultrasensitive optical detection and imaging [24, 28, 97, 98].

3.   Advantages of nanodiagnostics over conventional diagnostics

Nanodiagnostics is providing new ways with increased sensitivity and specificity for sample analysis for early detection of disease biomarkers, for the detection of pathogens and cancer biomarkers and is rapidly emerging to meet the demand of clinical diagnostics for determining disease status, pathology and identification of causative organisms by using hand held devices that are easy to use and are marketable [99]. Nanodiagnostics is being used in cancer diagnosis and therapy, surgery, molecular imaging, implant technology, bio-detection of molecular disease markers, tissue engineering, and devices for drug, protein, and gene and radionuclide delivery [100]. Nanodiagnostics approach offers the ease of on-the-spot diagnosis since most of the complex procedures could be integrated onto a simple device. The avalanche techniques available with the Nanotechnology approach makes the acquisition and analysis of biological information easily, quickly and inexpensively. This diagnostic approach also finds applications in preventive medicine, since with such excellent medical diagnosis, therapy becomes more specific and individualized that converges therapy and diagnostics to the so called theranostics. Here nanotechnology methods as well as medicines concomitantly serve both diagnostic and therapeutic purposes [101]. An example is the use of nanomaterials that could specifically recognize individual genes of impaired function and repair them [102].

4.   Nanodiagnostics to manage cancer

Nanoparticles can be used as novel intravascular or cellular probes for both diagnostic and therapeutic purposes for target-specific drug/gene delivery [103]. Ultra small super paramagnetic iron oxide particles can be used for detecting cancer cells by MRI [104]. This type of cancer imaging will help to provide real-time monitoring during therapy. QDs coated with paramagnetic and silica nanoparticle to generate MRI probe, anti-epidermal growth factor receptor conjugated with gold nanoparticles, etc are also used in cancer diagnosis.

5.   Limitations or concerns associated with nanodiagnostics

Even though the field of nanodiagnostics is rapidly gaining more and more importance, there are a few concerns as well. The reactivity of nanoparticles is much higher than even larger volumes of the same substance and most possess known/unknown toxicity. For example, carbon nanotubes are potentially toxic to human beings. The field of nano diagnostics also raises certain ethical concerns. For example, if a nanochip can analyze the entire DNA sequence from a drop of blood, an individual's entire genetic makeup could be accessed by laboratories or hospitals, which is morally not acceptable. The use of MEMS devices [MEMS are Microelectromechanical Systems employed in diagnostics as capsule like pills to visualize the gastrointestinal tract for diagnosis is also of concern, since introduction of these devices within the body might leave residual nanoparticles that may be toxic and become harmful to the digestive system [100].

Future prospects

Nanodiagnostics assays where nanoparticles are conjugated to recognition moieties such as oligonucleotides or antibodies for the detection of target biomolecules and such conjugates are utilized in immunoassays, DNA diagnostics, immunohistochemistry, cellular imaging, bio separation of specific cell populations, etc for the detection of infectious diseases, tumours, neurological diseases and bio-terrorism agents [15]. Future usage of nanodiagnostics will be more effective in blood or urine analysis to provide a rapid and sensitive assessment of health. Nanodevices such as nanorobotics may be of use for combined diagnosis and therapeutics. To conclude, Nanodiagnostics and Nanotheranostics will surely extend the frontiers of current diagnostics and therapeutics, and will enable the development of personalized medicine. [105]

References

[1] Jain K K. Nanotechnology in clinical laboratory diagnostics. Clin Chim Acta 358 (2005) [37–54]. https://doi.org/10.1016/j.cccn.2005.03.014

[2] Jain K K. Nanodiagnostics: application of nanotechnology in molecular diagnostics. Expert Rev Mol Diagn 3 (2003) [153–161]. https://doi.org/10.1586/14737159.3.2.153

[3] West J L, Halas N J. Engineered nanomaterials for biophotonics applications: improving sensing, imaging, and therapeutics. Annu Rev Biomed Eng 5 (2003) [285–292]. https://doi.org/10.1146/annurev.bioeng.5.011303.120723

[4] Hassan M.E. Azzazy, Mai M.H. Mansour, and Steven C. Kazmierczak, Nanodiagnostics: A New Frontier for Clinical Laboratory Medicine, Clinical Chemistry 52 (2006) [1238–1246]. https://doi.org/10.1373/clinchem.2006.066654

[5] Groneberg D A; Giersig M, Welte T, Pison U, Nanoparticle-Based Diagnosis and Therapy, Current Drug Targets 7 (2006) [643-648]. https://doi.org/10.2174/138945006777435245

[6] Elghanian R, Storhoff J J, Mucic R C, Letsinger R L, Mirkin C A. Selective colorimetric detection of polynucleotides based on the distance-dependent optical properties of gold nanoparticles. Science 277 (1997) [1078–1081]. DOI: 10.1126/science.277.5329.1078

[7] Liu J and Lu Y. A colorimetric lead biosensor using DNAzyme-directed assembly of gold nanoparticles. J. Am. Chem. Soc. 125 (2003) [6642–6643]. https://doi.org/10.1021/ja034775u

[8] Alivisatos P. The use of nanocrystals in biological detection. Nat Biotechnol 22 (2004) [47–52]. https://doi.org/10.1038/nbt927

[9] Rosi N L, Mirkin C A. Nanostructures in biodiagnostics. Chem Rev 105 (2005) [1547–1562]. https://doi.org/10.1021/cr030067f

[10] Dwaine F. Emerich and Christopher G. T, The pinpoint promise of nanoparticle-based drug delivery and molecular diagnosis, Biomolecular Engineering 23 (2006) [171-184]. https://doi.org/10.1016/j.bioeng.2006.05.026

[11] Link S, and El-Sayed M, Spectral properties and relaxation dynamics of surface plasmon electronic oscillations in gold and silver nanodots and nanorods. J. Phys. Chem. B 103, (1999) [8410-8426]. https://doi.org/10.1021/jp9917648

[12] Kelly K L, Coronado E, Zhao L.L, and Schatz G C, The optical properties of metal nanoparticles: the influence of size, shape, and dielectric environment. J. Phys. Chem. B 107 (2003) [668-677]. https://doi.org/10.1021/jp026731y

[13] Jensen T R, Schatz G C, and Van Duyne R P, Nanosphere lithography: surface plasmon resonance spectrum of a periodic array of silver nanoparticles by ultraviolet-visible extinction spectroscopy and electrodynamic modeling. J. Phys. Chem. B 103 (1999) [2394-2401]. https://doi.org/10.1021/jp984406y

[14] Jiang X, Jiang J, Jin Y, Wang E, and Dong S, Effect of colloidal gold size on the conformational changes of adsorbed cytochrome c: probing by circular dichroism, UV-visible, and infrared spectroscopy. Biomacromolecules, (2005) [46-53]. https://doi.org/10.1021/bm049744l

[15] Azzazy HME, Mansour MMH, In vitro diagnostic prospects of nanoparticles, Clinica Chimica Acta 403 (2009) [1-8]. https://doi.org/10.1016/j.cca.2009.01.016

[16] Brigger I, Dubernet C, Couvreur P, Nanoparticles in cancer therapy and diagnosis. Adv Drug Deliv Rev 54 (2002) [631-651]. https://doi.org/10.1016/j.addr.2012.09.006

[17]   Cuenca AG, Jiang H, Hochwald SN, Delano M, Cance WG, Grobmyer SR: Emerging implications of nanotechnology on cancer diagnostics and therapeutics. Cancer 107 (2006) [459-466]. https://doi.org/10.1002/cncr.22035

[18]   113.   Zhang Z, Yang X, Zhang Y, Zeng B, Wang S, Zhu T, Roden RB, Chen Y, Yang R: Delivery of telomerase reverse transcriptase small interfering RNA in complex with positively charged single walled carbon nanotubes suppresses tumor growth. Clin Cancer Res 12 (2006) [4933-4939]. https://doi.org/10.1158/1078-0432.CCR-05-2831

[19]   Bhattacharya R, Senbanerjee S, Lin Z, Mir S, Hamik A, Wang P, Mukherjee P, Mukhopadhyay D, Jain MK: Inhibition of vascular permeability factor/vascular endothelial growth factor mediated angiogenesis by the Kruppel-like factor KLF2. J Biol Chem 280 (2005) [28848-28851]. https://doi.org/10.1074/jbc.C500200200

[20]   Mukherjee P, Bhattacharya R, Wang P, Wang L, Basu S, Nagy JA, Atala A, Mukhopadhyay D, Soker S: Antiangiogenic properties of gold nanoparticles. Clin Cancer Res 11 (2005) [3530-3534]. https://doi.org/10.1158/1078-0432.CCR-04-2482

[21]   O'Neal DP, Hirsch LR, Halas NJ, Payne JD, West JL: Photo-thermal tumor ablation in mice using near infrared-absorbing nanoparticles. Cancer Lett 209 (2004) [171-176]. https://doi.org/10.1016/j.canlet.2004.02.004

[22]   Paciotti GF, Myer L, Weinreich D, Goia D, Pavel N, McLaughlin RE, Tamarkin L: Colloidal gold: a novel nanoparticle vector for tumor directed drug delivery. Drug Deliv 11 (2004) [169-183]. https://doi.org/10.1080/10717540490433895

[23]   Kam NW, Liu Z, Dai H: Carbon nanotubes as intracellular transporters for proteins and DNA: an investigation of the uptake mechanism and pathway. Angew Chem Int Ed Engl 45 (2006) [577-581]. https://doi.org/10.1002/anie.200503389

[24]   Chan W C, Nie S. Quantum dot bioconjugates for ultrasensitive nonisotopic detection. Science 281 (1998) [2016–2018]. DOI: 10.1126/science.281.5385.2016

[25]   Wu X, Liu H, Liu J, Haley K N, Treadway J A, Larson J P, et al. Immunofluorescent labeling of cancer marker Her2 and other cellular targets with semiconductor quantum dots. Nat Biotechnol 21 (2003) [41–46]. https://doi.org/10.1038/nbt764

[26]   Azzazy H M E, Mansour M M H. and Kazmierczak S C, Nanodiagnostics: A New Frontier for Clinical Laboratory Medicine, Clinical Chemistry 52 (2006) [1238–1246]. https://doi.org/10.1373/clinchem.2006.066654

[27]   Michalet X, Pinaud F F, Bentolila L A, Tsay J M, Doose S, Li J J, et al. Quantum dots for live cells, in vivo imaging, and diagnostics. Science 307 (2005) [538–44]. DOI: 10.1126/science.1104274

[28]   Bruchez M, Moronne M, Gin P, Weiss S and Alivisatos AP, Semiconductor nanocrystals as fluorescent biological labels, Science 281 (1998) [2013–2015]. DOI: 10.1126/science.281.5385.2013

[29]   Fortina P, Kricka L J, Surrey S, Grodzinski P. Nanobiotechnology: the promise and reality of new approaches to molecular recognition. Trends Biotechnol 23 (2005) [168–173]. https://doi.org/10.1016/j.tibtech.2005.02.007

[30]   41.Han M, Gao X, Su J Z, Nie S. Quantum-dot-tagged microbeads for multiplexed optical coding of biomolecules. Nat Biotechnol 19 (2001) [631–635]. https://doi.org/10.1038/90228

[31]   Larson D R, Zipfel W R, Williams R M, Clark S W, Bruchez M P, Wise F W, et al. Water-soluble quantum dots for multiphoton fluorescence imaging in vivo. Science 300 (2003) [1434–1436]. DOI: 10.1126/science.1083780

[32]   Crut A, Geron-Landre B, Bonnet I, Bonneau S, Desbiolles P, Escude C. Detection of single DNA molecules by multicolor quantum-dot end-labeling. Nucleic Acids Res 33 (2005) e98. https://doi.org/10.1093/nar/gni097

[33]   Dubertret B 1, Skourides P, Norris D J, Noireaux V, Brivanlou A H, Libchaber A. In vivo imaging of quantum dots encapsulated in phospholipid micelles. Science 298 (2022) [1759-1762]. DOI: 10.1126/science.1077194

[34]   Harma H, Soukka T, Lovgren T. Europium nanoparticles and time-resolved fluorescence for ultrasensitive detection of prostate- specific antigen. Clin Chem 47 (2001) [561–8]. https://doi.org/10.1093/clinchem/47.3.561

[35]   Jin T, Fujii F, Sakata H, Tamura M, Kinjo M. Amphiphilic p-sulfonatocalix[4]arene-coated CdSe/ZnS quantum dots for the optical detection of the neurotransmitter acetylcholine. Chem Commun (Camb) (2005) [4300–4302]. https://doi.org/10.1039/B506608E

[36]   Akerman M E, Chan W C, Laakkonen P, Bhatia S N, Ruoslahti E. Nanocrystal targeting in vivo. Proc Natl Acad Sci U S A 99 (2002) [12617–12621]. https://doi.org/10.1073/pnas.152463399

[37]   Yeh H C, Ho Y P, Shih I, Wang T H. Homogeneous point mutation detection by quantum dot-mediated two-color fluorescence coincidence analysis. Nucleic Acids Res 34 (2006) e35. https://doi.org/10.1093/nar/gkl021

[38]   Schmitt J, Decher G, Dressick W J, Brandow S L, Geer R E, Shashidar R, Calvert J M, AdV. Mater. 9 (1997) [61-65]. https://doi.org/10.1002/adma.19970090114

[39]   Mauriz E. Clinical Applications of Visual Plasmonic Colorimetric Sensing. Sensors (Basel) 20. (2020) [6214]. https://doi.org/10.3390/s20216214

[40]   de la Fuente A J M, Barrientos A G, Rojas T C, Rojo J, Canada J, Fernamdez A and Penades S, Angew. Chem., Int. Ed., 40 (2001) [2258–2260].

[41]   Tsai C, Yu T and Chen C T, Gold nanoparticle-based competitive colorimetric assay for detection of protein–protein interactions Chem. Commun. (2005) [4273–4275]. DOI: 10.1039/B507237A

[42]   Storhoff J J, Lazarides A A, Mucic R C, Mirkin Chad A, Letsinger R L and Schatz G C What Controls the Optical Properties of DNA-Linked Gold Nanoparticle Assemblies? J. Am. Chem. Soc. 122 (2000) [4640-4650]. https://doi.org/10.1021/ja993825l

[43]   Liu, J and Lu Y. Stimuli-responsive disassembly of nanoparticle aggregates for light-up colorimetric sensing. J. Am. Chem. Soc. 127(2005) [12677–12683]. https://doi.org/10.1021/ja053567u

[44]   Liu J and Lu Y Fast colorimetric sensing of adenosine and cocaine based on a general sensor design involving aptamers and nanoparticles. Angew. Chem. Int. Ed. Engl. 45 (2006) [90]. https://doi.org/10.1002/anie.200502589

[45]   Huang C C, Huang Y F, Cao Z, Tan W, and Chang H T. Aptamer modified gold nanoparticles for colorimetric determination of platelet derived growth factors and their receptors. Anal. Chem. 77 (2005) [5735–5741]. https://doi.org/10.1021/ac050957q

[46]   Guarise C, Pasquato L, Filippis V De, and Scrimin P, Gold nanoparticles-based protease assay, PNAS 103 (2006) [3978–3982]. https://doi.org/10.1073/pnas.0509372103

[47]   Chah S, Hammond M R, and Zare R N, Gold Nanoparticles as a Colorimetric Sensor for Protein Conformational Changes, Chemistry & Biology12 (2005) [323-328]. DOI 10.1016/j.chembiol.2005.01.013

[48]   Nath N and Chilkoti A, Interfacial phase transition of an environmentally responsive elastin biopolymer adsorbed on functionalized gold nanoparticles studied by colloidal surface plasmon resonance. J. Am. Chem. Soc. 123 (2001], [8197-8202]. https://doi.org/10.1021/ja015585r

[49]   Mirkin C A, Letsinger R L, Mucic R C and Storhoff J J, A DNA-based method for rationally assembling nanoparticles into macroscopic materials. Nature 382 (1996) [607-609].

[50]   Goodrich G P, Helfrich M R, Overberg J J and Keating C D, Effect of macromolecular crowding on DNA:Au nanoparticle bioconjugate assembly. Langmuir 20 (2004) [10246-10251]. https://doi.org/10.1021/la0484341

[51]   Maxwell D J, Taylor J R and Nie S, Self-assembled nanoparticle probes for recognition and detection of biomolecules. J. Am. Chem. Soc. 124 (2002) [9606-9612]. https://doi.org/10.1021/ja025814p

[52]   Lu Y and Liu J, Accelerated color change of gold nanoparticles assembled by DNAzymes for simple and fast colorimetric Pb^{2+} detection. J. Am. Chem. Soc. 126 (2004) [12298-12305]. https://doi.org/10.1021/ja046628h

[53]   Daniel, M C, and Astruc D, Gold nanoparticles: assembly, supramolecular chemistry, quantum-size-related properties, and applications toward biology, catalysis, and nanotechnology. Chem. Rev. 104 (2004) [293-346]. https://doi.org/10.1021/cr030698+

[54]   Wang J, Nanoparticle-based electrochemical DNA detection, Analytica Chimica Acta 500 (2003) [247-257]. https://doi.org/10.1016/S0003-2670(03)00725-6

[55]   Dykman LA and Khlebtsov NG, Gold Nanoparticles in Biology and Medicine: Recent Advances and Prospects. Acta Naturae 3 (2011) [34–55].

[56]   Andrea C, Robert M, Wolfgang F, Elghanian R, Storhoff J J, Mucic R C, Letsinger R L, Mirkin C A. Expert Review of Molecular Diagnostics 2 (2002) [187-193]. https://doi.org/10.1586/14737159.2.2.187

[57]   Freeman R G, Grabar K C, Allison K J, Bright R M, Davis J A, Guthrie A P, Hommer M B, Jackson M A, Smith P C, Walter D G, Natan M J. Science 267 (1995) [1629-1632]. DOI: 10.1126/science.267.5204.1629

[58]   Sun L, Zhang Z, Wang S, Zhang J, Li H, Ren L, Weng J, Zhang Q. Effect of pH on the Interaction of Gold Nanoparticles with DNA and Application in the Detection of Human p53 Gene Mutation Nanoscale Res Lett 4 (2009) [216–220]. https://doi.org/10.1007/s11671-008-9228-z

[59]   Kalogianni DP, Koraki T, Christopoulos TK, Ioannou PC, Nanoparticle-based DNA biosensor for visual detection of genetically modified organisms. Biosensors and Bioelectronics 21 (2006) [1069-1076]. https://doi.org/10.1016/j.bios.2005.04.016

[60]   Baptista PV, Koziol-Montewka M, Paluch-Oles J, Doria G, Franco R, Gold-Nanoparticle-Probe–Based Assay for Rapid and Direct Detection of Mycobacterium tuberculosis DNA in Clinical   Samples   Clinical   Chemistry   52   (2006). https://doi.org/10.1373/clinchem.2005.065391

[61]   Georganopoulou D G, Chang L, Nam J M, Thaxton C S, Mufson E J, Klein W L, et al. Nanoparticle-based detection in cerebral spinal fluid of a soluble pathogenic biomarker for Alzheimer's disease. Proc Natl Acad Sci U S A 102 (2005) [2273–2276]. https://doi.org/10.1073/pnas.0409336102

[62]   Huang X, Prashant K J, El-Sayed I H and El-Sayed M A, Special Focus: Nanoparticles for Cancer Diagnosis & Therapeutics – Review, Gold nanoparticles: interesting optical properties and recent applications in cancer diagnostics and therapy, Nanomedicine 2 (2007) [681-693]. https://doi.org/10.2217/17435889.2.5.681

[63]   El-Sayed I H, Huang X, and El-Sayed M A, Surface Plasmon Resonance Scattering and Absorption of anti-EGFR Antibody Conjugated Gold Nanoparticles in Cancer Diagnostics: Applications in Oral Cancer, Nano Lett. 5 (2005], [829–834]. https://doi.org/10.1021/nl050074e

[64]   Huang X, El-Sayed I H, Qian W and El-Sayed M A, Cancer Cell Imaging and Photothermal Therapy in the Near-Infrared Region by Using Gold Nanorods, J. Am. Chem. Soc. 128 (2006) [2115–2120]. https://doi.org/10.1021/ja057254a

[65]   Seol Y, Carpenter A E, Perkins T T, Gold nanoparticles: enhanced optical trapping and sensitivity coupled with significant heating, Optics Letters 31 (2006). https://doi.org/10.1364/OL.31.002429

[66]   Raj C R, Okajima T and Ohsaka T, Gold nanoparticle arrays for the voltammetric sensing of dopamine, Journal of Electroanalytical Chemistry 543 (2003) [127-133]. https://doi.org/10.1016/S0022-0728(02)01481-X

[67]   Simoniana A L, Good T A, Wang S S, and Wild J R Nanoparticle-based optical biosensors for the direct detection of organophosphate chemical warfare agents and pesticides, Analytica Chimica Acta 534 (2005) [69–77]. https://doi.org/10.1016/j.aca.2004.06.056

[68]   Chen X, Tam U C, Czlapinski J L, Lee G S, Rabuka D, Zettl A and Bertozzi C R, Interfacing carbon nanotubes with living cells J. Am. Chem. Soc. 128 (2006) [6292–6293]. https://doi.org/10.1021/ja060276s

[69]   Yu X, Munge B, Patel V, Jensen G, Bhirde A, Gong JD, Kim SN, Gillespie J, Gutkind JS, Papadimitrakopoulos F, Rusling JF, Carbon nanotube amplification strategies for highly sensitive immunodetection of cancer biomarkers J. Am. Chem. Soc. 128 (2006) [11199–11205]. https://doi.org/10.1021/ja062117e

[70]   Josephson L, Tung C H, Moore A and Weissleder R. High-efficiency intracellular magnetic labeling with novel superparamagnetic-tat peptide conjugates, Bioconjugate Chem. 10 (1999) [186–191]. https://doi.org/10.1021/bc980125h

[71]   Bulte JWM, Douglas T, Witwer B, Zhang SC, Strable E, Lewis BK, Zywicke H, Miller B, Van Gelderen P, Moskowitz BM, Duncan ID, Frank JA, Magnetodendrimers allow endosomal magnetic labeling and in vivo tracking of stem cells, Nat. Biotechnol. 19 (2001) [1141–1147]. https://doi.org/10.1038/nbt1201-1141

[72]   Gref R, Minamitake Y, Peracchia M T, Trubetskoy V, Torchilin V and Langer R. Biodegradable long-circulating polymeric nanospheres Science 263 (1994) [1600-1603]. DOI: 10.1126/science.8128245

[73]   Neuberger T, Schopf B, Hofmann H, Hofmann M and von Rechenberg B. Superparamagnetic nanoparticles for biomedical applications: Possibilities and limitations of a new drug delivery system, Journal of Magnetism and Magnetic Materials 293 (2005) [483-496]. https://doi.org/10.1016/j.jmmm.2005.01.064

[74]   Lee C, Jeong H, Kim S, Kim E, Kim D W, Lim S T, Jang K Y, Jeong Y Y, Nah J and Sohn M   SPION-loaded chitosan–linoleic acid nanoparticles to target hepatocytes, International Journal of Pharmaceutics 371 (2009) [163-169]. https://doi.org/10.1016/j.ijpharm.2008.12.021

[75]   Romanus E, Hückel M, Gro C, Prass S, Weitschies W, Brauer R and Weber P. Journal of Magnetism and Magnetic Materials 252 (2002) [387-389]. https://doi.org/10.1016/S0304-8853(02)00645-5

[76]   Fortina P, Kricka L J, Graves D J, Park J, Hyslop T, Tam F, Halas N, Surrey S and Scott A. Waldman Applications of nanoparticles to diagnostics and therapeutics in colorectal cancer, Trends in Biotechnology 25 (2007) [145-152]. https://doi.org/10.1016/j.tibtech.2007.02.005

[77]   Liu C H, Ren J Q, Yang J, Liu C, Mandeville J B, Rosen B R, Bhide P G, Yanagawa Y, Liu P K.   DNA-Based MRI Probes for Specific Detection of Chronic Exposure to Amphetamine in Living Brains J. Neurosci 29 (2009) [10663-10670]. https://doi.org/10.1523/JNEUROSCI.2167-09.2009

[78]   Berry CC, Possible exploitation of magnetic nanoparticle–cell interaction for biomedical applications J. Mater. Chem. 15 (2005) [543–547].

[79]   Gould P. Nanoparticles probe biosystems, Materials today 7 (2004) [36-43]. DOI10.1016/S1369-7021(04)00082-3

[80]   Nam J M, Thaxton C S, Mirkin C A. Nanoparticle-based bio-bar codes for the ultrasensitive detection of proteins. Science 301 (2003) [1884–1886]. DOI: 10.1126/science.1088755

[81]   Nam J M, Stoeva S I, Mirkin C A. Bio-bar-code-based DNA detection with PCR-like sensitivity. J Am Chem Soc 126 (2004) [5932–5933]. https://doi.org/10.1021/ja049384+

[82]   Taton T A, Mirkin C A, Letsinger R L. Scanometric DNA array detection with nanoparticle probes. Science 289 (2000) [1757–1760]. DOI: 10.1126/science.289.5485.1757

[83]   Fradinger E A, Bitan G. En route to early diagnosis of Alzheimer's disease: are we there yet? Trends Biotechnol 23 (2005) [531–533]. https://doi.org/10.1016/j.tibtech.2005.09.002

[84]   Oh B K, Nam J, Lee S W, Mirkin C A. A fluorophore-based biobarcode amplification assay for proteins. Small 2 (2006) [103–108]. https://doi.org/10.1002/smll.200500260

[85]   Haes A J and Van Duyne R P. A Nanoscale Optical Biosensor: Sensitivity and Selectivity of an Approach Based on the Localized Surface Plasmon Resonance Spectroscopy of Triangular Silver Nanoparticles, J. Am. Chem. Soc. 124 (2002) [10596–10604]. https://doi.org/10.1021/ja020393x

[86]   Abraham A M, Kannangai R, Sridharan G, Nanotechnology: A new frontier in virus detection in clinical practice, Indian Journal of Medical Microbiology 26 (2008) [297-301]. https://doi.org/10.1016/S0255-0857(21)01804-1

[87]   Valanne A, Huopalahti S, Soukka T, Vainionpaa R, Lovgren T and Harma H. A sensitive adenovirus immunoassay as a model for using nanoparticle label technology in virus diagnostics Journal of Clinical Virology 33 (2005) [217-223]. https://doi.org/10.1016/j.jcv.2004.11.007

[88]   Tripp R A, Alvarez R, Anderson B, Jones L, Weeks C, Chen W. Bioconjugated nanoparticle detection of respiratory syncytial virus infection. Int J Nanomed 2 (2007) [117-124]. https://doi.org/10.2147/IJN.S2.1.117

[89]   Tang S, Zhao J, Storhoff J J, Norris P J, Little R F, Yarchoan R, et al . Nanoparticle-Based biobarcode amplification assay (BCA) for sensitive and early detection of human immunodeficiency type 1 capsid (p24) antigen. J Acquir Immune Defic Syndr 46 (2007) [231-237]. DOI: 10.1097/QAI.0b013e31814a554b

[90]   McKendry R, Zhang J, Arntz Y, Strunz T, Hegner M, Lang H P, et al. Multiple label-free bio detection and quantitative DNA-binding assays on a nanomechanical cantilever array. Proc Natl Acad Sci U S A 99 (2002) [9783–9788]. https://doi.org/10.1073/pnas.152330199

[91]   Taylor J R, Fang M M and Nie S. Probing Specific Sequences on Single DNA Molecules with Bioconjugated Fluorescent Nanoparticles Anal. Chem. 72 (2000) [1979-1986]. https://doi.org/10.1021/ac9913311

[92]   Ahrens E T, Flores R, Xu H, Morel P A. In vivo imaging platform for tracking immunotherapeutic cells. Nat Biotechnol. 23 (2005) [983–987]. https://doi.org/10.1038/nbt1121

[93]   Neamtu J, Jitaru I, Malaeru T, Georgescu G, Kappel W and Alecu V V Synthesis and Properties of Magnetic Nanoparticles with Potential Applications in Cancer Diagnostic, NSTI Nanotechnology Conference & Trade Show 2005.

[94]   Alivisatos P, Johnsson K P, Peng X, Wilson T E, Loweth C J, Bruchez M and Schultz P G, Organization of nanocrystal molecules using DNA, Nature (London) 382 (1996) [609–611]. https://doi.org/10.1038/382609a0

[95]   Boal A K, Ilhan F, DeRouchey J E, Thurn-Albrecht T, Russell T P and Rotello V M, Self-assembly of nanoparticles into structured spherical and network aggregates, Nature (London) 404 (2000) [746– 748]. https://doi.org/10.1038/35008037

[96]   Templeton A C, Wuelfing M P and Murray R W, Monolayer protected cluster molecules, Acc. Chem. Res. 33 (2000) [27–36]. https://doi.org/10.1021/ar9602664

[97]   Mattoussi H, Mauro J M, Goldman E R, Anderson G P, Sundar V C, Mikulec F V and Bawendi M G, Self-assembly of CdSe- ZnS quantum dots bioconjugates using an engineered recombinant protein, J. Am. Chem. Soc. 122 (2000) [12142–12150]. https://doi.org/10.1021/ja002535y

[98]   Mitchell P, Turning the spotlight on cellular imaging — Advances in imaging are enabling researchers to track more accurately the localization of macromolecules in cells, Nat. Biotechnol. 19 (2001) [1013– 1017]. https://doi.org/10.1038/nbt1101-1013

[99]   Baptista, PV, Nanodiagnostics: Leaving the Research Lab to Enter the Clinics? Diagnosis 1 (2014) [305-309]. https://doi.org/10.1515/dx-2014-0055

[100] Alharbi KK, Al-sheikh YA, Role and implications of nanodiagnostics in the changing trends of clinical diagnosis, Saudi Journal of Biological Sciences 21 (2014) [109-117]. https://doi.org/10.1016/j.sjbs.2013.11.001

[101] Jackson, T, Patani, B. and Ekpa, D, Nanotechnology in Diagnosis: A Review. Advances in Nanoparticles 6 (2017) [93-102]. DOI: 10.4236/anp.2017.63008

[102] Lu, Z.R., Ye, F. and Vaidya, A, Polymer Platforms for Drug Delivery and Biomedical Imaging. Journal of Controlled Release 122 (2007) [269-277]. https://doi.org/10.1016/j.jconrel.2007.06.016

[103] Singh S, Hicham F, Singh B, Nanotechnology, Nanotechnology based drug delivery systems J. Occupat. Med. Toxcol. 2 (2007) [16]. https://doi.org/10.1186/1745-6673-2-16

[104] Lee S, Kwon I C, Kim K, Multifunctional Nanoparticles for Cancer Theragnosis X. Chen (Ed.], Nanoplatform-Based Molecular Imaging, John Wiley & Sons, Inc., Hoboken, NJ, USA (2011). https://doi.org/10.1002/9780470767047.ch22

[105] Moffatt S, Nanodiagnostics: a revolution in biomedical nanotechnology. MOJ Proteomics Bioinform.3 (2016) [34-36]. DOI: 10.15406/ mojpb.2016.03.00080

Nanoparticles in Healthcare: Applications in Therapy, Diagnosis and Drug Delivery   Materials Research Forum LLC
Materials Research Foundations 160 (2024) 163-184   https://doi.org/10.21741/9781644902974-7

Chapter 7

Green Synthesis of Nanoparticles via Plant Extract: A New Era in Cancer Therapy

Kirti Rani Saad[1]*

[1]Assistant Professor, Mahajana Education Society, SBRR Mahajana First Grade College PG. Wing (autonomous), Pooja Bhagavat Memorial Mahajana Education Centre, KRS Road, Metagalli, Mysuru, Karnataka

*kirti.saad@yahoo.in

Abstract

There are many potential applications for the production of green nanoparticles (NPs) in medical and environmental sciences. Green synthesis is focused on minimizing the use of harmful chemicals. Scientists are working to create unique approaches for treatment and diagnostics of cancer in light of development of the newest technology. Primarily plant-based approaches have been chosen as the best course of action due to benefits of biogenic approaches over traditional synthesis, including their simplicity, speed, energy efficiency, one-pot operations, safety, economics, and biocompatibility. The finest sources of biogenic NPs are secondary metabolites that operate as stabilizers or reducers to form NPs. Here, we discuss the role of plant mediated NPs for cancer therapy.

Keywords

Cancer Therapy, Green Synthesis, Nanotechnology, Nanoparticles, Plant Extract, Plant Metabolites

Contents

1.   Introduction

The design, development, and use of parts with a size between 10 and 1000 nanometres is referred to as "nanotechnology." Currently, engineering, biology, chemistry, and physics are all included in the multidisciplinary field of nanotechnology [1]. Nanotechnology is the study of, observation of, and manipulation of objects on an atomic or molecular scale, frequently less than 100nm. At this size, a variety of phenomena, such as a high surface-to-volume ratio, a weakening of gravitational influences, quantum effects, etc., take place [2]. Metal nanoparticles (MNPs) have drawn the interest of researchers due to their potential applications in a number of fields, including biomedicine, sensitive diagnostic tests, radiotherapy enhancement, thermal ablation, gene delivery, non-toxic carriers for drug and gene delivery, optical imaging, and labelling of biological systems [3]. These minuscule particles are known as nanoparticles (NPs), and they are typically created using the top down and bottom up approaches (Fig. 1).

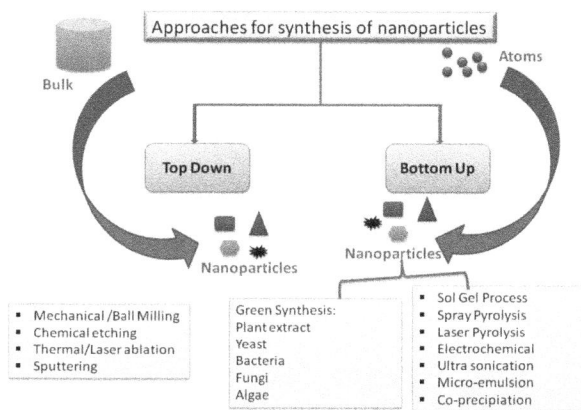

Figure 1. Different approaches for synthesis of metal nanoparticles

Metallic nanoparticles are tiny metal atoms with sizes between 1 and 100 nanometres (nm). In the creation of nanoparticles (NPs), materials such as gold (Au), silica (Si), zinc (Zn), aluminium (Al), copper (Cu), silver (Ag), iron (Fe), platinum (Pt), manganese (Mn), titanium (Ti), cerium (Ce), or thallium (TI) have all been used [4,5]. Metal nanoparticles are among the most efficient ones for biomedical applications due to their use as an imaging resource and their multifunctional theranostic properties, such as their antibacterial, anticancer, and drug carrier features [3]. The therapy and detection of cancer has been advanced by the development of several metallic nanoparticles. The most well-known anti-cancer nanoparticles are zinc (Zn), gold (Au), platinum (Pt) and silver (Ag), due to their size- and shape-dependent tuneable plasmonic properties [1].

While they have tremendous usage in many different industries, several nanoparticle materials have shown toxicity at the nanoscale dimension. This has piqued the interest of many scientists and researchers, who have expressed significant interest in their unique properties. Utilising microorganisms, plants, and other naturally occurring processes to address the toxicity issue, green chemistry and nanotechnology work together to produce eco-friendly nanoparticles [6]. Green synthesis is considered to be a feasible technique for the production of NPs because it uses inexpensive and non-hazardous raw materials. This environmentally friendly method of making NPs offers low toxicity, site-specific delivery, and great therapeutic efficacy. Additionally, it has no negative effects on the environment or human health [1].

While living organisms like fungi and bacteria can be used to produce nanoparticles, the synthesis platform using plants offers a suitable and environmentally acceptable method because it doesn't require the use of numerous expensive, toxic, and hazardous chemicals as growth media. Additional advantages of plant-derived nanoparticles include quick production, enhanced stability, and cost effectiveness [7] (Fig. 2). Unlike other species, plants are capable of producing nanoparticles of different sizes and forms. Contrary to bacteria and fungi, which require lengthy incubation time to decrease metal ions, phytochemicals can do this job extremely quickly, eliminating the need for an expensive and time-consuming downstream processing step [8].

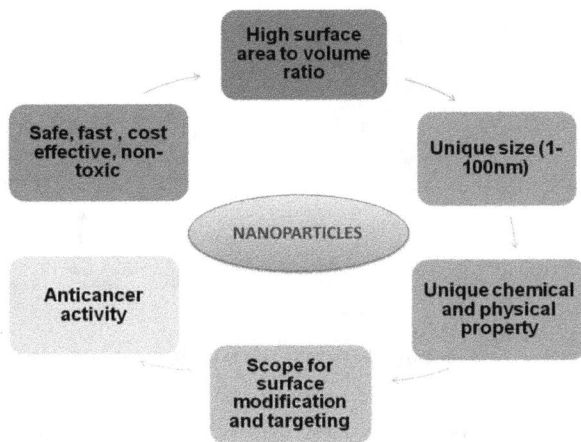

Figure 2. Advantages of green synthesis of nanoparticles

Using plants for green synthesis mitigates some biosafety concerns that are brought about by using green synthesis resources like fungi and bacteria. Hence, the most suitable platforms for green production of metal nanoparticles are those based on plants [9]. In comparison to huge biological molecules like enzymes and receptors, which range in size from 100 to 10,000 times smaller than human cells, nanoparticles are typically smaller than a few hundred nanometres. These nanoparticles have the potential to interact in previously unheard-of ways with biomolecules both outside and inside of bodily cells, which could revolutionise how cancer is detected and treated. We have yet to discover an effective drug or carrier for a drug that can be employed as an effective chemotherapeutic agent for many forms of cancer [10].

Cancer is a critical public health issue since it is the second leading cause of death in the world and one of the most common causes of illness and mortality. It was the reason for 8.8 million fatalities in 2015. Nearly 70% of cancer fatalities occur in developing and middle-income countries. By 2030, estimates indicate that there would be 21.6 million new cases annually, up from 14.1 million in 2012. Additionally, cancer has a large financial impact. Around US$ 1.16 trillion was anticipated to be the total annual economic cost of cancer in 2010, affecting economies at all income levels as well as causing individual and family financial disaster [11].

Cancer is a term used to describe a group of conditions that modify cellular microenvironment in a range of pathological and metabolic ways. It develops by a number of signaling mechanisms, such as metastasis, angiogenesis, and cell growth. In cancer cells, abnormal metabolic processes include genetic expression, alterations in respiratory chains, aerobic glycolysis, and the depletion of mitochondrial DNA [12]. Gene mutations are the main cause of cancer because they set off a chain of molecular events that result in the growth of tumours. A malignant state is characterised by unchecked cell division and consequent invasions of healthy cells and tissues [9].

Tumor tissues' leaky vasculature and poor lymphatic drainage make them more permeable to NPs than normal tissues. This made the produced NPs easier to penetrate tumour tissues and destroy malignant cells because they were more permeable to tumour tissues. Additionally, by shielding metal nanoparticles from macrophage-mediated absorption, hydrophilic molecules on their surfaces boost the solubility and half-life of the particles and prolong their time in the bloodstream [13].

Effective targeting, delayed release, prolonged half-life, and lesser toxicity are some of nanomedicine's key benefits. When phytotherapy and nanotechnology are combined in a clinical setting, patients may experience a stronger pharmacological response and a better clinical outcome [14].

In this chapter, we emphasise the reported plant sources for the synthesis of different nanoparticles. Plants are known to have a variety of healing chemicals that have been used traditionally as medicines since ancient times. Because of their enormous diversity, plants are constantly being researched for a wide range of applications in the medicinal, agricultural, industrial, etc. According to current research of plants that produce nanoparticles, a wide variety of biomolecules, such as tannins, alkaloids, quinines, terpenoids, flavonoids, and phenols, are known to mediate the synthesis of nanoparticles.

2.   Biogenic nanoparticles

Plant organs or plant extract are necessary for the generation of NPs. In reality, plant biomolecules such as enzymes, terpenoids, alkaloids, sugar, polysaccharides, proteins, and flavonoids, can function as stabilising, capping, as well as reducing agents [15] (Fig. 3). Green bio-nanomaterials use metals like copper, titanium, gold, iron, and silver. Because incident light is associated with the metal's plasmon excitation, which produces multiple times the amount of scattering of light than any other molecule, metal nanoparticles are attractive candidates for biological detection methods.

The green synthesis method is also cost-effective, harmless, eco-friendly, and suitable for developing a biological process [16]. The three main prerequisites for the synthesis of nanoparticles are the selection of a good reducing agent, an eco-friendly or green solvent, and a secure stabilising material. Nanoparticles have been produced by a variety of synthetic processes, such as biosynthetic, chemical and physical processes (Fig. 3). Generally speaking, the chemical procedures used are very expensive and contain hazardous and dangerous materials that are to blame for a number of environmental issues. The biosynthetic pathway provides a safe, biocompatible, and green way to make nanoparticles for biomedical applications. This synthesis can involve plants, algae, bacteria, fungi, and other living things. Because the extract from plant

Nanoparticles in Healthcare: Applications in Therapy, Diagnosis and Drug Delivery   Materials Research Forum LLC
Materials Research Foundations 160 (2024) 163-184   https://doi.org/10.21741/9781644902974-7

parts like seeds, stems, roots, leaves and fruits, contains phytochemicals that function as stabilising and reducing agents, these parts of plants have been used to create various nanoparticles [6].

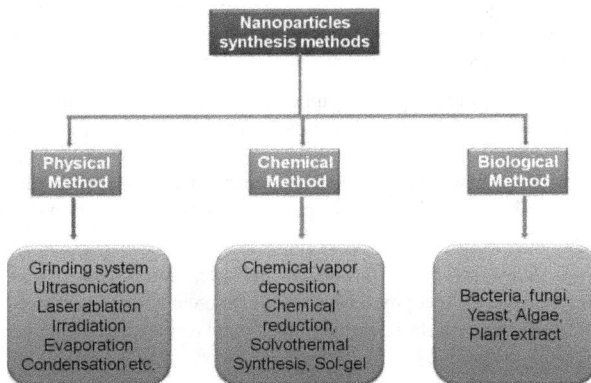

Figure 3. Various methods for synthesis of nanoparticles

2.1   Role of plants in green synthesis of nanoparticles

Numerous plant species along with its parts, including nitrogenous base amino acids, polyphenols, and sugars, have been found to contain antioxidant chemicals. When nanoparticles are made, these chemicals act as capping agents. Plant extracts can produce stable metal and metal oxide nanoparticles (NPs), which normally do not change even after a month. Plants' inherent capacity to absorb poisonous and dangerous substances has been reevaluated thanks to green synthesis of various metallic nanoparticles.

Heavy metals can assemble in different regions of plants in varying amounts. In a "one-pot" manufacturing process, several plants can be used to decrease and stabilise metallic nanoparticles. Many researchers have employed green synthesis techniques to synthesise the particles utilising plant leaf extracts in order to more completely investigate the different applications of metal/metal oxide nanoparticles. Examples of biomolecules found in plants that have a remarkable capacity to transform metal ions into nanoparticles include carbohydrates, proteins, and coenzymes [17].

Under a variety of reaction conditions, plant leaf extract and metal precursor solutions are mixed to create nanoparticles (Fig. 4). The factors that affect the plant leaf extract's circumstances, such as the kinds of phytochemicals present, their concentrations in metal salts, their pH levels, and their temperatures, are acknowledged to affect not only the production and stability of nanoparticles but also their rate of synthesis. The phytochemicals present in plant extracts have the capacity to reduce metal ions in a much shorter time than bacteria and fungi, which need an extended incubation period. Therefore, it is believed that plant leaf extracts constitute a superior and secure source for the production of metal and metal oxide nanoparticles. In addition, plant leaf extract serves as both a reducing and a stabilising agent during the production of nanoparticles. The main phytochemicals present in plants that are involved in the bioreduction of nanoparticles are sugar, flavones, amides, terpenoids, , aldehydes, ketones, and carboxylic acids [19].

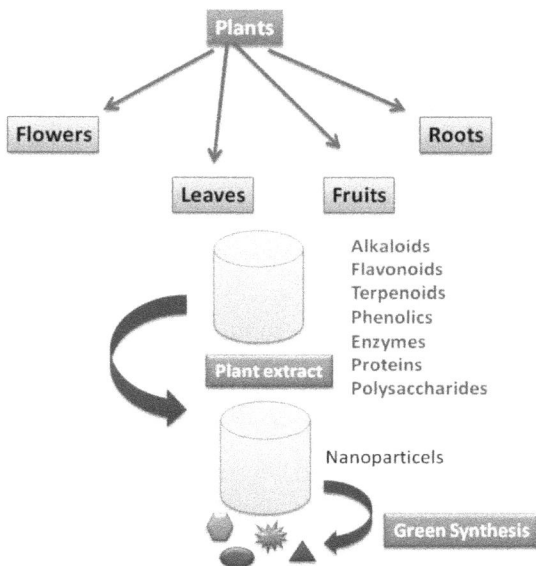

Figure 4. Diagram illustrating the process to produce nanoparticles in a sustainable manner employing different biological entities.

2.2   Methods for synthesis of nanoparticles

Numerous physical and chemical synthesis techniques have been generally accepted for the biosynthesis of NPs. The vast majority of those techniques were not well regarded because they required the employment of hazardous substances as well as elevated temperatures during the synthesis process. These might be dangerous for both people and the environment. The creation of extracellular nanoparticles is the technique used most frequently in biosynthesis. There have been reports of the environmentally friendly manufacture of gold nanoparticles using plant tissues, bacteria, fungi, actinomycetes, etc. The environmentally benign method is to create NPs from plants via green synthesis [20]. In order to produce the extract, several plant parts (such as the stem, leaf, root, and bark) are diced into small pieces and used as sources in the biosynthesis of NPs from the plant. The extract can be cleaned up using centrifugation and filtration. Metallic oxide and plant extract are typically combined for metal salt solution at room temperature (Fig. 5). Phenolic, flavonoids, alkaloids, proteins, cellulose and polysaccharides compounds are just a few of the organic components and secondary metabolites found in plant extracts that are used to create nanoparticles. They can function as stabilising agents and include the bio reduction of metallic ions to NPs. Plant extracts contain proteins that have functionalized amino groups (-NH$_2$) that can participate actively in the reduction of NPs. Functional groups like -C=O- , -C-O-C-, -C=C-, and -C-O-, present in phytochemicals like phenols, flavones, anthracenes, and alkaloids and make it easier for NPs to form. [1].

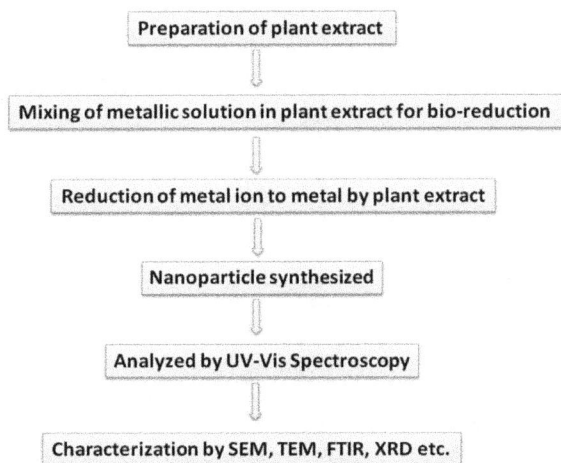

```
┌─────────────────────────────────────┐
│    Preparation of plant extract      │
└─────────────────────────────────────┘
                  ⇩
┌─────────────────────────────────────────────────────┐
│ Mixing of metallic solution in plant extract for bio-reduction │
└─────────────────────────────────────────────────────┘
                  ⇩
┌─────────────────────────────────────────────┐
│  Reduction of metal ion to metal by plant extract │
└─────────────────────────────────────────────┘
                  ⇩
┌───────────────────────────────┐
│   Nanoparticle synthesized     │
└───────────────────────────────┘
                  ⇩
┌───────────────────────────────┐
│  Analyzed by UV-Vis Spectroscopy │
└───────────────────────────────┘
                  ⇩
┌─────────────────────────────────────┐
│ Characterization by SEM, TEM, FTIR, XRD etc. │
└─────────────────────────────────────┘
```

Figure 5. Flow chat representing the steps involved in biosynthesis of nanoparticles

Several phytochemicals replace the toxicity of substances like sodium borohydride (NaBH$_4$) in these phenomena, acting as reducing and stabilizing/capping agents for the extracellular synthesis of NP. As a result, no external stabilizing/capping agents are necessary (NaBH$_4$). A metal ion's monovalent or divalent oxidation state is reduced in the bio reduction pathway to a zero-valent state. Then, new crystals are formed by the decreased metal atoms. Finally, the extract-containing metallic salt solution is converted into a metal ion, and an efficient, one-pot, one-step technique is used to produce NP within minutes to hours and characterised by various methods (Fig. 6). Although plant extract contains a wide range of phytochemicals, no specific mechanism for this manufacturing process has been identified.

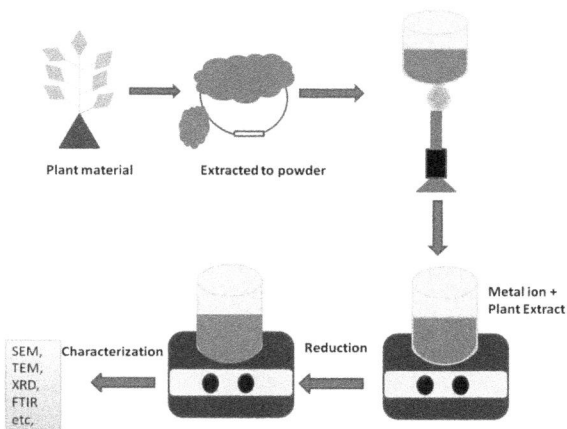

Figure 6. Steps involved in green synthesis of NPs from a plant.

2.3   Types of nanoparticles

2.3.1   Silver (Ag)

Silver nanoparticles (AgNPs) are being used more and more in a range of industries, including medicine, food, health care, consumer goods, and industrial applications, as a result of their unique physical and chemical properties. High electrical conductivity, thermal, optical, and biological properties are a few of them. The ecologically friendly synthesis of silver nanoparticles requires a reducing biological agent and a solution of silver metal ions. The easiest and most affordable method of producing AgNPs is by reducing and stabilising Ag ions with a variety of biomolecules, including vitamins, proteins, amino acids, polysaccharides, alkaloids, saponins, phenolics, and terpenes. Almost all plants can be utilised with the ability to prepare AgNPs [21].

2.3.2   Copper (Cu)

Numerous methods use L-ascorbic acid as a reducing agent in the synthesis, and the synthesised Cu NPs were highly stable. This can reduce the need for numerous dangerous chemicals [22].

2.3.3   Gold (Au)

Among all metallic nanoparticles, gold has garnered the most attention because of its special qualities, including its greater biocompatibility, strong scattering and absorption, tunable surface plasmon resonance, low toxicity, simple synthesis techniques, and ease of surface functionalization. Numerous chemical substances present in biogenic complexes operate as reducing agents and interact with the gold metal ion to reduce it and produce nanoparticles, which are the final product. Some biomolecules contained in plant extracts, such as proteins, phenols, flavonoids, etc., considerably aid in the reduction of metal ions and the coating of gold nanoparticles, according to numerous studies [23].

2.3.4 Titanium oxide (TiO₂)

Due to the stability of their chemical structure, optical, electrical, and physical properties, TiO_2 NPs have a wide range of uses. These features give TiO_2 a wide range of applications because it may be found in three different crystalline forms: anatase, rutile, and brocite. As a result of its photocatalytic activity, it is typically favoured. Titanium oxide nanoparticles have intriguing morphologies and surface chemistry, which is why they are so important. Green synthesis of TiO_2 nanoparticles from plants is a better choice for toxic-free synthesis [24].

2.3.5 Palladium (Pd) and platinum (Pt)

Platinum and palladium are both silvery-white, pricey metals with a high density. Due to their environmentally friendly, long-lasting, and cost-effective nature, both types of plant-derived nanoparticles have garnered the attention of several researchers. A valuable high-density metal is palladium. The production of Pd NPs has received substantial research due to its unique ligand-free catalysis [25].

2.3.6 Zinc oxide (ZnO)

Due to their numerous uses in the biomedical field as well as in optics and electronics, zinc oxide nanoparticles have attracted significant interest from researchers and scientists during the last 4–5 years. Due to their low cost, simple, and safe technique of manufacture, ZnO nanoparticles are of tremendous interest. Numerous biomedical uses for these nanoparticles, including antifungal, antibacterial, drug delivery, antidiabetic, and anticancer, were also discovered [17].

2.4 Quantification methods

After synthesis, precise particle characterisation is necessary because a particle's physicochemical properties may have a significant influence on those properties' biological properties. The specific characteristics of nanomaterials, such as size, shape, size distribution, surface area, form, solubility, aggregation, etc., must be examined before toxicity or biocompatibility can be determined. Numerous analytical techniques, including ultraviolet visible spectroscopy (UV-vis spectroscopy), X-ray diffractometry (XRD), transmission electron microscopy (TEM), dynamic light scattering (DLS), Fourier transform infrared spectroscopy (FTIR), scanning electron microscopy (SEM), atomic force microscopy (AFM), and others, have been used to evaluate the synthesised nanomaterials (Fig. 7) [26].

Nanoparticles in Healthcare: Applications in Therapy, Diagnosis and Drug Delivery   Materials Research Forum LLC
Materials Research Foundations 160 (2024) 163-184   https://doi.org/10.21741/9781644902974-7

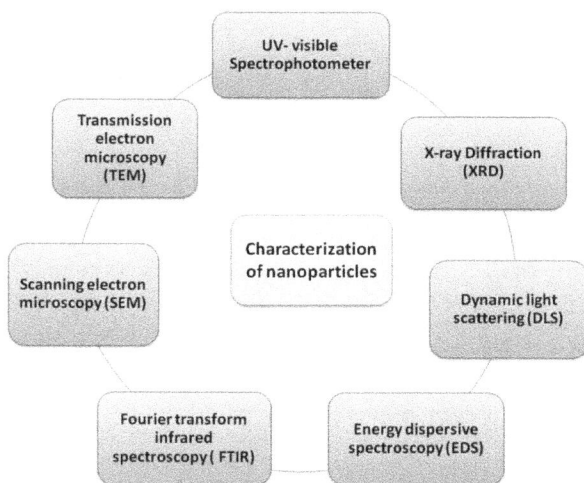

Figure 7. Various methods for characterization of nanoparticles

2.4.1 UV–visible spectroscopy

UV-vis spectroscopy is a very reliable and helpful method for the initial characterization of synthesised nanoparticles. It is also used to monitor the stability and formation of NPs. Additionally, UV-vis spectroscopy may characterise the particles in colloidal suspensions without a calibration because it is quick, simple, straightforward, sensitive, selective for different kinds of NPs, and only needs a short period of time for measurement. The arrangement of NPs from UV-visible spectroscopy may be investigated due to their surface plasmon reverberation assimilation band, which is caused by the consolidated wavering of conduction band electrons on the surface of metal NPs in reverberation with light wave [26]. Absorption spectroscopy is discussed in the context of UV-vis spectroscopy. Different metal nanoparticles (NPs) with sizes ranging from 2 nm to 100 nm are frequently depicted using light frequencies between 200 and 800 nm.

2.4.2 Fourier-transform infrared spectroscopy (FTIR)

With FTIR, it is possible to obtain accuracy, consistency, and an excellent signal-to-noise ratio. FTIR spectroscopy is frequently used to ascertain if biomolecules have a role in the formation of nanoparticles. Additionally, FTIR has been utilised for investigating nanoscaled products in order such as validating biological molecules that have been covalently grafted onto, carbon nanotubes, gold, graphene, silver, and carbon nanotubes nanoparticles, or interactions that occur among an enzyme and the substrate during a catalytic reaction. It is also a non-intrusive technique. FTIR spectrometers are superior to dispersive ones in terms of data collection speed, signal-to-noise ratio, sample heating, and signal intensity among other factors [27].

FTIR is a relevant, useful, non-invasive, affordable, and straightforward approach to identify the properties of biological groups or metabolites existing on the outermost layers of NPs, which may be in charge of decreasing and stabilising NPs, as well as details regarding capping and stabilising NPs.

2.4.3 Transmission electron microscopy (TEM)

TEM is a useful, popular, and important tool for obtaining accurate measurements of the particle size, size distribution, and shape of nanomaterials. TEM directly sees the image produced from the transmitted electron. The structure and chemical behaviour of the nanoparticles can be observed using a strong electron beam with excellent resolution. The magnification of TEM is primarily determined by the distance between the objective lens and the specimen in relation to the distance between the objective lens and its image plane [26,28].

2.4.4 Scanning electron microscopy (SEM)

SEM can be used to determine the surface anatomy of nanoparticles, including their shape, size, and size distribution. In SEM imaging, an electrical current passes via electromagnetic coils and lenses to create a focused electron beam that collides with the surface of a specimen to produce electrons known as secondary electrons. A highly accurate reconstruction of the sample surface morphology is created using the data on the produced electrons. A test sample's surface is scanned by a SEM, which records the backscattered light. Due to their high electrical conductivity, metal nanoparticles like silver and gold are simple to scan with a SEM. This microscopy equipment has a substantial benefit in that samples can be immediately placed on a black surface without worrying about unwanted incoming beam scattering [29].

2.4.5 X-ray diffraction spectroscopy (XRD)

X-ray diffraction (XRD), a widely used analytical technique, has been used, among other things, to analyse crystal and molecular structures, qualitatively and quantitatively resolve chemical species, assess the degree of crystallinity, detect isomorphous substitutions, and measure the dimensions of particles. Any crystal that receives an X-ray reflection will produce different diffraction patterns, which will represent the physico-chemical characteristics of the crystal structures. Typically, diffracted beams come from powder specimens and reflect the structural and physico-chemical properties of the sample. Thus, XRD can be used to examine the structural properties of a wide range of materials, including polymers, inorganic catalysts, glasses, superconductors, biomolecules, and more. The purity or impurity of the sample materials can also be seen in the diffracted patterns. Additional techniques for NP characterization include dynamic light scattering (DLS) and energy dispersive spectroscopy (EDS). The components of metal NPs are identified using EDS, which separates the characteristic X-rays of individual metals into an energy spectrum. The dynamic light scattering analysis of the incident photons is used to determine the surface charge and hydrodynamic radius of the NPs [28].

2.4.6 Dynamic light scattering (DLS)

DLS gives the measurement of the diameter of the particles in the formulation that are dispersed throughout the liquid. It establishes the NPs colloidal suspension's size. The foundation of DLS is the scattering of light theory. NPs that are made utilising phytoconstituents are frequently characterised using DLS. The colloidal suspension's dispersed particles scatter the light, producing an image of the particles and allowing for the estimation of their dimension distribution in the range of 0.3 to 10 m [30].

2.5   Role of nanoparticles in cancer treatment

In the twenty-first century, cancer is recognised as a severe medical concern. When healthy cells in the body deviate from their normal state and multiply uncontrolled, they may develop into cancer. The increasing death rate and the prevalence of cancer are two major concerns for scientists and medical professionals. The creation of highly efficient cancer treatments and diagnostics has been one of the most challenging and complex issues in health and medicine this century. Some of the most researched nanostructures for therapeutic application in cancer treatment and detection include microbially synthesised nanoparticles, polymeric nanoparticles, metallic nanoparticles, magnetic nanoparticles, carbon nanotubes, and liposomes [31].

Despite tremendous advancements in medicine research and delivery, the numerous adverse effects, low accuracy, and sensibility, which frequently harm healthy tissues and organs, have severely limited the use of medications. On the other hand, the growing number of tumour cells that are tolerant to conventional cancer treatments like chemotherapy, radiotherapy, and surgery, as well as the fact that these treatments frequently have negative side effects and inadequate treatment outcomes, highlight the urgent need for advancement and growth of tumor-fighting therapeutic research [32]. These medicines offer the ability to successfully target tumours and kill cancer cells while having a low risk of unpleasant side effects. Cancer nanomedicines, also referred to as "magic bullets," are now routinely employed in the treatment of cancer due to their outstanding efficacy and safety. Plant extract-mediated nanoparticle synthesis has demonstrated promising anticancer effects against a range of carcinoma cells. The green synthesised nanoparticles reduce the multiplication of carcinoma cells by halting cell cycle, which is how they work against cancer [33]. (Table-1) Presents the plant and its part associated with NPs formulation and showed anticancer activity [34-59].

Apoptosis is a normal process that occurs in all living things. It starts with different processes of destruction of DNA, mitochondrial breakdown, apoptotic gene activation, the production of an apoptosome, and finally cell contraction. These end up being the most important targets for cancer therapy. Recent research indicates that the main effects of NPs on DNA synthesis, cellular oxidative stress, and the production of reactive oxygen species (ROS). ROS are necessary for maintaining healthy cellular equilibrium, which is crucial for maintaining the survival of cells. As a by-product of cellular metabolism, free radicals (ROS) are involved in the signaling pathways for cellular transduction. An excessive amount of intracellular ROS disrupts DNA, lipids, and proteins as a mechanism of NPs-induced toxicity [60].

Table 1: The anticancer properties of synthesised NPs derived from various plant extracts.

Plant	Nanoparticles	Plant parts	References
Cynara scolymus	Silver Nanoparticles	Leaf extract	[34]
Olea europaea	Silver Nanoparticles	Leaf	[35]
Hypericum Perforatum L	Silver Nanoparticles	Aerial part	[36]
Zinnia elegans	Silver Nanoparticles	Leaf extract	[37]
Azadirachta indica	Silver Nanoparticles	Fruit extract	[38]
Mangifera indica	Silver Nanoparticles	Seed extract	[39]
Carica papaya	Silver Nanoparticles	Peel extract	[40]
Zingier officinale	Silver Nanoparticles	Leaf extract	[41]
Cocos nucifera L.	Silver Nanoparticles	Fruit extract	[42]
Perilla frutescens	Silver Nanoparticles	Leaf extract	[43]
Ziziphus nummularia	Silver Nanoparticles	Leaf extract	[44]
Gloriosa superba	Silver Nanoparticles	Tuber extract	[45]
Ocimum americanum	Silver Nanoparticles	Leaf extract	[46]
Solanum incanum	Silver Nanoparticles	Leaf extract	[47]
Eriobotrya japonica	Silver Nanoparticles	Leaf extract	[48]
Eucalyptus camaldulensis	Silver Nanoparticles	Leaf extract	[49]
Cyperus conglomeratus	Silver Nanoparticles	Root extract	[50]
Caesalpinia pulcherrima	Silver Nanoparticles	Leaf extract	[51]
Ruta graveolens	Silver Nanoparticles	Leaf extract	[52]
Cucumis prophetarum	Silver Nanoparticles	Leaf extract	[53]
Fumaria parviflora	Silver Nanoparticles	Leaf extract	[54]
Teucrium polium	Silver Nanoparticles	Leaf extract	[7]
Azadirachta indica	Zinc Oxide Nanoparticles	Leaf extract	[55]
Borago officinalis	Gold Nanoparticles	Leaf extract	[2]
Moringa oleifera	Nickel oxide Nanoparticles	Leaf extract	[56]
Ziziphus spina-christi	Gold Nanoparticles	Leaf extract	[57]
Cordia myxa L.	Gold Nanoparticles	Leaf extract	[57]
Eclipta Alba	Gold Nanoparticles	Leaf extract	[58]
Butea monosperma	Gold Nanoparticles	Leaf extract	[59]

According to De Matteis et al. [61], the NPs-triggered impacts are mostly driven by metal ions as a result of the acidic pH causing NP ionisation easier. Additionally, NPs have the ability to produce ROS, which triggers biological pathways that lead to cell death by inducing extreme oxidative stress. As a result, dysfunctions such as a lack of mitochondrial membrane and structural instability coexist with the stress brought on by NPs in mitochondria. Cytochrome C may subsequently be liberated from the mitochondria and enter the cytoplasm, resulting in cell death. Additionally, Rageh et al. [62] demonstrated that AgNPs could cause DNA damage in cancer cells in order to exercise their anti-tumor effects. (Fig. 8) suggested a potential mechanism via which NPs may function.

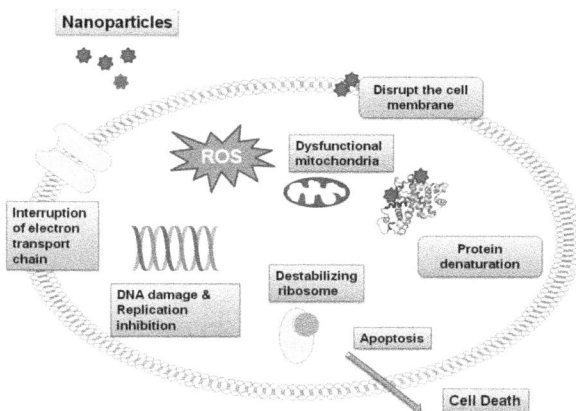

Figure 8. Possible anticancer mechanism of green synthesized NPs

Conclusions

Because these processes are expensive and produce extremely dangerous products, there is currently an urgent need to reduce the risk associated with environmental damage from the numerous chemicals used in the physical and chemical methods of manufacturing NPs. The "green synthesis" is one of the alternatives techniques for producing NPs that has been found. The use of plant material in the green synthesis of nanoparticles is a fascinating and rapidly developing area of nanotechnology that provides considerable environmental benefits that support a sustainable environment and ongoing development in the field of nanoscience. The hyper-accumulation and reduction of metallic ions are commonly found in plants. Analyses of various plant extracts have confirmed the existence of various phytochemical types that serve as precursors for the stabilisation and reduction of nanoparticles. For the synthesis of environmentally friendly nanoparticles (NPs) that might be less harmful to human health, plants are a desirable source because of these characteristics. The most current developments in bioactive metallic NPs' green manufacture, characterization, and cancer therapeutic applications are covered in this chapter. It also discusses how these environmentally friendly methods utilise secondary metabolites originating from plants, such as terepnoids, alkaloids, quinones, and others, as reducing and capping agents. These harmless plant components are naturally converted into secondary metabolites. Recent research has focused on using plant-derived nanoparticles to selectively identify and kill cancer cells while sparing healthy, non-cancerous cells. The emphasis of this section is on the unique capabilities of NPs produced using different plant materials for cancer detection and treatment. The mechanism that works of phytosynthesized nanoparticles employed in cancer-fighting applications is also emphasised.

References

[1] Bharadwaj, K. K., Rabha, B., Pati, S., Sarkar, T., Choudhury, B. K., Barman, A., & Mohd Noor, N. H. Green synthesis of gold nanoparticles using plant extracts as beneficial prospect for cancer theranostics. Molecules, 2021, 26(21), [6389]. https://doi.org/10.3390/molecules26216389

[2] Ealia, S. A. M., & Saravanakumar, M. P. A review on the classification, characterisation, synthesis of nanoparticles and their application. In IOP conference series: materials science and engineering, 2017, Vol. 263, No. 3, p. 032019. IOP Publishing. https://doi.org/10.1088/1757-899X/263/3/032019

[3] Singh, J., Singh, T., & Rawat, M. Green synthesis of silver nanoparticles via various plant extracts for anti-cancer applications. Nanomedicine, 2017, 7(3), [1-4].

[4] Ali, A., Shah, T., Ullah, R., Zhou, P., Guo, M., Ovais, M., & Rui, Y. Review on recent progress in magnetic nanoparticles: Synthesis, characterization, and diverse applications. Frontiers in chemistry, 2021, 9, [629054]. https://doi.org/10.3389/fchem.2021.629054

[5] Jeevanandam, J., Kulabhusan, P. K., Sabbih, G., Akram, M., & Danquah, M. K. Phytosynthesized nanoparticles as a potential cancer therapeutic agent. 3 Biotech, 2020, 10, [1-26]. https://doi.org/10.1007/s13205-020-02516-7

[6] Jadoun, S., Arif, R., Jangid, N. K., & Meena, R. K. Green synthesis of nanoparticles using plant extracts: A review. Environmental Chemistry Letters, 2021, 19, [355-374]. https://doi.org/10.1007/s10311-020-01074-x

[7] Hashemi, S. F., Tasharrofi, N., & Saber, M. M. Green synthesis of silver nanoparticles using *Teucrium polium* leaf extract and assessment of their antitumor effects against MNK45 human gastric cancer cell line. Journal of Molecular structure, 2020, 1208, [127889]. https://doi.org/10.1016/j.molstruc.2020.127889

[8] Goel, A., & Bhatia, A. K. Phytosynthesized nanoparticles for effective cancer treatment: a review. Nanoscience & Nanotechnology-Asia, 2019, 9(4), [437-443]. https://doi.org/10.2174/2210681208666180724100646

[9] Ovais, M., Khalil, A. T., Raza, A., Khan, M. A., Ahmad, I., Islam, N. U., & Shinwari, Z. K. Green synthesis of silver nanoparticles via plant extracts: beginning a new era in cancer theranostics. Nanomedicine, 2016, 12(23), [3157-3177]. https://doi.org/10.2217/nnm-2016-0279

[10] Raghunandan, D., Ravishankar, B., Sharanbasava, G., Mahesh, D. B., Harsoor, V., Yalagatti, M. S., & Venkataraman, A. Anti-cancer studies of noble metal nanoparticles synthesized using different plant extracts. Cancer nanotechnology, 2011, 2, [57-65]. https://doi.org/10.1007/s12645-011-0014-8

[11] Khan, T., & Gurav, P. PhytoNanotechnology: enhancing delivery of plant based anti-cancer drugs. Frontiers in pharmacology, 2018, 8, [1002]. https://doi.org/10.3389/fphar.2017.01002

[12] Abdel-Fattah, W. I., & Ali, G. W. On the anti-cancer activities of silver nanoparticles. J Appl Biotechnol Bioeng, 2018, 5(2), [1-4]. https://doi.org/10.15406/jabb.2018.05.00116

[13] Yao, Y., Zhou, Y., Liu, L., Xu, Y., Chen, Q., Wang, Y., & Shao, A. Nanoparticle-based drug delivery in cancer therapy and its role in overcoming drug resistance. Frontiers in molecular biosciences, 2020, 7, [193]. https://doi.org/10.3389/fmolb.2020.00193

[14] Kashyap, D., Tuli, H. S., Yerer, M. B., Sharma, A., Sak, K., Srivastava, S., & Bishayee, A. Natural product-based nanoformulations for cancer therapy: Opportunities and challenges. In Seminars in cancer biology, 2021, (Vol. 69, pp. 5-23. Academic Press. https://doi.org/10.1016/j.semcancer.2019.08.014

[15] Jain, N., Jain, P., Rajput, D., & Patil, U. K. Green synthesized plant-based silver nanoparticles: Therapeutic prospective for anticancer and antiviral activity. Micro and Nano Systems Letters, 2021, 9(1), [5]. https://doi.org/10.1186/s40486-021-00131-6

[16] Karmous, I., Pandey, A., Haj, K. B., & Chaoui, A. Efficiency of the green synthesized nanoparticles as new tools in cancer therapy: insights on plant-based bioengineered nanoparticles, biophysical properties, and anticancer roles. Biological trace element research, 2020, 196, [330-342]. https://doi.org/10.1007/s12011-019-01895-0

[17] Hano, C., & Abbasi, B. H. Plant-based green synthesis of nanoparticles: Production, characterization and applications. Biomolecules, 2021, 12(1), [31]. https://doi.org/10.3390/biom12010031

[18] Singh, J., Dutta, T., Kim, K. H., Rawat, M., Samddar, P., & Kumar, P. 'Green'synthesis of metals and their oxide nanoparticles: applications for environmental remediation. Journal of nanobiotechnology, 2018, 16(1), [1-24]. https://doi.org/10.1186/s12951-018-0408-4

[19] Marslin, G., Siram, K., Maqbool, Q., Selvakesavan, R. K., Kruszka, D., Kachlicki, P., & Franklin, G. Secondary metabolites in the green synthesis of metallic nanoparticles. Materials, 2018, 11(6), [940]. https://doi.org/10.3390/ma11060940

[20] Akintelu, S. A., & Folorunso, A. S. A review on green synthesis of zinc oxide nanoparticles using plant extracts and its biomedical applications. BioNanoScience, 2020, 10(4), [848-863]. https://doi.org/10.1007/s12668-020-00774-6

[21] Fierascu, R. C., Ortan, A., Avramescu, S. M., & Fierascu, I. Phyto-nanocatalysts: Green synthesis, characterization, and applications. Molecules, 2019, 24(19), [3418]. https://doi.org/10.3390/molecules24193418

[22] Akintelu, S. A., Folorunso, A. S., Folorunso, F. A., & Oyebamiji, A. K. Green synthesis of copper oxide nanoparticles for biomedical application and environmental remediation. Heliyon, 2020, 6(7), e04508. https://doi.org/10.1016/j.heliyon.2020.e04508

[23] Begum, S. J., Pratibha, S., Rawat, J. M., Venugopal, D., Sahu, P., Gowda, A., ... & Jaremko, M. Recent advances in green synthesis, characterization, and applications of bioactive metallic nanoparticles. Pharmaceuticals, 2022, 15(4), [455]. https://doi.org/10.3390/ph15040455

[24] Shah, M., Fawcett, D., Sharma, S., Tripathy, S. K., & Poinern, G. E. J. Green synthesis of metallic nanoparticles via biological entities. Materials, 2015, 8(11), [7278-7308]. https://doi.org/10.3390/ma8115377

[25] Rokade, S. S., Joshi, K. A., Mahajan, K., Patil, S., Tomar, G., Dubal, D. S., & Ghosh, S. Gloriosa superba mediated synthesis of platinum and palladium nanoparticles for induction of apoptosis in breast cancer. Bioinorganic chemistry and applications, 2018. https://doi.org/10.1155/2018/4924186

[26] Zhang, X. F., Liu, Z. G., Shen, W., & Gurunathan, S. Silver nanoparticles: synthesis, characterization, properties, applications, and therapeutic approaches. International journal of molecular sciences, 2016, 17(9), [1534]. https://doi.org/10.3390/ijms17091534

[27] Roy, S., & Das, T. K. Plant mediated green synthesis of silver nanoparticles-A. Int. J. Plant Biol. Res, 2015, 3, [1044-1055].

[28] Jain, N., Jain, P., Rajput, D., & Patil, U. K. Green synthesized plant-based silver nanoparticles: Therapeutic prospective for anticancer and antiviral activity. Micro and Nano Systems Letters, 2021, 9(1), [5]. https://doi.org/10.1186/s40486-021-00131-6

[29] Alharbi, N. S., & Alsubhi, N. S. Green synthesis and anticancer activity of silver nanoparticles prepared using fruit extract of *Azadirachta indica*. Journal of Radiation Research and Applied Sciences, 2022, 15(3), [335-345]. https://doi.org/10.1016/j.jrras.2022.08.009

[30] Sargazi, S., Laraib, U., Er, S., Rahdar, A., Hassanisaadi, M., Zafar, M. N., & Bilal, M. Application of green gold nanoparticles in cancer therapy and diagnosis. Nanomaterials, 2022, 12(7), [1102]. https://doi.org/10.3390/nano12071102

[31] Ijaz, I., Gilani, E., Nazir, A., & Bukhari, A. Detail review on chemical, physical and green synthesis, classification, characterizations and applications of nanoparticles. Green Chemistry Letters and Reviews, 2020, 13(3), [223-245]. https://doi.org/10.1080/17518253.2020.1802517

[32] Soto, K. M., Mendoza, S., López-Romero, J. M., Gasca-Tirado, J. R., & Manzano-Ramírez, A. Gold nanoparticles: Synthesis, application in colon cancer therapy and new approaches-review. Green Chemistry Letters and Reviews, 2021, 14(4), [665-678]. https://doi.org/10.1080/17518253.2021.1998648

[33] Tadele, K. T., Abire, T. O., & Feyisa, T. Y. Green synthesized silver nanoparticles using plant extracts as promising prospect for cancer therapy: a review of recent findings. J. Nanomed, 2021, 4, [1040].

[34] Erdogan, O., Abbak, M., Demirbolat, G. M., Birtekocak, F., Aksel, M., Pasa, S., & Cevik, O. Green synthesis of silver nanoparticles via *Cynara scolymus* leaf extracts: The characterization, anticancer potential with photodynamic therapy in MCF7 cells. PloS one, 2019, 14(6), e0216496. https://doi.org/10.1371/journal.pone.0216496

[35] Felimban, A. I., Alharbi, N. S., & Alsubhi, N. S. Optimization, Characterization, and Anticancer Potential of Silver Nanoparticles Biosynthesized Using *Olea europaea*. International Journal of Biomaterials, 2022. https://doi.org/10.1155/2022/6859637

[36] Alahmad, A., Feldhoff, A., Bigall, N. C., Rusch, P., Scheper, T., & Walter, J. G. *Hypericum perforatum* L.-mediated green synthesis of silver nanoparticles exhibiting antioxidant and anticancer activities. Nanomaterials, 2021, 11(2), [487]. https://doi.org/10.3390/nano11020487

[37] Haque, S., Norbert, C. C., Acharyya, R., Mukherjee, S., Kathirvel, M., & Patra, C. R. Biosynthe-sized Silver Nanoparticles for Cancer Therapy and In Vivo Bioimaging. Cancers, 2021, 13, [6114]. https://doi.org/10.3390/cancers13236114

[38] Alharbi, N. S., Alsubhi, N. S., & Felimban, A. I. Green synthesis of silver nanoparticles using medicinal plants: Characterization and application. Journal of Radiation Research and Applied Sciences, 2022, 15(3), [109-124]. https://doi.org/10.1016/j.jrras.2022.06.012

[39] Donga, S., & Chanda, S. Facile green synthesis of silver nanoparticles using *Mangifera indica* seed aqueous extract and its antimicrobial, antioxidant and cytotoxic potential (3-in-1 system). Artificial Cells, Nanomedicine, and Biotechnology, 2021, 49(1), [292-302]. https://doi.org/10.1080/21691401.2021.1899193

[40] Devanesan, S., Jayamala, M., AlSalhi, M. S., Umamaheshwari, S., & Ranjitsingh, A. J. A. Antimicrobial and anticancer properties of *Carica papaya* leaves derived di-methyl flubendazole mediated silver nanoparticles. Journal of Infection and Public Health, 2021, 14(5), [577-587]. https://doi.org/10.1016/j.jiph.2021.02.004

[41] Wang, Y., Chinnathambi, A., Nasif, O., & Alharbi, S. A. Green synthesis and chemical characterization of a novel anti-human pancreatic cancer supplement by silver nanoparticles containing *Zingiber officinale* leaf aqueous extract. Arabian Journal of Chemistry, 2021, 14(4), [103081]. https://doi.org/10.1016/j.arabjc.2021.103081

[42] Das, G., Shin, H. S., Kumar, A., Vishnuprasad, C. N., & Patra, J. K. Photo-mediated optimized synthesis of silver nanoparticles using the extracts of outer shell fibre of *Cocos nucifera* L. fruit and detection of its antioxidant, cytotoxicity and antibacterial potential. Saudi Journal of Biological Sciences, 2021, 28(1), [980-987]. https://doi.org/10.1016/j.sjbs.2020.11.022

[43] Reddy, N. V., Li, H., Hou, T., Bethu, M. S., Ren, Z., & Zhang, Z. Phytosynthesis of silver nanoparticles using *Perilla frutescens* leaf extract: characterization and

evaluation of antibacterial, antioxidant, and anticancer activities. International journal of nanomedicine, 2021, 16, [15]. https://doi.org/10.2147/IJN.S265003

[44] Padalia, H., & Chanda, S. Synthesis of silver nanoparticles using *Ziziphus nummularia* leaf extract and evaluation of their antimicrobial, antioxidant, cytotoxic and genotoxic potential (4-in-1 system). Artificial Cells, Nanomedicine, and Biotechnology, 2021, 49(1), [354-366]. https://doi.org/10.1080/21691401.2021.1903478

[45] Murugesan, A. K., Pannerselvam, B., Javee, A., Rajenderan, M., & Thiyagarajan, D. Facile green synthesis and characterization of *Gloriosa superba* L. tuber extract-capped silver nanoparticles (GST-AgNPs) and its potential antibacterial and anticancer activities against A549 human cancer cells. Environmental Nanotechnology, Monitoring & Management, 2021, 15, [100460]. https://doi.org/10.1016/j.enmm.2021.100460

[46] Manikandan, D. B., Sridhar, A., Sekar, R. K., Perumalsamy, B., Veeran, S., Arumugam, M., & Ramasamy, T. Green fabrication, characterization of silver nanoparticles using aqueous leaf extract of *Ocimum americanum* (Hoary Basil) and investigation of its in vitro antibacterial, antioxidant, anticancer and photocatalytic reduction. Journal of Environmental Chemical Engineering, 2021, 9(1), [104845]. https://doi.org/10.1016/j.jece.2020.104845

[47] Lashin, I., Fouda, A., Gobouri, A. A., Azab, E., Mohammedsaleh, Z. M., & Makharita, R. R. Antimicrobial and in vitro cytotoxic efficacy of biogenic silver nanoparticles (Ag-NPs) fabricated by callus extract of *Solanum incanum* L. Biomolecules, 2021, 11(3), [341]. https://doi.org/10.3390/biom11030341

[48] Jabir, M. S., Hussien, A. A., Sulaiman, G. M., Yaseen, N. Y., Dewir, Y. H., Alwahibi, M. S., & Rizwana, H. Green synthesis of silver nanoparticles from *Eriobotrya japonica* extract: a promising approach against cancer cells proliferation, inflammation, allergic disorders and phagocytosis induction. Artificial cells, nanomedicine, and biotechnology, 2021, 49(1), [48-60]. https://doi.org/10.1080/21691401.2020.1867152

[49] Çetintaş, Y., Nadeem, S., Sakalli, E., Eliuz, E., & Ozler, M. Green synthesis, antimicrobial and anticancer activities of AgNPs prepared from the leaf extract of *Eucalyptus camaldulensis*. Mugla Journal of Science and Technology, 2020, 6(1), [146-155]. https://doi.org/10.22531/muglajsci.714696

[50] Al-Nuairi, A. G., Mosa, K. A., Mohammad, M. G., El-Keblawy, A., Soliman, S., & Alawadhi, H. Biosynthesis, characterization, and evaluation of the cytotoxic effects of biologically synthesized silver nanoparticles from *cyperus conglomeratus* root extracts on breast cancer cell line MCF-7. Biological trace element research, 2020, 194, [560-569]. https://doi.org/10.1007/s12011-019-01791-7

[51] Deepika, S., Selvaraj, C. I., & Roopan, S. M. Screening bioactivities of *Caesalpinia pulcherrima* L. swartz and cytotoxicity of extract synthesized silver nanoparticles on

HCT116 cell line. Materials Science and Engineering:, C, 2020, 106, [110279].
https://doi.org/10.1016/j.msec.2019.110279

[52] Ghramh, H. A., Ibrahim, E. H., Kilnay, M., Ahmad, Z., Alhag, S. K., Khan, K. A., ...
& Asiri, F. M. Silver nanoparticle production by *Ruta graveolens* and testing its
safety, bioactivity, immune modulation, anticancer, and insecticidal
potentials. Bioinorganic Chemistry and
Applications, 2020. https://doi.org/10.1155/2020/5626382

[53] Hemlata, Meena, P. R., Singh, A. P., & Tejavath, K. K. Biosynthesis of silver
nanoparticles using *Cucumis prophetarum* aqueous leaf extract and their antibacterial
and antiproliferative activity against cancer cell lines. ACS omega, 2020, 5(10),
[5520-5528]. https://doi.org/10.1021/acsomega.0c00155

[54] Sattari, R., Khayati, G. R., & Hoshyar, R. Biosynthesis and characterization of silver
nanoparticles capped by biomolecules by *fumaria parviflora* extract as green
approach and evaluation of their cytotoxicity against human breast cancer MDA-MB-
468 cell lines. Materials Chemistry and Physics, 2020, 241, [122438].
https://doi.org/10.1016/j.matchemphys.2019.122438

[55] Rani, N., Rawat, K., Saini, M., Yadav, S., Shrivastava, A., Saini, K., & Maity, D.
Azadirachta indica leaf extract mediated biosynthesized rod-shaped zinc oxide
nanoparticles for in vitro lung cancer treatment. Materials Science and Engineering:
B, 2022, 284, [115851]. https://doi.org/10.1016/j.mseb.2022.115851

[56] Kiran, M. S., Kumar, C. R., Shwetha, U. R., Onkarappa, H. S., Betageri, V. S., &
Latha, M. S. Green synthesis and characterization of gold nanoparticles from *Moringa
oleifera* leaves and assessment of antioxidant, antidiabetic and anticancer
properties. Chemical Data Collections, 2021, 33, [100714].
https://doi.org/10.1016/j.cdc.2021.100714

[57] Abed, A. S., Khalaf, Y. H., & Mohammed, A. M. Green synthesis of gold
nanoparticles as an effective opportunity for cancer treatment. Results in Chemistry,
2023, 5, [100848]. https://doi.org/10.1016/j.rechem.2023.100848

[58] Mukherjee, S., Sushma, V., Patra, S., Barui, A. K., Bhadra, M. P., Sreedhar, B., &
Patra, C. R. Green chemistry approach for the synthesis and stabilization of
biocompatible gold nanoparticles and their potential applications in cancer
therapy. Nanotechnology, 2012, 23(45), [455103]. https://doi.org/10.1088/0957-
4484/23/45/455103

[59] Patra, S., Mukherjee, S., Barui, A. K., Ganguly, A., Sreedhar, B., & Patra, C. R.
Green synthesis, characterization of gold and silver nanoparticles and their potential
application for cancer therapeutics. Materials Science and Engineering: C, 2015, 53,
[298-309]. https://doi.org/10.1016/j.msec.2015.04.048

[60] Jabeen, S., Qureshi, R., Munazir, M., Maqsood, M., Munir, M., Shah, S. S. H., &
Rahim, B. Z. Application of green synthesized silver nanoparticles in cancer

treatment—a critical review. Materials Research Express, 2021, 8(9), [092001]. https://doi.org/10.1088/2053-1591/ac1de3

[61] De Matteis, V., Malvindi, M. A., Galeone, A., Brunetti, V., De Luca, E., Kote, S., & Pompa, P. P. Negligible particle-specific toxicity mechanism of silver nanoparticles: the role of Ag+ ion release in the cytosol. Nanomedicine: Nanotechnology, Biology and Medicine, 2015, 11(3), [731-739]. https://doi.org/10.1016/j.nano.2014.11.002

[62] Rageh, M. M., El-Gebaly, R. H., & Afifi, M. M. Antitumor activity of silver nanoparticles in *Ehrlich carcinoma*-bearing mice. Naunyn-Schmiedeberg's archives of pharmacology, 2018, 391, [1421-1430]. https://doi.org/10.1007/s00210-018-1558-5

Nanoparticles in Healthcare: Applications in Therapy, Diagnosis and Drug Delivery   Materials Research Forum LLC
Materials Research Foundations 160 (2024) 185-199   https://doi.org/10.21741/9781644902974-8

Chapter 8

Metal Doped Nanoparticles: Advances in Synthesis and their Applications in Wound Healing

Laxmikant R. Patil[1], Shivalingsarj V. Desai[1]*, Veeranna S. Hombalimath[1]

[1]Department of Biotechnology, KLE Technological University, Hubballi, Karnataka, India-580031

*desaisv@kletech.ac.in

Abstract

Nanoparticles are proven to possess significant and versatile applications in various fields which are attributed to their distinct physicochemical properties. Top-down and bottom-up approaches are generally employed for their synthesis which includes various physical (micro-wave assisted, sono-chemical, laser ablation and combustion), chemical (chemical reduction, co-precipitation, and sol gel) and biological methods. Doping of nanoparticles with elements like silver, copper, cobalt, iron, zinc, rare earth elements and various transition metal elements offers synergistic functional and biological properties. Due to their excellent anti-microbial property which includes both antibacterial and antifungal, they are used as dressing agents in wound healing laden with medicines for effective drug delivery to the site of infection, thus facilitating faster and effective healing.

Keywords

Nanoparticles, Dopants, Wound Healing, Antimicrobial, Healing Mechanism

Contents

1.   Introduction

Over the past several years investigations in the field of nanotechnology have increased phenomenally and thus has added significantly to the existing body of knowledge. Nanoscale particles are being used extensively in various fields, particularly in medicine, owing to their distinct properties [1]. Nanomaterial-based medical advancements leverage from their small size, rendering them less noticeable but accessible to the body. Additionally, nanoparticles can be modified in various ways, giving them greater versatility and adaptability than non-nano particles. Researchers are working on altering the properties of nanoparticles and more so with their doping (altering the particles to ameliorate the applications) to improve their diverse usage in multiple areas. Various types of nanoparticles with their characteristic distinct features are employed in drug deliver investigations which include gold nanoparticles, carbon nanotubes, albumin bonds, liposomes, quantum dots, dendrimers, and nanoparticles like Titanium Dioxide (TiO_2) and Zinc Oxide (ZnO). These are commonly used for anti-cancer, anti-bacterial and drug delivery purposes and considered safe substances for various biological applications. Nanoparticles in doped form are being used extensively employed in addressing the wounds whose healing has been considered critical especially with aged and diabetic people[2].

The effectiveness of using nanoparticles for biomedical applications like wound healing is influenced by factors like encapsulation efficacy of drugs, loading capacity and surface area which are affected by the physical and chemical properties of the nanoparticle. Doping is a method of intentionally inserting foreign elements into a crystal lattice of another element to alter its characteristics and it is commonly used to enhance the physicochemical, electrical, optical, and biological characteristics of nanoparticles. Metals such as silver, manganese, cobalt, nickel rare earth and transition elements are employed as dopants to modulate electrical structure of materials and to improve its biological activities such as anti-bacterial, anti-cancer properties and wound healing.

This article provides a detailed information on different methods of synthesizing the doped nanoparticles, various dopants used, their characterization methods and
bio-medical application regarding wound healing [3] [4] [5] [6]

2.   Methods of synthesising doped nanoparticles

Different methods have been used to dope nanoparticles onto different substrates. These include various physical, chemical, and biological approaches. Two main methods are used for creating nanosized materials: the top-down strategy, which involves breaking down larger precursor particles into smaller units, and the bottom-up approach, which involves creating nanosized materials with smaller units through bottom-up approach.

The top-down strategy uses decomposition techniques, while the bottom-up approach employs solvothermal approaches which include sol-gel, sonication, laser ablation, dip-coating, arc-discharge, combustion, and chemical breakdown methods to produce doped nanoparticles.

A list of the various methods used to prepare doped nanoparticles along with their dopants and substrate precursors are represented in Table 1. [7] [8] [9]

Table 1. Synthesis methods of various doped nanoparticles

S. No.	Materials Synthesised	Dopant Precursor	Substrate Precursor	Method of Synthesis
01	Ag-ZnO NPs	Silver Nitrate	Zinc Acetate Dehydrate	Sol-Gel
02	Ga-Doped ZnO NPs	Gallium Nitrate Hydrate	Zinc Nitrate Hexahydrate	Combustion
03	Aluminium doped ZnO NPs	Al-target	Zn-target	Laser Ablation
04	Nickel doped Cobalt ferrite NPs	Nickel Nitrate	Cobalt Nitrate and Ferric Nitrate	Microwave-Assisted Bioengineered Method
05	Cu-doped ZnO NPs	Copper Chloride Dihydrate	Zinc Nitrate Hexahydrate	Biological Reduction
06	Ni-doped ZnS NPs	Nickel (II) Acetate Tetrahydrate	Zinc Acetate	Sonochemical Technique

2.1   Sol gel method

The sol-gel method is a wet chemical process for producing various nanostructures, particularly metal oxide nanoparticles. It involves dissolving a molecular precursor, often a metal alkoxide, in alcohol or water, and then heating and stirring it through hydrolysis/alcohol-lysis to create a gel. This technique consists of several stages like polycondensation, aging, densification, crystallization, gelation, hydrolysis and drying which can be utilized to create particles, xerogels, aerogels, glass, and ceramics, depending on the final processing step.

In the doping process, the sol-gel method typically involves blending dopant and substrate materials, followed by stirring for a specified period, rinsing with water, ethyl alcohol and drying at elevated temperatures. Dopant inclusion can be achieved by mixing the dopant ion solution with the precursor prior to blending with the base material.

This technique has several advantages, including low-temperature chemistry, economics, ease of experimentation and regulated synthesis. Additionally, the sol-gel method can modify form and size, has a high surface-to-volume ratio, and is a repeatable process [10] [11]

2.2   Laser ablation

The process of laser ablation involves using laser irradiation to remove material from a surface, which can generate nanoparticles through nucleation and growth of laser-vaporized species in a background gas. This technique is commonly used for doping nanoparticles and synthesizing thin films of nanoparticles.

Different materials have been doped using laser ablation in various mediums such as Polyvinyl Alcohol/Chitosan blend, polyethylene oxide/polyvinyl pyrrolidone blend, and Milli-Q water. The composition, size and characteristics of the nanoparticles can be regulated by changing the count of laser pulses and magnitude of the laser output. Laser ablation in liquids with low laser intensity and small number of laser pulses can change the characters of synthesized nanoparticles [12]

2.3   Biological methods (green synthesis)

The use of leaf extracts in a biological technique is a popular method for creating doped nanoparticles due to its simplicity, eco-friendliness, and non-toxicity. This approach eliminates the need for hazardous chemicals and solvents. Different plant extracts are used as reagents to create various types of doped nanoparticles, which are shown to be less toxic compared to undoped nanoparticles. In most of the cases, doped nanoparticles are known to show better biological activity than pure nanoparticles. The technique is promising for creating a range of nanoparticles for various biomedical applications [13].

2.4   Sono chemical method

Sono chemical synthesis is the generation of molecules through chemical reaction by the application of powerful ultrasound radiations. It has gained significant attention due to the high energy produced by gas bubble cavitation, which can surpass the limitations of traditional techniques such as pressure and temperature. The merits of the Sono chemical approach of synthesis over traditional methods are in terms of high energy and pressure in short time, improved phase purity, faster reaction time, higher surface ratio and uniform size distribution. However low yield of the product is a limitation of the process [14] [15].

2.5   Combustion method

This involves heating substrate and dopant precursors together in fuels like urea, followed by dissolving in de-ionised water and agitation until completely soluble. The resulting mixture is heated to approximately 80°C before being placed in a pre-heated muffle furnace at approximately 500°C. This process causes combustion and ignites a flame, allowing for the acquisition of a sample. Combustion techniques have been used to create silver and gold doped ZnO nanoparticles.

This has an advantage of low energy requirement but presence of large volume of carbon in the final product is a disadvantage. Microwave-assisted combustion has also been used to produce doped nanoparticles[16] [17]

3.   Dopant materials

3.1   Silver

The element silver is commonly used as a dopant material due to its antimicrobial properties, particularly in biomedical applications. It has been traditionally employed to prevent bacterial spoilage of food. Studies suggest that silver ions penetrate bacterial cells and bind to DNA, leading to denaturation, or they disturb respiratory chains in microorganisms. To enhance the antibacterial properties, silver is often doped with compounds such as zinc oxide. Researchers have developed various Ag-doped nanoparticles with synergistic effects that show improved antibacterial activity against bacteria like *Staphylococcus aureus* and *Escherichia coli*. Silver-doped nanoparticles are also being studied for their anticancer properties. [18] [19] [20] [21] [22] [23] [24] [25]

3.2   Copper

Copper is used as an anti-infective agent due to its potent antibacterial effect. This effect is attributed to the release of ions that damage the bacterial wall and cause oxidative stress and genotoxicity, leading to the bacteria's death. Copper doping has been used for various biomedical applications, particularly as antibacterial and anticancer agent. Several studies have demonstrated the enhanced antibacterial and anticancer properties of copper-doped nanoparticles compared to their undoped counterparts. For instance, copper doped MgO, ZnO, $CdSO_4$, and Ag_2S nanoparticles have shown superior antibacterial activity against various bacterial strains [26] [27] [28] [29] [30].

3.3   Cobalt

Cobalt is known to be a potent antibacterial agent. Cobalt-doped zinc oxide nanoparticles were more effective than undoped ones in their antibacterial activity. Similarly, cobalt-doped titanium dioxide nanoparticles were found to have a significant influence on their antibacterial properties. Doped cobalt is found to have anti-cancer properties [31] [32].

3.4   Iron

Iron doping has been extensively studied for its antibacterial properties. Recent studies have shown that iron-doping in MgO nanoparticles increases their antibacterial effect against *E. coli*. Similarly, iron-doping in titanium dioxide films and ZnO nanoparticles also enhanced their antibacterial activity. Fe-doped cobalt oxide nanoparticles were found to be biocompatible and can serve as a possible vehicle for drug delivery in various biomedical applications [33].

3.5   Zinc

Zinc oxide is an inorganic substance that has been effectively utilized in various sectors, such as biomedicine, energy harvesting and gas sensing. The characters of doped ZnO are greatly influenced by the doping element, which includes ion size, electronegativity, and coordination state. The choice of dopant plays a major role in determining its functionality.

3.6   Rare Earth Elements

The exploitation of Rare Earth (RE) or Lanthanide elements is prevalent in modifying the electronic band structure of nanomaterials through doping. For instance, the addition of Cerium has been demonstrated to increase the photocatalytic properties of ZnO nano rods beyond the typical reference material, Titanium Dioxide (TiO_2). These elements have been extensively researched for their potential to increase the photocatalytic and optical characters of nanomaterials and, in some instances, to improve their electromechanical properties [34] [35].

3.7   Transition metal Elements

Transition metals, such as cobalt, chromium, iron, manganese, and copper, have been commonly used as doping agents for ZnO to enhance its properties or create new features, making it interesting for nanomedicine. Doping of these metal elements contributes to ferromagnetic behaviour, optical and electrical properties.

Among the transition metals, iron is one of the most used, and its doping affects both the electrical and structural properties. Manganese has also been extensively used, and its doping modifies the magnetic, structural, and optical properties of the semiconductor that is formed. Copper has been used to increase the antimicrobial property of ZnO, while chromium and cobalt have been found to enhance the piezoelectric response and increase the bandgap, respectively. Gold and silver have also been studied as doping agents, with varying effects on ZnO's properties. Overall, doping ZnO with transition metals has the potential to create multifunctional nanomaterials with new applications as nanomedicine [36] [37].

3.8   Other elements

In addition to using rare earth and transition metal elements as dopants, other elements have also been utilized for this purpose. For example, aluminium has been commonly used to produce flexible detectors. Magnesium is another popular option for doping ZnO, which has been found to improve the optical and electromechanical characters of pure ZnO. Mg doping also enhances the photoluminescence characters of zinc oxide in the visible light spectrum, as well as the degradation due to photocatalytic activity of rhodamine B and antibacterial activity due to a lesser bandgap which gets reflected in increased photocatalytic activity compared to pure zinc oxide [38].

4.   Characterization of doped nanoparticles

Various methods are employed to analyse the physicochemical properties of nanoparticles (NPs), including particle size analysis, Brunauer–Emmett–Teller (BET), TEM, SEM, infrared (IR), photoelectron spectroscopy (XPS) and X-ray diffraction.

The structural properties of nanocomposites, such as the doped nanomaterials can be investigated using XRD, FT-IR, and SEM/TEM analysis. These elucidate the structural and spatial properties of the nanomaterials along with their dopant characteristics [39] [40].

5.   Wound healing applications of doped nanoparticles

The use of doped nanoparticles has various biomedical applications, including their anti-microbial properties, ability to act as anti-cancer agents, antioxidant, antimicrobial, drug delivery, photo-catalysts, biosensors, bio-imaging and nanomedicines and wound healing. The present article deals with the role of doped nanoparticles in the process of wound healing and its underlying mechanism.

Skin, apart from being a major sensory organ, acts as frontline defense barrier and protects the body against mechanical, physical, and thermal injuries, pathogens, UV rays and obnoxious chemicals. It acts as thermoregulator and plays a role in preventing the loss of moisture. Skin when breaks down manifests in the form of wounds. Wounds are the result of the damage or loss of

Nanoparticles in Healthcare: Applications in Therapy, Diagnosis and Drug Delivery　　Materials Research Forum LLC
Materials Research Foundations 160 (2024) 185-199　　　　　　https://doi.org/10.21741/9781644902974-8

epithelial tissue [41] [42]. This damage to the tissue could be due to various external factors which include surgical interventions, incisions, chemicals, pressure, heat, shear, stress, and microbial infections [43][44]

5.1 Classification of wounds

Wounds can be classified into different categories based on different criteria. Figure1 shows the classification of wounds in general.

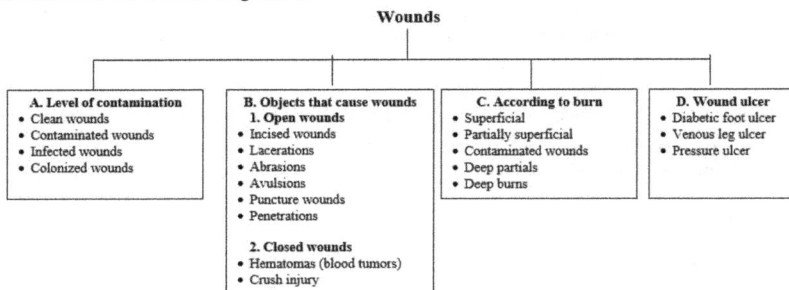

Wounds

A. Level of contamination	B. Objects that cause wounds	C. According to burn	D. Wound ulcer
• Clean wounds • Contaminated wounds • Infected wounds • Colonized wounds	**1. Open wounds** • Incised wounds • Lacerations • Abrasions • Avulsions • Puncture wounds • Penetrations **2. Closed wounds** • Hematomas (blood tumors) • Crush injury	• Superficial • Partially superficial • Contaminated wounds • Deep partials • Deep burns	• Diabetic foot ulcer • Venous leg ulcer • Pressure ulcer

Figure 1. Classification of wounds

5.2 Acute and chronic wound healing

Acute wounds generally heal within a span of 2-3 weeks. It is a normal physiological phenomenon which occurs in four successive stages namely remodelling, proliferation, inflammation and hemostasias.

The healing phases overlap each other on account of different physiological and anatomical features such as signalling of intracellular components characterized by the release of pro-inflammatory and anti-inflammatory release of cytokines [45].

The sequential and temporal factors play a significant role during the healing process. The immunological response of the body normally takes care of the infections associated with the wound healing. The innate (skin) and adaptive immunity play role in the healing process of acute wounds. When the infection is too severe due to extensive microbial colorization, the immune system fails to circumvent, and the wounds progresses towards chronic stage which takes longer span to heal [46].Under these circumstances, therapeutic intervention becomes necessary. Chronic wounds are predominantly caused by bacterial infections and characterized by elevated degree of colonization, severe inflammation, reduced oxygenation of tissues and delayed restoration of epithelial tissues. [47] [48]. Although wound healing is a natural process that takes part Suo moto, many traditional formulations and therapies which include dressings, creams, ointments, bandages, growth factors and surgery, are employed to accelerate the process [49].

Various factors affect the wound healing process which include local factors (infections, venous sufficiency, and oxygenation) that directly influence the wound characteristics and systemic factors (level of sex hormones, chronic diseases, stress, ischemia, gender (Age, nutrition, or underlying comorbidities) related to the patient's overall healing process.

5.3   Conventional approaches in wound healing

While wound healing is a natural process, conventionally, natural plant extracts like turmeric, neem and Aloe have been used as healing agents. Wound dressings made of cotton and wool have been employed which act as protective barrier at the site of wounds to prevent the entry of pathogens[50].

Exposed wounds lack protective barrier on their surface which is compensated by wound dressings thus shielding it from external contamination. The wound dressing in addition has therapeutic properties which hastens the process of wound healing. In this context, advanced dressings which are potent of creating a protective barrier have an edge over traditional dressing material like cotton and wool which merely act as passive barrier. Hence various combinations of natural and synthetic materials have been developed such as hydro fibre mats, hydrocolloids, hydrogels, and sponges.

These strategies came with their limitations of variations in treatment outcomes amongst patients.

However, silver-based therapy, vascular surgery, growth factor therapy, bioengineered substitutes, bioengineered skin substitutes, artificial and biomaterial wound dressings, and stem therapy emerged as alternate modes of wound healing [51] [52]

5.4   Metallic nanoparticles as wound healing agents

In this context nanotechnology has emerged as a novel tool to address the wound healing process, which is both effective, fast, and economically feasible. Nanotechnology is a multidisciplinary field which deals with synthesis, assembly, and application of materials of nanoscale (1-100 nm). Materials are known to behave differently with versatility and altered functions under those dimensions, which has been employed for various applications[52]. Metallic nanoparticles of inorganic origin are proven to be more effective than their organic counterparts when it comes to functional aspects of the nanoparticles, which is attributed to their unique physicochemical characteristics [53] These include:

a). their size and shape that determine biological efficacy by enhancing penetration ability, (either through cell membranes or through macrophage mediated phagocytosis) and cellular responses. [54]

b) their high surface-area- to volume ratio leading to enhanced and synergistic healing characters.

c) their enhanced mechanical strength helps in effective drug release making them excellent candidates for drug delivery against microbial infections. [55]

d) their high reactivity which enables deeper penetration into the wounds and facilitate better organic association with the biological factors.

e) potent antimicrobial properties and low toxicity profiles [56].

Various structures can be utilized to create nanomaterials for tissue regeneration, including nanospheres, nano capsules, nano emulsions, nanocarriers, and nano colloids. One type of nanomaterial used for wound therapy is nanoparticles, which can be categorised into two categories. The first category consists of metallic/metal oxide nanomaterials and non-metallic nanomaterials with inherent wound healing properties, while the second category involves using nanoparticles as delivery mechanisms for therapeutic agents.

The significant types of nanomaterials used for wound treatment are represented by nanoparticles (inorganic-metal & non-metals and organic- polymeric & non-polymeric) nanocomposites (porous materials, colloids, copolymers, and gels), coatings and scaffolds (hydrogels, nanofibers, films, coatings).

Several metals that are successfully employed in their nanoscale form for wound healing include silver, copper, gold, and iron. The doped nanomaterials that are employed in wound healing include Poly (3-hydroxy octanoate-co-3-hydroxy-10-undecenoate) encapsulated gold clusters, Poly-propylene polyethylene glycol membranes impregnated with gold and gold-cobalt oxide nanoparticles, graft copolymers containing gold nanoparticles and Polymer-based silver nanoparticles [57].

5.5   Limitations of doped nanoparticles in wound healing

Nanoparticles have great potential for promoting wound healing, but it's important to note that they come directly in contact with the tissues that are wounded, thus making their biological safety crucial. Nanoparticles have been known to cause skin irritation and allergies, exacerbate skin inflammation and irritation, and even lead to DNA damage and possible cell cancerization. The harmfulness of nanoparticles is known to predominantly depend on factors like concentration, stability, size, shape, and charge on the surface.

Nanoparticles can also enter the bloodstream, causing haemolysis and hence their physicochemical properties need to modify by coating them with biologically active substances [58].

Conclusion

Doped nanoparticles play a significant role in various biomedical applications including wound healing. The nanocomposites being synthesized by different methods are known to possess distinct physicochemical characters which qualifies them as potent candidates for the effective application in wound healing process. The process, which is affected by various intrinsic and extrinsic factors, with the most significant being the presence of microorganisms has many challenges to overcome. Several methods have been developed to minimize scarring, promote optimal healing, and maintain a moist environment while fighting infections. However, treating chronic wounds remains still a challenge, due to microbial biofilms and prevalence of multidrug-resistant microorganisms. Nanoparticles are emerging as promising solutions to this problem of wound therapy. Extensive research is being conducted on new formulations of active nanoparticles due to their significant potential in treating wounds. This potential is further enhanced by the ability of nanoparticles to carry various traditional and potential drugs, which increases the effectiveness of treating skin wounds.

References

[1] Wang, E. C., & Wang, A. Z. (2014). Nanoparticles and their applications in cell and molecular biology. Integrative biology, 6(1), 9-26. https://doi.org/10.1039/c3ib40165k

[2] Espitia, P. J. P., Soares, N. D. F. F., Coimbra, J. S. D. R., de Andrade, N. J., Cruz, R. S., & Medeiros, E. A. A. (2012). Zinc oxide nanoparticles: synthesis, antimicrobial activity and food packaging applications. Food and bioprocess technology, 5, 1447-1464. https://doi.org/10.1007/s11947-012-0797-6

[3] Rekha, K., Nirmala, M., Nair, M. G., & Anukaliani, A. (2010). Structural, optical, photocatalytic and antibacterial activity of zinc oxide and manganese doped zinc oxide nanoparticles. Physica B: Condensed Matter, 405(15), 3180-3185. https://doi.org/10.1016/j.physb.2010.04.042

[4] Zhang, X., Song, H., Yu, L., Wang, T., Ren, X., Kong, X., ... & Wang, X. (2006). Surface states and its influence on luminescence in ZnS nanocrystallite. Journal of luminescence, 118(2), 251-256. https://doi.org/10.1016/j.jlumin.2005.07.003

[5] Laurent, S., Forge, D., Port, M., Roch, A., Robic, C., Vander Elst, L., & Muller, R. N. (2008). Magnetic iron oxide nanoparticles: synthesis, stabilization, vectorization, physicochemical characterizations, and biological applications. Chemical reviews, 108(6), 2064-2110. https://doi.org/10.1021/cr068445e

[6] Jamkhande, P. G., Ghule, N. W., Bamer, A. H., & Kalaskar, M. G. (2019). Metal nanoparticles synthesis: An overview on methods of preparation, advantages and disadvantages, and applications. Journal of drug delivery science and technology, 53, 101174. https://doi.org/10.1016/j.jddst.2019.101174

[7] Rajendran, V., Deepa, B., & Mekala, R. (2018). Studies on structural, morphological, optical and antibacterial activity of Pure and Cu-doped MgO nanoparticles synthesized by co-precipitation method. Materials Today: Proceedings, 5(2), 8796-8803. https://doi.org/10.1016/j.matpr.2017.12.308

[8] Naik, M. M., Naik, H. B., Kottam, N., Vinuth, M., Nagaraju, G., & Prabhakara, M. C. (2019). Multifunctional properties of microwave-assisted bioengineered nickel doped cobalt ferrite nanoparticles. Journal of Sol-Gel Science and Technology, 91, 578-595. https://doi.org/10.1007/s10971-019-05048-6

[9] Othman, A. A., Osman, M. A., Ali, M. A., Mohamed, W. S., & Ibrahim, E. M. M. (2020). Sonochemically synthesized Ni-doped ZnS nanoparticles: structural, optical, and photocatalytic properties. Journal of Materials Science: Materials in Electronics, 31(2), 1752-1767. https://doi.org/10.1007/s10854-019-02693-z

[10] Sharma, N., Kumar, J., Thakur, S., Sharma, S., & Shrivastava, V. (2013). Antibacterial study of silver doped zinc oxide nanoparticles against Staphylococcus aureus and Bacillus subtilis. Drug Invention Today, 5(1), 50-54. https://doi.org/10.1016/j.dit.2013.03.007

[11] Rajendran, R., & Mani, A. (2020). Photocatalytic, antibacterial and anticancer activity of silver-doped zinc oxide nanoparticles. Journal of Saudi Chemical Society, 24(12), 1010-1024. https://doi.org/10.1016/j.jscs.2020.10.008

[12] Menazea, A. A., Ismail, A. M., Awwad, N. S., & Ibrahium, H. A. (2020). Physical characterization and antibacterial activity of PVA/Chitosan matrix doped by selenium nanoparticles prepared via one-pot laser ablation route. Journal of Materials Research and Technology, 9(5), 9598-9606. https://doi.org/10.1016/j.jmrt.2020.06.077

[13] Matinise, N., Fuku, X. G., Kaviyarasu, K., Mayedwa, N., & Maaza, M. J. A. S. S. (2017). ZnO nanoparticles via Moringa oleifera green synthesis: Physical properties & mechanism of formation. Applied Surface Science, 406, 339-347. https://doi.org/10.1016/j.apsusc.2017.01.219

[14] Bang, J. H., & Suslick, K. S. (2010). Applications of ultrasound to the synthesis of nanostructured materials. Advanced materials, 22(10), 1039-1059. https://doi.org/10.1002/adma.200904093

[15] Karunakaran, C., Gomathisankar, P., & Manikandan, G. (2010). Preparation and characterization of antimicrobial Ce-doped ZnO nanoparticles for photocatalytic detoxification of cyanide. Materials Chemistry and Physics, 123(2-3), 585-594. https://doi.org/10.1016/j.matchemphys.2010.05.019

[16] Pathak, T. K., Kroon, R. E., & Swart, H. C. (2018). Photocatalytic and biological applications of Ag and Au doped ZnO nanomaterial synthesized by combustion. Vacuum, 157, 508-513. https://doi.org/10.1016/j.vacuum.2018.09.020

[17] Sathish, P., Dineshbabu, N., Ravichandran, K., Arun, T., Karuppasamy, P., SenthilPandian, M., & Ramasamy, P. (2021). Combustion synthesis, characterization and antibacterial properties of pristine ZnO and Ga doped ZnO nanoparticles. Ceramics International, 47(19), 27934-27941. https://doi.org/10.1016/j.ceramint.2021.06.224

[18] Klueh, U., Wagner, V., Kelly, S., Johnson, A., & Bryers, J. D. (2000). Efficacy of silver-coated fabric to prevent bacterial colonization and subsequent device-based biofilm formation. Journal of Biomedical Materials Research: An Official Journal of The Society for Biomaterials, The Japanese Society for Biomaterials, and The Australian Society for Biomaterials and the Korean Society for Biomaterials, 53(6), 621-631. https://doi.org/10.1002/1097-4636(2000)53:6<621::AID-JBM2>3.0.CO;2-Q

[19] Jansen, B., Rinck, M., Wolbring, P., Strohmeier, A., & Jahns, T. (1994). In vitro evaluation of the antimicrobial efficacy and biocompatibility of a silver-coated central venous catheter. Journal of biomaterials applications, 9(1), 55-70. https://doi.org/10.1177/088532829400900103

[20] Stensberg, M. C., Wei, Q., McLamore, E. S., Porterfield, D. M., Wei, A., & Sepúlveda, M. S. (2011). Toxicological studies on silver nanoparticles: challenges and opportunities in assessment, monitoring and imaging. Nanomedicine, 6(5), 879-898. https://doi.org/10.2217/nnm.11.78

[21] Ghosh, S., Goudar, V. S., Padmalekha, K. G., Bhat, S. V., Indi, S. S., & Vasan, H. N. (2012). ZnO/Ag nanohybrid: synthesis, characterization, synergistic antibacterial activity and its mechanism. Rsc Advances, 2(3), 930-940. https://doi.org/10.1039/C1RA00815C

[22] Abdulkadhim, W. K. (2021, September). Synthesis titanium dioxide nanoparticles doped with silver and Novel antibacterial activity. In Journal of Physics: Conference Series (Vol. 1999, No. 1, p. 012033). IOP Publishing. https://doi.org/10.1088/1742-6596/1999/1/012033

[23] Li, P., Li, J., Wu, C., Wu, Q., & Li, J. (2005). Synergistic antibacterial effects of β-lactam antibiotic combined with silver nanoparticles. Nanotechnology, 16(9), 1912. https://doi.org/10.1088/0957-4484/16/9/082

[24] Thiel, J., Pakstis, L., Buzby, S., Raffi, M., Ni, C., Pochan, D. E., & Shah, S. I. (2007). Antibacterial properties of silver-doped titania. Small, 3(5), 799-803. https://doi.org/10.1002/smll.200600481

[25] Bahadur, J., Agrawal, S., Panwar, V., Parveen, A., & Pal, K. (2016). Antibacterial properties of silver doped TiO 2 nanoparticles synthesized via sol-gel technique. Macromolecular Research, 24, 488-493. https://doi.org/10.1007/s13233-016-4066-9

[26] Vincent, M., Duval, R. E., Hartemann, P., & Engels-Deutsch, M. (2018). Contact killing and antimicrobial properties of copper. Journal of applied microbiology, 124(5), 1032-1046. https://doi.org/10.1111/jam.13681

[27] Rajendran, V., Deepa, B., & Mekala, R. (2018). Studies on structural, morphological, optical and antibacterial activity of Pure and Cu-doped MgO nanoparticles synthesized by co-precipitation method. Materials Today: Proceedings, 5(2), 8796-8803. https://doi.org/10.1016/j.matpr.2017.12.308

[28] Samavati, A., Ismail, A. F., Nur, H., Othaman, Z., & Mustafa, M. K. (2016). Spectral features and antibacterial properties of Cu-doped ZnO nanoparticles prepared by sol-gel method. Chinese Physics B, 25(7), 077803. https://doi.org/10.1088/1674-1056/25/7/077803

[29] Garg, A., Singh, A., Sangal, V. K., Bajpai, P. K., & Garg, N. (2017). Synthesis, characterization and anticancer activities of metal ions Fe and Cu doped and co-doped TiO 2. New Journal of Chemistry, 41(18), 9931-9937. https://doi.org/10.1039/C7NJ02098H

[30] Rishikesan, S., & Basha, M. A. M. (2020). Synthesis, Characterization and Evaluation of Antimicrobial, Antioxidant & Anticancer Activities of Copper Doped Zinc Oxide Nanoparticles. Acta Chimica Slovenica, 67(1). https://doi.org/10.17344/acsi.2019.5379

[31] Naik, E. I., Naik, H. B., Sarvajith, M. S., & Pradeepa, E. (2021). Co-precipitation synthesis of cobalt doped ZnO nanoparticles: Characterization and their applications for biosensing and antibacterial studies. Inorganic Chemistry Communications, 130, 108678 https://doi.org/10.1016/j.inoche.2021.108678

[32] Shi, D., Yang, H., & Xue, X. (2020). Preparation, characterization and antibacterial properties of cobalt doped titania nanomaterials. Chinese Journal of Chemical Engineering, 28(5), 1474-1482. https://doi.org/10.1016/j.cjche.2020.03.017

[33] Hong, X., Yang, Y., Li, X., Abitonze, M., Diko, C. S., Zhao, J., & Zhu, Y. (2021). Enhanced anti-Escherichia coli properties of Fe-doping in MgO nanoparticles. RSC advances, 11(5), 2892-2897. https://doi.org/10.1039/D0RA09590G

[34] Malik, R., Tomer, V. K., Mishra, Y. K., & Lin, L. (2020). Functional gas sensing nanomaterials: A panoramic view. Applied Physics Reviews, 7(2), 021301. https://doi.org/10.1063/1.5123479

[35] Rivera, V. F., Auras, F., Motto, P., Stassi, S., Canavese, G., Celasco, E., & Cauda, V. (2013). Length-dependent charge generation from vertical arrays of high-aspect-ratio ZnO nanowires. Chemistry-A European Journal, 19(43), 14665-14674. https://doi.org/10.1002/chem.201204429

[36] Djerdj, I., Jagličić, Z., Arčon, D., & Niederberger, M. (2010). Co-doped ZnO nanoparticles: minireview. Nanoscale, 2(7), 1096-1104. https://doi.org/10.1039/c0nr00148a

[37] Yang, Y. C., Song, C., Wang, X. H., Zeng, F., & Pan, F. (2008). Cr-substitution-induced ferroelectric and improved piezoelectric properties of Zn 1− x Cr x O films. Journal of Applied Physics, 103(7), 074107. https://doi.org/10.1063/1.2903152

[38] Namgung, G., Ta, Q. T. H., Yang, W., & Noh, J. S. (2018). Diffusion-driven Al-doping of ZnO nanorods and stretchable gas sensors made of doped ZnO nanorods/Ag nanowires bilayers. ACS applied materials & interfaces, 11(1), 1411-1419. https://doi.org/10.1021/acsami.8b17336

[39] Zhao, Y., Li, C., Liu, X., Gu, F., Du, H. L., & Shi, L. (2008). Zn-doped TiO2 nanoparticles with high photocatalytic activity synthesized by hydrogen-oxygen diffusion flame. Applied Catalysis B: Environmental, 79(3), 208-215. https://doi.org/10.1016/j.apcatb.2007.09.044

[40] Fujishima, A., Rao, T. N., & Tryk, D. A. (2000). Titanium dioxide photocatalysis. Journal of photochemistry and photobiology C: Photochemistry reviews, 1(1), 1-21. https://doi.org/10.1016/S1389-5567(00)00002-2

[41] Maklebust, J., & Sieggreen, M. (2001). Pressure ulcers: Guidelines for prevention and management. Lippincott Williams & Wilkins.

[42] Flanagan, M. (2013). Wound healing and skin integrity: principles and practice. John Wiley & Sons.

[43] Ovais, M., Ahmad, I., Khalil, A. T., Mukherjee, S., Javed, R., Ayaz, M., ... & Shinwari, Z. K. (2018). Wound healing applications of biogenic colloidal silver and gold nanoparticles: recent trends and future prospects. Applied microbiology and biotechnology, 102, 4305-4318. https://doi.org/10.1007/s00253-018-8939-z

[44] Tejiram, S., Kavalukas, S.L., Shupp, J.W., Barbul, A., 2016. 1 - Wound healing, in: Ågren, M.S. (Ed.), Wound Healing Biomaterials. Woodhead Publishing, pp. 3-39. https://doi.org/10.1016/B978-1-78242-455-0.00001-X https://doi.org/10.1016/B978-1-78242-455-0.00001-X

[45] Chen, H., Lan, G., Ran, L., Xiao, Y., Yu, K., Lu, B., ... & Lu, F. (2018). A novel wound dressing based on a Konjac glucomannan/silver nanoparticle composite sponge effectively kills bacteria and accelerates wound healing. Carbohydrate polymers, 183, 70-80. https://doi.org/10.1016/j.carbpol.2017.11.029

[46] Ather, S., Harding, K. G., & Tate, S. J. (2009). Advanced Textiles for Wound Care. Woodhead Publishing, Duxford, UK, doi: http://dx. doi. org/10.1533/9781845696306.1, 3, 3-19. https://doi.org/10.1533/9781845696306.1.3

[47] Mihai, M. M., Dima, M. B., Dima, B., & Holban, A. M. (2019). Nanomaterials for wound healing and infection control. Materials, 12(13), 2176. https://doi.org/10.3390/ma12132176

[48] Hamdan, S., Pastar, I., Drakulich, S., Dikici, E., Tomic-Canic, M., Deo, S., & Daunert, S. (2017). Nanotechnology-driven therapeutic interventions in wound healing: potential uses and applications. ACS central science, 3(3), 163-175. https://doi.org/10.1021/acscentsci.6b00371

[49] Medici, S., Peana, M., Nurchi, V. M., & Zoroddu, M. A. (2019). Medical uses of silver: history, myths, and scientific evidence. Journal of medicinal chemistry, 62(13), 5923-5943. https://doi.org/10.1021/acs.jmedchem.8b01439

[50] Mihai, M. M., Dima, M. B., Dima, B., & Holban, A. M. (2019). Nanomaterials for wound healing and infection control. Materials, 12(13), 2176. https://doi.org/10.3390/ma12132176

[51] Hampton, S., 2015. Selecting wound dressings for optimum healing. Nurs Times 111, 20-23. https://doi.org/10.12968/bjcn.2015.20.Sup6.S10

[52] Bansod, S. D., Bawaskar, M. S., Gade, A. K., & Rai, M. K. (2015). Development of shampoo, soap and ointment formulated by green synthesised silver nanoparticles functionalised with antimicrobial plants oils in veterinary dermatology: treatment and prevention strategies. IET nanobiotechnology, 9(4), 165-171. https://doi.org/10.1049/iet-nbt.2014.0042

[53] Mihai, M. M., Dima, M. B., Dima, B., & Holban, A. M. (2019). Nanomaterials for wound healing and infection control. Materials, 12(13), 2176. https://doi.org/10.3390/ma12132176

[54] Lin, P. C., Lin, S., Wang, P. C., & Sridhar, R. (2014). Techniques for physicochemical characterization of nanomaterials. Biotechnology advances, 32(4), 711-726. https://doi.org/10.1016/j.biotechadv.2013.11.006

[55] Parani, M., Lokhande, G., Singh, A., & Gaharwar, A. K. (2016). Engineered nanomaterials for infection control and healing acute and chronic wounds. ACS Applied Materials & Interfaces, 8(16), 10049-10069. https://doi.org/10.1021/acsami.6b00291

[56] Negut, I., Grumezescu, V., & Grumezescu, A. M. (2018). Treatment strategies for infected wounds. Molecules, 23(9), 2392. https://doi.org/10.3390/molecules23092392

[57] Erol, A., Rosberg, D. B. H., Hazer, B., & Göncü, B. S. (2020). Biodegradable and biocompatible radiopaque iodinated poly-3-hydroxy butyrate: synthesis, characterization and in vitro/in vivo X-ray visibility. Polymer Bulletin, 77, 275-289. https://doi.org/10.1007/s00289-019-02747-6

[58] Lai, X., Wang, M., Zhu, Y., Feng, X., Liang, H., Wu, J., & Shao, L. (2021). ZnO NPs delay the recovery of psoriasis-like skin lesions through promoting nuclear translocation of p-NFκB p65 and cysteine deficiency in keratinocytes. Journal of hazardous materials, 410, 12456 https://doi.org/10.1016/j.jhazmat.2020.124566

Nanoparticles in Healthcare: Applications in Therapy, Diagnosis and Drug Delivery   Materials Research Forum LLC
Materials Research Foundations 160 (2024) 200-246   https://doi.org/10.21741/9781644902974-9

Chapter 9

Role of Nanomaterials in Diagnosis, Drug Delivery and Treatment of Neurodegenerative Diseases

Justin Lalu Perumal[1], George Kanjooparambil Rajeev[1], Kamakshi Gnanasekaran[1],
Rakesh Gunasekhar[1], Revathi Ravind[1], Richa Sunil Sikligar[1], Smithi Vibha Toppo[1],
Shinomol George Kunnel[1]*

[1]Faculty of Life Sciences, Kristu Jayanti College (Autonomous), Bengaluru, India

*shinojesu@gmail.com, shinomol@kristujayanti.com

Abstract

Neurodegenerative diseases affect the neuronal system, with irreversible loss of neurons. Many factors play a role in the development of NDD and till now no conclusive cure is available for any of these diseases. As early diagnosis and specific drug delivery being the most challenging part of NDD, along with the absence of cure has led the scientists to explore the amazing nanotechnology and nanomaterials for diagnosis, delivery and cure for these diseases. *In vitro* and *in vivo* studies using nanomaterials have revealed promising results in these fields. Since the inception, nanotechnology has found its applications in healthcare, transforming the field and making the omnipresence in Nano medicine with many discoveries replacing the conventional systems revolutionizing the healthcare sector.

Keywords

Neurodegenerative Diseases, Green Nanoparticles, Gold Nanoparticles, Drug Delivery, Blood Brain Barrier

Contents

1.   Introduction

Neurodegenerative diseases (NDD) primarily affect the cells of the nervous system, which function for coordination in humans. This type of diseases now affects a large portion of the human population with some complex NDD do not have a specific or permanent cure. However, various cutting-edge technologies, including the use of nanoparticles, can alleviate symptoms leading to better quality of life [1]. Some of these NDD are Alzheimer's disease (AD), Parkinson's disease (PD), Lewy body dementia (LBD), Frontotemporal dementia, Vascular dementia, Spinocerebellar ataxia, prion disease, Huntington's disease, Amyotrophic lateral sclerosis (ALS) etc. [2]. With most of these diseases associated with the aging population, there is high requirement to have therapeutic approaches to treat NDD.The current advancements in nanotechnology could help us to overcome the limitations of Blood Brain Barrier (BBB), Blood-Cerebrospinal fluid barrier (BCFB) and the P-glycoproteins and help in much local delivery of drugs too. In the present chapter we will emphasis on how nanoparticles or nanocarriers will aid in the delivery of drugs as a non-invasive method of drug delivery, diagnosis and therapy.

*Alzheimer's Disease (AD):*AD is the maximum customary form of dementia. It slowly starts with mild memory loss and later leads to loss of speech and response to the environment. It entails elements of the mind that manage thought, memory, and language. Almost at the age of 75, four percent of the total population is affected, but by age 85 the average climbs to 50 percent and chances increases if there is family history [3,4]. AD symptoms develop mainly due to the abnormal build-up of protein which mainly involves Amyloid and Tau- both of these are responsible for the formation of plaques inside and around neurons. As major part of the brain is affected, it decreases the production of the most common neurotransmitters [5]. These later decreases the level of acetylcholine in the brain. Hence different parts of brain are affected that results in memory loss and problems with vision or language. This leads to shrinking of brain and reduced connections among neurons leading to serious affects in regions of memory, cortex, hippocampus and later regions accountable for language, reasoning, and social behaviour [6,7]. There are various risk factors involved including age, family history and genetics [8].

Parkinson's disease (PD): This progressive NDD caused by the degeneration of neurons in the part of the brain that controls movement called the substantia nigra causes the neurons die or weaken, losing the ability to produce the neurotransmitter- dopamine .Studies have ascertained that symptoms of PD develop in patients who have 80 percent or greater loss of dopamine-producing cells. Normally, dopamine works in a delicate balance with other neurotransmitters to help coordinate the millions of nerve and muscle cells involved in movement. In the absence of sufficient dopamine, this balance is disturbed, causing tremors, stiffness, slowness of movement and disorders of balance and co-ordination [9]. Scientist haven't yet discovered the real cause behind the disease. However, oxidative damage, environmental toxins, genetic factors and accelerated aging have been theorized as possible causes of the disease [10,11].

Lewy body dementia (LBD) involves a build-up of abnormal proteins called Lewy bodies and results in dementia. These protein deposits forms α-synuclein which inhibits the neurotransmitter release when over expressed as they localize in the nerve terminal. These deposits built-up mass and end ups in blocking the messages between the cells of brain which involve thinking and movement. When they occur with PD or AD in a same person such conditions is known as "Mixed Dementia"[12]. LBD can be divided into two main types: dementia with Lewy bodies (DLB) and Parkinson's disease dementia (PDD). Both types have similar symptoms, but PDD characteristically starts with movement problems and later progresses to dementia, while DLB typically starts with cognitive or behavioural symptoms and later develops movement problems [13]. LBD can have similar symptoms like that of PD and AD. The common symptom is progressive loss of memory. Memory loss may not affect at the early stages, unlike Alzheimer's. People suffering from LBD may also have fluctuations in cognitive ability, like calculating numbers/figures or a sudden change in attention, alertness, and orientation. In the early stages visual hallucinations are also common [14]. LBD can be challenging to diagnose because its symptoms overlap with those of other NDD. However, there is currently no cure for LBD [15,16].

Frontotemporal Dementia (FTD) is due to progressive loss of nerve cell in the frontal lobes and / or in temporal lobe, leading to loss of function and impairment of cognitive and behavioural abilities and also involves difficulty with languages. It is sometimes referred to as frontotemporal lobar degeneration (FTLD). The disease typically affects people between the ages of 45 and 65, although it also affect at younger age and it is more common in men than women [17]. There are several types of FTD, each of which distresses diverse parts of the brain and can result in various symptoms. The most common types include 1). Behavioural variant FTD: This type is typical by changes in behaviour, personality, and social skills. People with this type of FTD may exhibit apathy, disinhibition, lack of empathy, and changes in eating habits 2). Semantic variant primary progressive aphasia: This type of FTD affects language skills, making it difficult for people to understand and use words. People with this type of FTD may have trouble with naming objects, understanding the meaning of words, and using correct grammar. 3). Nonfluent/agrammatic variant primary progressive aphasia:This type of FTD also affects language skills, but it primarily affects the ability to speak fluently and form grammatically correct sentences. People with this type of FTD may have trouble with word order, grammar, and pronouncing words correctly. Other types of FTD comprise progressive supranuclear palsy, corticobasal degeneration, and FTD with motor neuron disease [18,19].

Vasculardementia (VD) occurs when impairmentin blood vessels that supply blood to the brain occur due to a stroke or transient ischemic attacks resulting in blocks or reduce the flow of blood inside the brain. The resulting damage to the brain can lead to cognitive impairment, including memory loss, difficulty with language, and changes in personality. These mainly coexist with changes including Alzheimer's or dementia with lewy bodies [20]. Symptoms of VD can vary depending on the severity of the damage and the areas of brain affected. It can also induce thinking difficulties and leads into conditions like confusion, disorientation, difficulty with reasoning and problem-solving, and trouble with balance and coordination. Increase in number of strokes can eventually reduce the flow of blood inside the brain [21]. VD is the second most reason of dementia after AD, blameable for up to 20% of all cases of dementia. The threat of developing VD is greater in persons who have a history of stroke or cardio vascular diseases. They can be identified by several neurocognitive testing and presently no cure for VD is available[22].

Spinocerebellar ataxia (SCA) is a genetic disorder that mainly affects the people over the age of 18 in the region of brain responsible for managing movement and balance *ie* cerebellum [23]. Symptoms of SCA can vary depending on the specific type of the disorder and the severity of the condition. Common symptoms include clumsiness, unsteadiness, difficulty with speech and swallowing, and changes in eye movements. As the disease progresses, individuals with SCA may also experience muscle wasting, difficulty with fine motor skills, and impaired cognition. There are many different types of SCA, each caused by a different genetic mutation. The most common type is SCA type 3, also known as Machado-Joseph disease, which is caused by a mutation in the ATXN3 gene. There is currently no cure for SCA and treatment focuses on managing symptoms and improving quality of life [24].

Prion disease: Prions are special contagious agents that are made up entirely of protein. They are responsible for causing a variety of neurological disorders in humans as well other animals which includes Creutzfeldt-Jakob disease (CJD), variant CJD, mad cow disease and chronic wasting disease [25]. Unlike other contagious agents like viruses and bacteria, prions do not contain DNA or RNA. Instead, they are made up of an abnormal form of a protein called PrP (prion protein). This abnormal PrP, known as PrPSc, is resistant to normal cellular processes that would typically break down and eliminate proteins that are no longer needed. When PrPSc comes into contact with normal PrP in the body, it can cause the normal PrP to misfold into the abnormal PrPSc [26].This sets off a chain reaction, leading to the accumulation of PrPSc inside the brain and other tissues, which results in the death of cells and the development of neurodegenerative disease.Prions are particularly challenging to study and treat because they are incredibly resistant to heat, radiation, and chemical agents that would normally destroy other infectious agents. This makes it difficult to sterilize equipment and prevent the spread of prion disease [27].

Huntington's disease (HD) is a kind of genetic disorder which usually distresses the brain, leading to the progressive loss of brain cells, cognitive impairment and motor functions. HD is caused due to mutation in the huntingtin gene, which results in the production of a defective protein that accumulates inside the brain and leads to neuronal damage. Symptoms of HD typically begin to appear in middle age, although the age of onset can vary widely. Early symptoms may include mood changes, irritability, and problems with coordination and movement [28,29]. HD patients experience trouble with language, experience involuntary movements known as chorea, and cognitive decline as the disease advances. There is no permanent cure for the disease, but treatment can reduce the intensity of the symptoms [30] Fig: 1

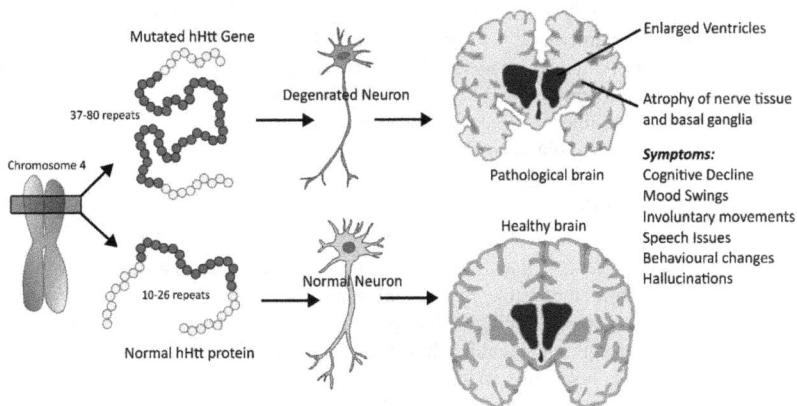

Fig.1- Huntington's disease.

Amyotrophic lateral sclerosis (ALS) also known as Lou Gehrig's disease, is a progressive neurodegenerative disease that causes problems and results in death of cells in the brain and spinal cord. It causes weakness in muscle and atrophy, which can lead to difficulty with movement, speech, and breathing. A combination of genetic and environmental factors results in the onset of the disease and about 10% of cases are inherited, while the majority are sporadic. The genes associated with ALS include C9orf72, SOD1, TARDBP, FUS, and UBQLN2. Environmental factors that have been linked to ALS include exposure to lead, pesticides, and other toxins. The symptoms usually vary from person to person, but typically include muscle weakness, cramping, and twitching and as the disease progresses, these symptoms can grow to other parts of the body, eventually leading to difficulty with speech, swallowing, and breathing. There is no permanent cure for ALS, but with the help of treatments, the intensity of the symptoms can be reduced and improve the quality of life [31].

1.1   Drug delivery across blood brain barrier

One of the major hurdles in the effectiveness of medicines is the ability to penetrate the blood brain barrier and work effectively. The blood brain barrier or BBB is a defensive system in the human body designed to block most compounds in the blood that can flow to the brain [32]. Only tiny molecules with a weight of under 400-600 Dalton or below 100 nanometers can actually cross the BBB [33]. Accumulation of toxic materials can affect the Central Nervous System (CNS) and other organs causing symptoms of progressive neurodegeneration like mental retardation and loss of motor functions. Severe exposure will almost certainly cause death. Damage to the BBB (Fig.2) can also cause the immune cells to attack neurons therefore great care must be taken to ensure that methods of drug delivery should avoid being invasive or damaging to the membrane (Table 1).

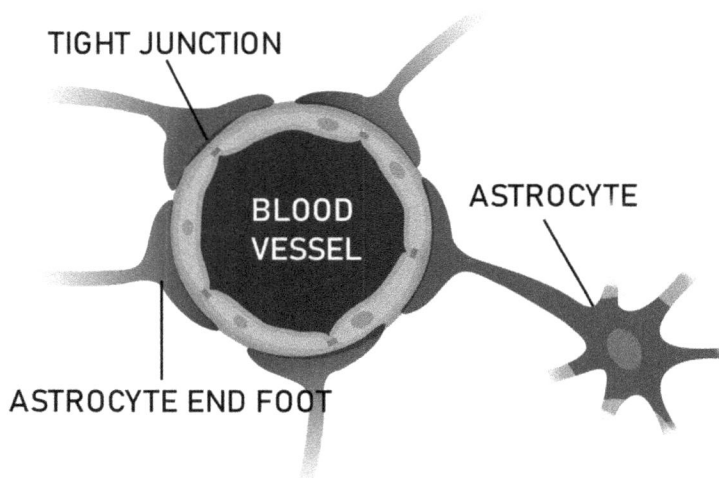

Fig.2-Blood brain Barrier

Table 1-Certain methods of delivering drugs across the blood brain barrier.

INVASIVE TECHNIQUES	NON-INVASIVE TECHNIQUES
Stereotactic Brain Surgery: Use of computer aided imaging to insert a device like a catheter	MR-Guided Focused Ultrasound: Ultrasound is emitted which temporarily opens BBB
Intercranial Injections: Direct Injection of drugs through a small hole in the skull. Used when non-invasive procedures fail.	Osmotic opening: Hyperosmotic agents are used to increase osmotic pressure and temporarily open BBB
Convection-enhanced delivery (CED): Using pressure to overpower the BBB and deliver	Nanoparticle-mediated delivery: Utilizes the small nanoparticle size to transmit drugs
Intracerebroventricular (ICV) infusion: Injection of drugs into the cerebrospinal fluid.	Transcranial Magnetic stimulation: Magnetic field induces a BBB opening

[34,35,36].

1.2   Uptake mechanism of nanoparticles into the brain

The BBB's purpose is to regulate the flow of factors, particles and debris that would damage or adversely affect the neurons in the brain, in other words; to prevent neurotoxic materials entering the brain [37]. Nanoparticles when implemented into an organism will be challenged when required to deal with subcellular components in cells. Cells may be able to intake nanoparticles through a process known as endocytosis [38]. The effectiveness of endocytosis is reliant on size, shape and chemistry of membrane [39]. Particles like nanocarriers can be transported from the site of deposition to auxiliary parts of the body like the brain with the help of the blood circulatory system [40].

Polymeric nanoparticles and amphiphilic molecules form liposomes- both artificially created and can target specific drug. While both vehicles have their own unique advantages and may be specific to each patient, it is generally known that nanoparticles have an extra edge in its ability to be controlled in its drug delivery, as well as better at targeting specific organs or locations [41,42]. It is imperative to note that polymeric nanoparticles are not the only type of nanomaterials used, however they and liposomes are the ones extensively used in brain-drug delivery systems [43]. The method of how nanoparticles are to be introduced into a body or an organism is dependent on the particular location of the target like an organ. The nanoparticles can reach the body at certain points for example:

Lungs: where rapid absorption into the blood stream can occur (and also to reach the BBB if necessary)

Intestine: Absorption through the epithelial tissue in the digestive tract, most commonly through oral means

Skin: Possible given the nanoparticles are small enough to infiltrate deep into the skin [44].

1.3   Nanomaterial stem-cell hybrid

The ability of stem cells (specifically pluripotent stem cells) to differentiate into nearly any type of cell in the body is extremely powerful and may open up opportunities for regenerative medicine, personalized medicine, new drug developments and even food/organ culturing [45]. However, while we can isolate stem cells that don't spontaneously differentiate, we struggle to fully utilize the potential of stem cells, they are complex and variable and thus prevents enough reproducible results to make it permissible for human use. Stem cell research has shown that there is definitely a potential risk to the formation of tumors which can cause more harm than good [46]. This is where nanomaterials come into picture, if used in conjunction with stem cells may allow stem cells to perform to their fullest potential. With the help of nanomaterials Tissue engineering can be improved further, as stem cells can now be grown on nanomaterial scaffolding that supports stem cell growth, transplantation and drug delivery [47]. Biosensing by sensitive hybrids can allow great performance by taking more control on how the stem cells will be deployed or triggered. There has been some research on how nanomaterials can be sensitive to certain chemical or physical triggers like electrochemiluminescence [48].

Mechanical Advantage: Nanotechnology since it is often made out of small molecular sized particles like the carbon atoms in graphene for example, also has the added advantage of strength which can be enhanced and modified with unique structures like nanotubes and gold nanoparticles. This may provide great help in medicine related to bone damage or other orthopedic issues. Regenerative tissue for cartilage, tendons and even nerve can be an absolute boon in surgical procedures [49].

Stem cell modification: The nanomaterial hybrid can also open the possibility of taking cultured stem cells and facilitating specific ways the stem cells can proliferate and differentiate with the influence of the nanomaterials on its growth thus its hybrid nature may affect the fate of a stem cell differently [50].

1.4   Nanocarriers for drug delivery in the central nervous system

Similarly, to the BBB, the CNS which consists of the brain and the spinal cord has another barrier called the Blood-cerebrospinal fluid barrier (BCFSB). These highly regulated environments protect the brain and spinal cord against pathogens and toxic substances but similarly also prevents the uptake of medical drugs and substances. The existence of nanocarriers may finally provide a solution to pass drugs through the barriers and begin affecting the brain and the CNS. These nanocarriers can not only be capable of crossing the barriers but also dispense the drugs in a safe and controlled manner all the while without being toxic to the body and protecting the drug molecule [51].

There are a multitude of particles that are considered nanocarriers:

Liposomes: spherical vesicles composed of a lipid bilayer.

Polymeric nanoparticles: made of biodegradable polymers like PLA, PEG, and PLGA

Solid Lipid nanoparticles: made of lipids and surfactants, these encapsulate drugs [51].

*Dendrimers:*highly branched, nanoscale polymers that can be functionalized with aiming ligands [52].

*Carbon nanotubes:*cylindrical, carbon-based nanomaterials with high surface area and unique electrical and mechanical properties [53] [Fig.3]

Gold nanoparticles are small, spherical particles made of gold that can be functionalized with targeting ligands and have unique optical properties [54] [Fig.4].

Iron oxide nanoparticles are magnetic nanoparticles that can be targeted and visualized using magnetic resonance imaging (MRI) and thus monitored in bodies [55].

Nanoemulsions: made out of lipids and surfactant which can encapsulate drugs and improve their bioavailability [56].

*Nanoemulsions:*made of a lipid and surfactant mixture that can form a stable emulsion, which can encapsulate drugs and improve their bioavailability [57].

1.5   Nanomaterials used in neurodegenerative diseases

Nanoparticles are typically insoluble polymers sized 10-1000 nanometres in range. They are designed in the treatment of NDD to cross the BBB to targets and overcome the disadvantage of typical drug delivery procedures to provide improved ways to repair cell structures. Nanoparticles can be designed to bind to specific biomarkers or proteins that are associated with NDD. These nanoparticles can then be detected using imaging techniques (like magnetic resonance imaging (MRI) or positron emission tomography (PET)), allowing for early and precise diagnosis of these diseases [58].

For nanoparticles to be ideal for medical usage and for brain drug delivery, certain criteria must be met such as

- Nontoxic
- Biodegradable
- Biocompatible
- Capable of targeting the brain and passing through the BBB
- Diameter below 100nm of particle
- Should be chemically stable in blood to prevent reaction and aggregation
- Prolonged circulation in blood to avoid the mononuclear phagocyte system
- Easily scalable and in expensive manufacturing costs
- Amenable to small molecules, peptides, proteins and nucleic acids
- Drug release profiles should be modifiable and controlled

Overall, nanocarriers provide a promising solution to transporting useful and even life-saving drugs through the BCFSB for the cure of CNS disorders, they can improve the efficacy of the medicine, reduce the potential risk of side effects and increase patient compliance through the reduced frequency in dosage. The advantages offered by nano carriers are almost similar to their process in transporting across the BBB as well [59,60].

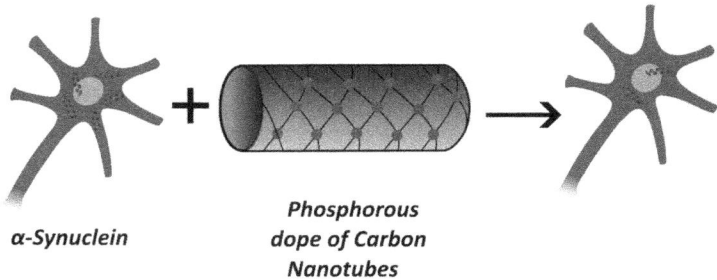

α-Synuclein   ***Phosphorous dope of Carbon Nanotubes***

Fig.3- Carbon nanotubes.

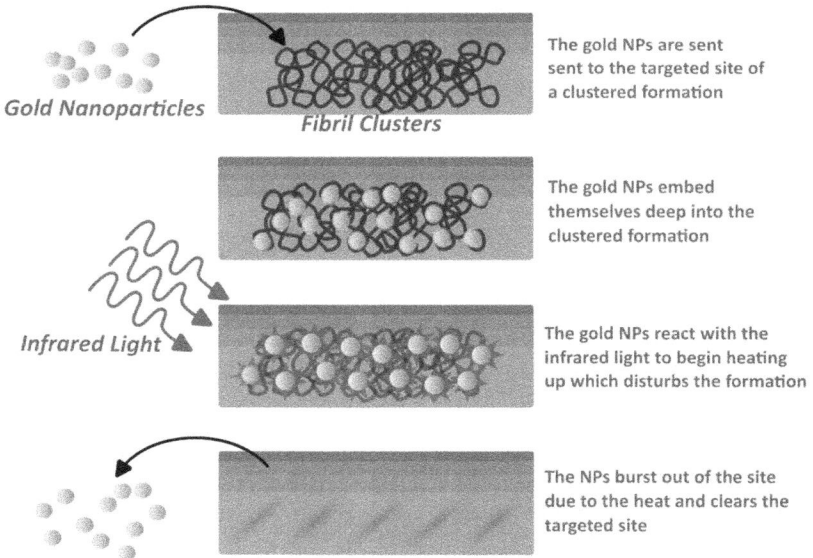

Fig.4- Gold nano particles

Nanomaterials used in treatment of NDD include Liposomes, lipid based nanomaterials, Solid lipid nanoparticles (SLN), Nanoparticles, Polymeric micelles, Nanogels, Exosomes, Inorganic nanomaterials, Quantum dots, Calcium phosphate nanoparticles, Silica Nanoparticles,

Nanoparticles made of organic materials, Polymeric nanoparticle, Nano emulsions, Hybrid nanoparticles, Dendrimers etc. [61].

Liposomes: Liposome nanoparticles are small vesicular structures that are typically between 50-1000 nanometres in size, made up of single or multiple lipid bilayers that enclose an aqueous partition and can be engineered to have specific physicochemical properties, such as charge and size [62]. Liposomes are designed to target specific tissues or organs, allowing for targeted drug delivery and reducing potential side effects. They can also be modified with various surface coatings, such as polyethylene glycol (PEG), to increase their circulation time in the body and reduce clearance by the immune system. In the case of ischemic stroke, modified liposomal nanoparticles are used to target injured brain regions and deliver neuroprotective agents [63]. Liposome based drug delivery system across BBB are now at the starting stages and there are not much studies proving the capacity of liposome as a drug carrier. Liposome have hydrophilic as well as hydrophobic layer, hence a hydrophilic drug can be encapsulated by liposome and can be used as a drug delivery molecule. Liposome can travel through BBB by covering itself with glutathione and glucose as nutrients. Liposomes can also be used as double functionalized *viz.*, curcumin with any peptide useful for Alzheimer's or any other NDD to enhance the crossing and reach the target site. A study showed that curcumin conjugated with HIV TAT peptide under observation using [3H]-sphingomyelin by mass spectrometry and radioactive assay showed 3 times increase in nanoliposome absorption after TAT functionalization in human brain capillary endothelial cells [64,65]. Under this category, Niosomes are structures that can carry hydrophilic as well as hydrophobic drugs together in the same system[66] [Fig.5].

Lipid-based nanoparticles (LNPs) are a type of nanoparticle that is composed of a lipid bilayer that surrounds an aqueous core. These nanoparticles are used for drug delivery and gene therapy applications due to their competence to encapsulate and shield nucleic acids and other therapeutic agents. LNPs have been shown to have low toxicity and high biocompatibility, making them a safe option for drug delivery. They also have the ability to protect therapeutic agents from degradation in the bloodstream, increasing their bioavailability and efficacy [67].

Solid lipid nanoparticles (SLN) are another type of nanoparticle system used for drug delivery. They are composed of a solid lipid core surrounded by a surfactant layer for stability. The size of SLNs typically ranges from 10 to 1000 nm, and they can be designed to have a specific drug release profile. SLNs have been investigated for their ability to deliver drugs that target the Aβ protein in AD and the accumulation of α-synuclein in PD which are key contributors to the development of the disease. SLNs have been presented as to improve the bioavailability and efficacy of these protein-targeting drugs and reduce the toxicity associated with these drugs. SLNs have some limitations, such as their tendency to aggregate and their low drug-loading capacity compared to other nanoparticle systems. Researchers are continuing to explore ways to improve the properties of SLNs for targeted brain drug delivery [68].

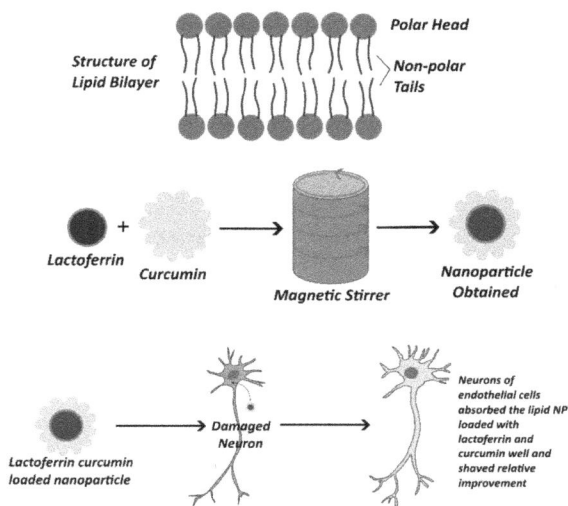

Fig.5- Lactoferin-curcumin nano particles based therapy

Polymeric Micelles are another type of nanoparticle system used for drug delivery composed of a self-assembled core structure of amphiphilic copolymers in a solvent. The hydrophobic core of the shell carries the drug enhancing the solubility and stability. They are seen as a potential way of drug delivery to the brain by increasing the bioavailability of drugs [69] and delivering gene therapies. Research has been undergoing to investigate the usage of these particles in brain tumor imaging.

Polymeric nanoparticles have been many research done on polymeric nanomaterials that have been proven to be having low toxic, good drug delivery capacity, drug encapsulation and biodegradability. Curcumin when treated with APP/PS1 mice that are good model for AD and other NDD have showed that it did not affect much or inhibit the amyloid beta fibrillation or reach the brain by crossing the BBB, but curcumin when loaded with poly(lactic-co-glycolic acid) enhanced the crossing of the drug from the BBB, heightened the drug loading capacity and were also seen penetrating the neural stem and neural spheres And the modified conjugated polymeric nanoparticles also had very low toxic level when the cytotoxicity assay MTT was done [70, 71].

Nanogels are nanosized hydrophilic cross-linked polymers of polyethyleneimine (PEI) and carbonyl diimidazole-activated polyethyleneglycol (PEG) used as drug delivery systems. They are shown to increase the stability of bioactive enzymes, reduce cytotoxic side effects, and possess an increased water-retaining capacity. They are seen as potential means for delivering oligonucleotides and DNA molecules for antisense and gene delivery to the CNS. [72].

Protein based nanoparticles: Protein conjugated nanoparticles have low toxicity, stability, surface modification, biodegradability and are biopolymer-based nanoparticles. Fibroin nanoparticles are one such example of protein-based nanoparticles. Fibroin nanoparticles can circumvent the limitations of low-molecular-weight medications. They are better for medication therapy because all fibroin nanoparticles loaded with small-molecule medicines have increased drug solubility and stability, avoided drug degradation, and reduced toxicity [73].

Exosomes are small extracellular vesicles having the potential of being used as prognostic biomarkers and therapeutic targets of NDD. Exosomes have been found carrying beta-amyloid (Aβ) and tau proteins in AD, alpha-synuclein (α-syn) protein, which is the main component of Lewy bodies in PD and in HD, exosomes have been found to contain mutant huntingtin protein (mHTT). These exosomes are taken up by neighbouring cells and cause the spread of the disease. Understanding the pathology of exosomes can help contribute to diagnosis and treatment in the future [74].

Inorganic Nanoparticles are broadly used for bio imaging and the detection of biomarkers in disease diagnosis and drug department to the CNS in the form of nanocarriers. Particular metal, metal oxide, and semiconductor nanoparticles (NP) are exploited based on their properties. Due to their reduced biodegradation and higher retention properties, they are utilized in long-term treatment methods. Gold NPs can be administered with antibodies or peptides that specifically bind to amyloid-βprotein associated with AD, and be used to image amyloid-β plaques in the brain. Cerium oxide nanoparticles have been shown to have antioxidant properties and may help to protect neurons from oxidative damage [75].

Quantum dots are nanosized semiconductor particles with electronic and optical properties. Their main applications are based on their Fluro luminescent properties: to visualize the structures of cells and tissues in the brain functionalized as probes, biosensing, and targeted drug and gene delivery across the BBB using antibodies or peptides [76].

Silica Nanoparticles (SiNP) are multi-functionalized using different moieties like Polyethylene glycol and surface modified to increase efficacy in crossing the BBB. In a study it was observed that SiNPs were found across the BBB in mice, and their transport efficiency is dependent on size and not affected by drug loading [77]. SiNPs bio-tagged with fluorescent probes have shown applications in cancer bioimaging [78]. Their versatile properties like optical transparency, surface functionality, drug-carrying capacity, biocompatibility and size tunability make them a prospective research domain for the treatment and diagnosis of NDD.

Calcium Phosphate Nanoparticles (CaPNs). They are widely used in bone regeneration and bone diseases by encapsulation of drugs and delivery to injury sites due to their biocompatibility and biodegradation properties. They show a great affinity towards nucleic acids, therapeutic drugs, and enzymes [79,80]. They are lower in toxicity compared to other inorganic nanoparticles and structurally stable than liposomes. CaPNs have been shown to reduce toxicity by aggregate formation and promote the clearance of causative proteins in NDD like AD and PD.

Dendrimers are nanosized polymers that can help as drug delivery vehicles, gene transfection, and molecular imaging agents [81]. Their properties include highly branched 3-dimensional structure, well-defined molecular weight and size, structural uniformity, surface functionality, and biocompatibility. Dendrimers can also be functionalized with peptides or antibodies that target specific protein aggregates, such as amyloid beta or alpha-synuclein. Dendrimers can be operationalized with peptides that target specific cell types, such as neurons or glial cells, and deliver drugs or other molecules to those cells. Additionally, dendrimers can be used as scaffolds to deliver multiple drugs or other molecules simultaneously, which can enhance their efficacy [82].

Nanoemulsions are a type of emulsion where the dispersed phase droplets have a size range in the nanometer scale. They possess properties like increased stability, larger surface area, improved bioavailability, and enhanced delivery of active ingredients making them a potential targeted drug delivery system with minimized adverse effects and toxic reactions in treating NDD like AD, PD, and Prion's disease [83].

Hybrid Nanoparticles are nanosized chemical conjugates of either organic, inorganic or both materials. They are multi functionalized to modify target environment and induce therapeutic benefits as seen in the case of mouse models in treatment of AD [84]. A cholesterol modified hepta-peptide hybrid nanoparticle was shown to deliver cholesterol across BBB for the treatment of HD which is characterized by inadequate biosynthesis of brain cholesterol [85]. The use of hybrid nanoparticles in NDD is a promising area of exploration that has the potential to revolutionize the identification and management of diseases in reference to their versatile tuneable properties.

Nanoparticles made of organic materials consist of Liposomes, LNPs, SLNs, Polymeric Micelles and Dendrimers.

1.6   Green extract nanoparticles

There are numerous potential uses for green nanoparticles in medicinal and environmental sciences. Green synthesis specifically tries to reduce the use of harmful chemicals. It is typically safe to employ organic resources like plants as these phytochemicals give the NPs chelation and stability. The conventional use of the green chemistry-based approach is concentrated on using medicinal plants or phytochemicals that have therapeutic properties in their purest form [86]. Green extract nanoparticles have been investigated as a potential treatment for NDD such as AD and PD. Green extracts, such as those from tea, are rich in polyphenols & bioactive compounds which have been shown to have antioxidant and anti-inflammatory properties that can help protect brain cells from damage. Epigallocatechin-3-gallate (EGCG), a major bioactive compound in green tea targets misfolded proteins such as amyloid β-peptide linked with AD [87,88].

Researchers have also found that the condition of patients with NDD was greatly improved by a number of nanoformulations derived from many natural compounds, including curcumin (Cur), quercetin (QC), resveratrol (RSV), piperine (PIP), *Ginkgo biloba,* and *Nigella sativa.* [89,90]. Among these, *Ginkgo biloba* (Gingkoaceae) containing certain glycosides, terpene trilactones, bilobalide as well as ginkgolic acid are used in the treatment of AD and Dementia. Apart from this, the active lipophillic element Thymoquinone (TQ) present in *Nigella sativa* (Ranunculaceae) have been found useful in many immunomodulatory, neurodegenerative & cognitive impairments medicinal activities [91,92].

The antioxidant capabilities of ayurvedic medicinal plant-*Convolvulus pluricaulis* was discovered to improve memory in nanoparticles created using iron oxide as a precursor. Scientists are still unsure of the molecular mechanism underlying these tiny particles. Traditional Chinese medicine makes extensive use of the *Pulicaria undulata* plant as an anti-epileptic, insect repellent, and antioxidant. Higher concentrations of silver nanoparticles (AgNPs) are detected in the brain, and their interactions with proteins prevent the development of fibrils by lowering protein conformation and self-association in silver nanoparticles from *Pulicaria undulata* (AgNPs) [86].

Phyto nanomaterials: Plant extracts when tested with different cell lines have always been proving that they have anti-bacterial, anti-oxidant, anti-inflammatory, anticholinesterase properties [93,94]. The outcomes demonstrated biogenic gold nanoparticles' considerable neuroprotection and biocompatibility. The acetylcholine esterase (AChE) was successfully inhibited, the amyloid β fibrillation process was slowed down, and at low concentrations, mature fibrils were destabilised [95]. One of the pharmacological effects of the spirocyclic alkaloid rhynchophylline (RIN), which is generated by Uncaria species, is neuroprotection. Hyperexcitability of hippocampus neurons were considerably suppressed by RIN [96]. Overall, compared to already marketed treatments for NDD, phyto-nanomedicines show promise as phytochemicals have a variety of qualities, including anti-inflammatory, antioxidant, and anticholinesterase actions, they may one day be employed to treat NDD. To fully understand the degree of these drugs' neuroprotective properties, the mechanisms by which they exert their protective effects, and whether combination therapy could be synergistic as neuroprotectants, more research is necessary.

1.7   Biodegradable nanoparticles

Colloidal particles having a gene of interest enclosed inside a polymeric matrix are known as biodegradable nanoparticles (NPs). These are typically 100 nm in diameter and made of biodegradable, biocompatible polymers that have received FDA approval, such as poly(D,L-lactide-co-glycolide) (PLGA), polylactide (PLA), chitosan & polycaprolactone. Among these PLGA based biodegradable nanoparticles have been exclusively used in the sustained & localized administration of various agents such as plasmid DNA, proteins and peptides and low molecular weight compounds. Due to their high bioavailability, superior encapsulation, controlled release, and less toxic features, biodegradable nanoparticles have been used often as drug delivery systems [97,98, 99]. FDA approved biodegradable nanotherapeutics have also been used in tumor treatments by enabling uniform delivery of drugs to all regions of tumors with sufficient quantities in a controlled manner, reducing the need for frequent dosing. These nanoparticles overcome all the physiological barriers & ensure proper delivery of the drug to the target site because of its small size allowing improved targeting of specific tissues & cells, reducing off-target effects & improving therapeutic efficacy [100].

Drug delivery using biodegradable nanoparticles in NDD: Drug delivery using biodegradable nanoparticles is an emerging field that has shown promise in the treatment of NDD such as AD, PD and HD. The use of biodegradable nanoparticles can help to overcome some of the challenges associated with traditional drug delivery methods, such as poor drug solubility, short half-life, and limited bioavailability. Treatment of NDD are so much restricted due to the limitations produced by the BBB, BCFP and P-glycoproteins. [101, 89]. The use of biodegradable nanoparticles for drug delivery has shown promise in preclinical studies of NDD. For example, PLGA nanoparticles loaded with Neurotrophic Factors (NTFs) have been shown to enhance the survival of neurons and improve motor function in a mouse model of PD. NTFs are given directly into patients' brains through expensive and risky intracranial surgery because some of them cannot cross the BBB [102]. Similarly, chitosan nanoparticles loaded with curcumin, a natural anti-inflammatory compound, have been shown to reduce neuroinflammation and improve cognitive function in a mouse model of AD [103,104].

Therapy using biodegradable nanoparticles: Nanomedicine, or the use of precisely manufactured materials at the nanoparticle size range to create innovative therapeutic and diagnostic modalities, is the use of nanotechnology to medicine. These particles are special in their physicochemical properties which could be used to overcome some of the limitations found in traditional therapeutic and diagnostic agents [105]. There are many potential applications for therapy using biodegradable nanoparticles. One area of active research is the treatment of cancer. Biodegradable nanoparticles can be engineered to selectively target cancer cells and release cytotoxic drugs, which can improve the efficacy of chemotherapy and reduce side effects [106]. The physicochemical characteristics of these particles can be adjusted to help carry substances via the various biological barriers needed to reach different cell subsets. They can be designed to exhibit triggered capability when at a specific position or activated by an external source, as well as to degrade under physiological conditions [107]. These particles can be used to selectively target cancer cells while sparing normal cells which is achieved by exploiting the unique properties of tumor tissues, such as their leaky vasculature, abnormal cell membranes, and enhanced permeability and retention effect. By designing nanoparticles that can exploit these properties, it is possible to deliver higher doses of drugs directly to tumor sites, which can improve efficacy and reduce toxicity. These can be tried for brain tumors and glioma's as well [108].

Bioinformatics driven by nanotechnology for early detection: To understand diseases and choose the best treatments, integrated bioinformatics technologies can collect and analyse enormous amounts of biological and laboratory data [109]. Nanotechnology when merged with bioinformatics it is called 'nano informatics'. The neurodegenerative disorder experiments are more expensive and delicate. Bioinformatics may be utilised for early detection, understanding the disease, and application, whereas nanotechnology can be used to investigate the diagnostic or treatment plan by using medicine delivery that is target specific [110].

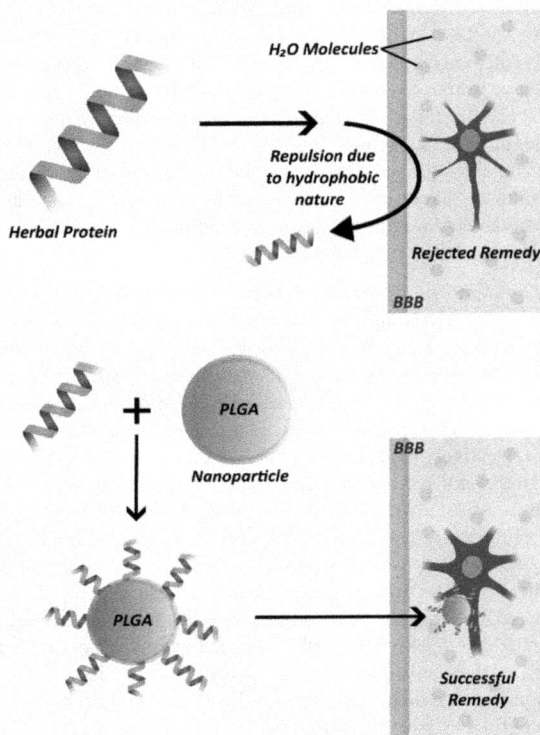

Fig.6- Herbal nanoparticle remedies for Alzheimer's Disease

1.8   Treatment of neurodegenerative diseases with nanoparticles

1.8.1   Treatment of Alzheimer's disease with nanoparticles

Lipid based Nanoparticles: Lipids are composed of one or more layers of phospholipid and thus are very stable structures, this property of them helps in acting as brain drug delivery particle when induced with nanoparticle. They are less toxic and have antigenic properties. Modified structure of liposome perforate easily and can convey the therapeutic substance at the target site. For focussed drug supply in the brain lipid structure modified with lactoferrin loaded curcumin was used which showed highest rate of brain capillary endothelial cell absorption. When tested in rats it showed improvement of with abnormal neuronal damage. Quercetin and rosamarinic acid with conjugated phosphatidic acid and apolipopoprotein loaded liposomes showed neurons recovery from Amyloid beta induced neurotoxicity both in-vitro and in-vivo by causing no harm to the blood brain barrier [111].

Polymeric nanoparticles are composed of Poly(lactic-co-golic acid), PEG, Polyamidoamine, Polyalkyl cyanoacrylate which are biodegradable polymers and biocompatible materials. Curcumin loaded Poly(lactic-co-golic acid) were treated in-vitro as well as in in-vivo conditions. In in-vitro condition, curcumin loaded poly(lactic-co-golic acid) were incorporated and were seen penetrating into neural stem cells and neuro spheres. In-vivo curcumin loaded poly(lactic-co-golic acid) could cross the BBB following intraperitoneal treatment, concentrated in the hippocampus in rats. Free curcumin and curcumin brimmed poly(lactic-co-golic acid) comparison study showed that the Curcumin loaded poly(lactic-co-golic acid) Nanoparticles level in brain increased by 2.8 times. Also, cell proliferation of hippocampus and neuronal cells was enhanced [112,113]. Galantamine, resveratrol, quercetin, etc. like already existing drugs when encapsulated with polymeric nanoparticles showed better results than the normal commercial drug when injected or consumed [114].

Inorganic NPs are ultra-small in size, have very low melting point if the inorganic material is gold, excellent biocompatibility, inert chemical property, can impregnate easily. Gold nanoparticles have been mostly used as drug delivery carriers. Gold nanoparticles were sent to the affected region and a beam of laser targeting the gold nanoparticle was sent. Laser light is used because it only targets the gold nanoparticle it does not affect the other organs of the body. Gold nanoparticles have low melting point thus when it gets hot the clusters near the nanoparticle also get affected by the heat and thus the zone is cleared. Thus, gold nanoparticles have been seen as a possible AD treatment. In in-vitro experiment showed that when nanoclusters were coated with L-glutathione inhibited the growth of Amyloid β nhibition [115]. Gold nanoparticle when conjugated with curcumin showed better inhibition of amyloid β fibrillation and breaking down of amyloid fibrils in varied concentrations when compared to the nanoparticle and curcumin reacting alone. In order to know the ability of a nanoparticle to use it for the treatment of AD it should be able to cross the BBB to reach the affected spot. Selenium nanoparticles, carbon nanoparticles, cerium oxide nanoparticles and Iron oxide nanoparticles are seen to be having the ability to cross the BBB, inhibit amyloid β fibrillation, breakdown four beta fibrils and protect the neurons and synapses [116]. For the treatment of AD, inorganic nanoparticles can serve as multipurpose drug delivery systems and anti-aging agents [111] [Fig.7].

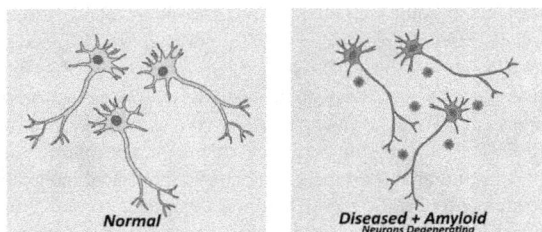

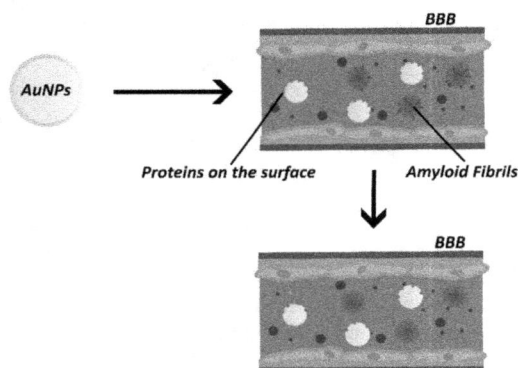

Fig.7- Gold nano particles based therapy for AD

Nanogels are a kind of 3-dimensional polymers at the nanoscale that are thermally and optically controllable and have the ability to enhance drug release [117]. They are made of polymers that have undergone chemical and/or physical crosslinking. Nanogels have superior biocompatibility, sustained release, high drug loading and colloidal stability when compared to other nanoparticles. Moreover, they can be surface-functionalized and successfully traverse the BBB [118]. Beta site amyloid precursor protein cleaving enzyme (BACE1) gene catalyses when the amyloid precursor protein is first broken down, it produces Aβ. BACE1 activity could stop the initial pathogenic occurrences in AD. A study stated that a particular siRNA strand was immobilised on magnetite nanoparticles to silence the BACE1 gene expression in order to track the capacity to track, regulate, and target particular organs, tissues, or cells. Immobilized siRNA nanoconjugates demonstrated mild and acute cytotoxicity in the MTT experiment, and after cellular incubation, the BACE1 gene was silenced, but with a very low yield [119]. Also, Co-immobilized OmpA protein was present on PEGylated nanoparticles. Nanoparticles that have been OmpA/PEGylated show improved endosomal escape [120]. Evidence depicts that immobilisation of siRNA on PEGylated magnetite appears to preserve the biological activity of the molecules which is evidenced by the efficient silencing of the BACE1 gene in HFF-1 cells. Owing to the existence of the extremely hydrophilic PEG molecules on the conjugates surface, this method offered a good means of enhancing stability and wettability while having no discernible effect on cytotoxicity. Because of their small size, siRNA conjugates will probably be endocytosed by brain cells when administered intravenously [119].

Curcumin loaded with PLGA-PEG and B6 peptide: Curcumin can correctly target signs of AD that is Amyloid β and tau. Curcumin might lessen amyloid synthesis, senile plaque development and aggregation, neurofibrillary tangles' development and hyperphosphorylation of tau [121,122,123]. Curcumin loaded PLGA nanoparticles were used because they proved to be more efficient than the curcumin alone and could at least nine times enhance the bioavailability [124]. A study used native curcumin, PLGA-PEG-B6/Cur and PLGA-PEG-B6 nanoparticles. When cytotoxic studies were done the results showed very low toxicity levels and also proved to be biocompatible. When the native curcumin, PLGA-PEG-B6/Cur, PLGA-PEG-B6 were tested *in vivo* on APP/PS1 mice, the mice pre-treated with PLGA-PEG-B6/Cur nanoparticles demonstrated more platform crossings and consumed more time scouring the target quadrant platform in comparison to the APP/PS1 and cur group. These groups swimming rates were the same, excluding out any physical influences on the cognitive exam. The spatial learning ability of APP/PS1 mice, a useful model for studies on AD is impaired. However, the cognitive impairment in Mice APP/PS1 may dramatically be improved by PLGA-PEG-B6/Cur nanoparticles, which could be further useful for AD. Hippocampal APP load was decreased in APP/PS1 mice by PLGA-PEG-B6/Cur nanoparticles and PLGA-PEG-B6 nanoparticles. In the hippocampus region APP/PS1 mice induced mass amyloid accumulation. As a result, when compared to the APP/PS1 control group, Curcumin and PLGA-PEG-/Cur nanoparticles slightly decreased the development of Amyloid plaques. Using PLGA-PEG-B6/Cur nanoparticles, the pathogenesis was significantly slowed down. Bielschowsky silver staining and immunostaining results showed that PLGA-PEG-B6/Cur nanoparticles may have reduced AD pathology in APP/PS1 mice, leading to further examination of A-related proteins and p-tau protein levels in animal brains. APP/PS1 mice had considerably higher levels of tau-phosphorylation and Aβ levels than WT mice. The fascinating idea that APP processing and secretase cleavage in APP/PS1 mice may be impacted by PLGA-PEG-B6/Cur nanoparticles is raised by the observation of the inhibition of BACE1, APP, and PS1. HT22 cells were used for in-vitro research to test the biosafety and absorption capacity of the PLGA-PEG-B6/Cur nanoparticles. The cell viability analysis of the relatively high concentration of PLGA-PEG-B6/Cur and PLGA-PEG-B6 nanoparticles revealed no discernible difference in cytotoxicity. PLGA-PEG-B6/Cur nanoparticles significantly improved the cellular absorption in comparison to native Cur, in a dose-dependent way. As a result, it was decided that PLGA-PEG-B6/Cur nanoparticles would make a relatively safe and promising treatment strategy [124].

1.8.2   Treatment of Parkinson's disease with nanoparticles

Encapsulating medications within nanoparticles and delivering them directly to the brain is one method of employing them to treat PD. By doing so, the BBB may be crossed, which may increase the effectiveness of medications for treating NDD. During *in vivo* testing, acute treatment was given to rats with the produced nanoparticles, and microdialysate samples were obtained to assess striatal dopamine levels. A surge in striatal dopamine levels and a robust ability to transport across the BBB were also observed. The Chitosan NPs' release profile was likewise shown to be fast and pulsatile. Dopamine-loaded cellulose acetate phthalate (CAP) nanoparticles were used in a related study. The findings showed that dopamine release peaked 3 days after injection and thereafter continued [125].

Several studies on potential treatments for PD have centred on re-establishing normal dopamine levels as a means of protecting the brain. Dopamine-encapsulating polymeric chitosan nanoparticles were created, and their ability to load and release the chemical was evaluated in vitro using the Madin Darby canine kidney II-multidrug resistance gene-1 (MDCKII-MDR1) cell line, a prototypical of the BBB. Also, it was proposed that the dopamine release profile formed from chitosan nanoparticles was advantageous to that obtained from CAP nanocarrier due to its rapid and pulsatile nature and lesser risk of possible neurotoxicity [126]. Chitosan's potential as a dopamine nanocarrier has also been demonstrated by its amazing capacity to relax the close bonds between endothelial cells in the BBB and improve the absorption of hydrophilic drugs [127]. Due to their lipophilic, bio acceptable, and biodegradable attributes, which enable a less risky therapeutic strategy, LNPs have shown to be the most prominent nanoparticle regarding advancement into clinical trials. After being used as a means of mRNA administration in the COVID-19 vaccine, these LNPs have gained popularity and demonstrated promise as therapeutic delivery systems [128]. Depending on their environment, LNPs can express a negative charge at physiological pH with little toxicity or a positive charge at low pH that permits nucleic acid complexation. These LNPs have better endosomal escape, circulation half-life, and cellular absorption [129]. The use of lipid-based NPs for immunisation and treatment has been highlighted. Its potential for use in treating neurological illnesses is increased by its inherent capacity to permeate the brain. These LNPs have evolved into effective nanocarriers as a result of their ideal size, surface charge, customizable surface area, and shape [130].

To increase therapeutic indices, various parameters must be considered while building a LNP for drug delivery. Some of the factors are:
1. Desorption of the adsorbed drug
2. Combination of erosion/diffusion processes
3. Drug solubility
4. Drug diffusion over nanoparticle matrix
5. Nanoparticle matrix degradation

Another study using LNPs coupled with the dopamine agonist ropinirole (RP) by [131], highlights the effectiveness of nanomedicine and pharmacological therapy. In terms of the drug's pharmacokinetics in the host, there was a three-fold increase with topical treatment, a one-fold increase with topical bioavailability in SLN and NLC complexes, and a more than two-fold increase with oral administration. Higher levels of glutathione, catalase, and dopamine are associated with decreased levels of lipid peroxidation, according to pharmacodynamic investigations [131].

In recent years, there has been a lot of interest in the ability of various nanoparticles to prevent the formation of α-synuclein amyloid. Only a few of the nanoparticles that have been researched for this purpose include gold nanoparticles, super paramagnetic iron oxide nanoparticles, QDs graphene, and graphene derivatives [132]. The strongest linkages between the CNTs and α-synuclein were seen in the phosphorus-doped CNTs. Doped-CNTs, in specific phosphorus-doped carbon nanotubes, are capable of successfully suppressing the synthesis of α -synuclein amyloid and may hold potential as a treatment for PD. Nonetheless, more *in vitro*, *in vivo* and clinical studies are necessary as direct contact with a misfolded α -synuclein protein can exacerbate the condition and cause the misfolding of more α –synucleins [133].

As free radical production, mitochondrial dysfunction, and oxidative stress are all associated with PD, antioxidants' capacity to safeguard mitochondria and lessen oxidative stress is constrained. In MPTP-mouse models, cerium oxide nanoparticles (CeONPs) are a prospective therapeutic for treating PD because they can retain striatal dopamine and stop the degeneration of dopaminergic neurons. In a mammalian (C57BL/6) model, PD can be effectively treated with low dosages (0.5– 5 l) of CeONPs. These findings support the idea that CeO2NP, a nanomedicine, can slow the progression of PD [134,135].

For the treatment of PD, numerous herbal remedies have lately been discovered. Due to its capacity to lessen tyrosine hydroxylase (TH)+ DA neuronal mortality, puerarin has been observed to shield against the death of DA neurons in animal models of 6-hydroxydopamine (6-OHDA)-mediated neuronal toxicity [136]. This protective effect could be attributed to activation of the protein kinase B/phosphatidylinositol 3-kinase signalling pathway, which stopped the build-up of p53 in nucleus and the subsequent caspase-3-mediated apoptotic death [137]. Although puerarin has the potential to be an effective treatment for PD, its unfavourable water solubility and low bioavailability have limited its application [138]. Researchers created nanoparticles that might be used to encapsulate puerarin in order to increase its bioavailability, *in vivo* half-life, and brain accumulation for the treatment of PD symptoms. Six-armed star-shaped poly(lactide-co-glycolide) (6-s-PLGA) NPs with 89.52 1.74% encapsulation efficiency, 42.97 1.58% drug loading, and a 48-h sustained drug release were made into spherical nanoparticles. Hydrophobic interactions played a significant role in nanoparticle synthesis and drug loading, while variations in the outdoor environment caused these nanoparticles to become more hydrophilic, resulting in drug release. In contrast to rats given free puerarin, oral puerarin - nanoparticles treatment to rats showed considerable changes in puerarin accumulation within the plasma and brain. It was discovered that puerarin - nanoparticles administration corrected disease-associated behavioural impairments and dopamine and its metabolite depletion in mice with MPTP-mediated neurotoxicity. These results suggested that puerarin - nanoparticles may be an effective method for increasing puerarin oral absorption and promoting its distribution to the brain, where it can help treat Parkinson's disease [138]. Several investigations have found the similar pattern of improved performance with fewer side effects for other phytochemicals as well, along with considerable reductions in parkinsonian symptoms.

1.8.3   Treatment of Lewy body dementia with nanoparticles

Amyloid accumulation in the brain has been linked to DLB. These findings, in accordance with Edison and colleagues, suggest that anti-amyloid medication may be beneficial for the majority of DLB patients [139]. A significant number of these diseases' treatments rely on Amyloid oligomers and fibrils being inhibited, delayed, or dissociated. Nanoparticles depicted significant effects on the Amyloid fibrillation process, although their effects on deteriorating memory and cognitive function have not been well studied *in vivo*. Although the amyloid β-fibrillation process has been demonstrated to be significantly impacted by nanoparticles, *in vivo* studies on the effects of nanoparticles on memory and cognitive decline have been scanty. The protein -lactalbumin was chosen as a suitable sample to assess the impact of AuNPs on the development of amyloid. It was discovered that increased protein adsorption to the surface of the nanoparticles, which prevents their structural alterations, causes AuNPs to limit the production of amyloid fibrils. This can be utilised as a powerful therapeutic drug to treat and prevent the development of amyloid illness [140]

For all Lewy body illnesses, levodopa-the gold-standard medication for treating PD can be utilised as the main symptomatic treatment for parkinsonian symptoms [141] which improves patient quality of life significantly but only temporarily. During treatment, L-DOPA exhibits tolerance and the emergence of induced dyskinesias [142]. In an experiment on rat in order to deal with PD using levodopa coated with nanoparticles resulting in reductions in oxidative stress brought on by PD progression, striatal dysfunction brought on by dopamine deprivation, and mobility difficulties without acute damage to the major organs. This efficient administration of homogenous, biocompatible L-DOPA-loaded nanoparticles into the brain's lymphatic system has significant therapeutic application for the treatment of PD [143].

Kim and colleagues looked into a novel strategy to target -synuclein fibrils utilising graphene quantum dots, a carbon-based nanomaterial (GQDs) [144]. Researchers demonstrate that GQDs can prevent -synuclein from fibrilizing and cause constructed fibrils to disaggregate. They offer convincing insight into the process by which GQDs engage with and separate these fibrils using structural biology techniques and molecular dynamics simulations. Synuclein preformed fibrils (PFFs) cause toxicity in primary neurons that includes deficiencies in mitochondrial respiration, synuclein aggregation, and cell-to-cell transmission of synuclein aggregates. Administration of GQDs in primary neurons greatly reversed these toxic effects [144]. The researchers demonstrated the GQDs' extensive brain penetration and proposed that they dispense in the brain *via* endocytosis and release; however, additional research is required to fully comprehend GQD metabolism as well as its absorption, distribution, in the brain and in other parts of the body. However, research in this area is ongoing, and it is possible that in the future, nanoparticles may be developed as a potential treatment for Lewy body dementia. As always, any potential treatment would need to undergo extensive testing in preclinical and clinical trials before it could be approved for use in patients.

1.8.4   Treatment of frontotemporal dementia with nanoparticles

FTD, which affects behaviour, psychology, communication, and motor skills, is accompanied by changes in the frontal as well as temporal lobes. One approach that has been studied in preclinical studies involves using nanoparticles to deliver drugs that selectively target proteins implicated with FTD. For instance, gold nanoparticles were used to deliver an antibody that targets the tau protein, which is involved in several NDD, where the antibody delivered by nanoparticles reduced tau pathology in a mouse model of FTD.Regarding ALS, it has been shown that curcumin interacts with reduced wt SOD1's aggregation-prone sites to prohibit it from aggregating in vitro. SOD1 aggregates that were curcumin-bound were less cytotoxic, unstructured, and smaller [145]. Positive results have been obtained for tau, a protein associated to FTD, where beta-sheet formation and aggregation were inhibited in the presence of curcumin [146]. In contrast, there was no indication of curcumin binding to tau clumps in post-mortem brain tissue sections from FTD patients [147]. Innovative approaches to get over these limitations and successfully provide drugs to the CNS include a nanotechnology-based drug delivery platform, which offers potential therapeutic ways for treating a number of prevalent neurological disorders, including FTD [138].

1.8.5 Treatment of vascular dementia with manomaterials

Resveratrol solid lipid nanoparticle: Resveratrol belongs to a phytoalexin superfamily and is a phenolic compound that occurs naturally. It has an array of pharmacological features including anti-tumor, anti-aging, antioxidant, anti-apoptotic, anti-inflammatory, cardiac, and neuroprotective properties. Resveratrol has been seen defensive for various NDD like stroke, PD, AD, Ischemia, HD etc. Resveratrol is used as it has been studied and known that it has the ability to inhibit the action of HIF-1 α. Resveratrol and other lipophilic substances that preferentially partition into lipidic nanoparticles are good candidates for inclusion. The submicron sized molecules known as solid lipid nanoparticles replace the liquid lipid with solid lipid. Solid lipid nanoparticles have enormous potential in pharmaceutical and neutraceuticals because they have special qualities such as compact size, big surface area and high drug loading capacity. By lowering the protein carbonyl levels, Resveratrol therapy has been shown to ameliorate 6-hydroxydopamine-induced oxidative damage [148]. Resveratrol has also been demonstrated to increase SOD activity, reducing oxidative stress in aged mice's brain regions brought on by D-galactose [149]. The outcomes show that Resveratrol -SLNs (R-SLNs) mitigate oxidative stress in VD rats to have positive benefits. In summary, the work shows that R-SLNs activate the Nrf2/HO1 pathway to reduce the oxidative stress and cognitive deficits caused by VD. The use of SLNs improved the effectiveness of Resveratrol. R-SLNs are therefore obviously a possible treatment approach for treating BCCAO-induced VD [150]. For now, this is the only study done for the treatment of VD using nanomaterials. Further research and study are going on and there are more types of nanomaterials that can be used for the treatment and can also be studied further as there are no much information on them.

1.8.6   Treatment of Prion disease with nanomaterials

Nano-emulsions (NE) are emulsions created at nanoscales to enhance drug distribution to the target area and lessen adverse reactions and dangerous reactions [151]. Also called as miniemulsions as its either oil in water type or water in oil type. There is a slight difference that differentiates miniemulsions and microemulsions. Microemulsions are generated through a spontaneous thermodynamic self-assembly process and are thermodynamically stable. Nanoemulsions, however, are kinetically high or low external shear forces are applied to create sturdy structures [152]. NE is administered intravenously and has proven to be effective, in part due to the small droplet size (less than 1 m). In addition, due to the advantages it has in terms of formulations for intranasal drug delivery systems using mucoadhesive substances have attracted the interest of various researchers due to their potential to speed up drug transport from the nose to the brain. The advantages of this specific nose-to-brain nanoemulsion delivery connection have been shown in a number of academic research. In TgMHu2ME199′K mice, pomegranate seed oil in NE form has been proven to delay clinical development, neurodegenerative pathological traits, and prions [153]. The E200K Prions protein mutation was associated with this mouse model. It was located showed the mouse model used to fight ND had a delayed start to action due to the injection of Pomegranate seed oil. As comparison to lesser doses of natural PSO, nano-PSO slowed the course of the disease more quickly. There were no side effects recorded, and investigation of brain samples revealed that Pomegranate seed oil formulation had no impact on the deposition of PK resistant Prions protein. The treated mice's brains displayed a strong neuroprotective effect [154]

Nanoimaging: Early stages of protein aggregation and misfolding occur too quickly and involve too few species to be experimentally probed using the majority of currently available methods. To explore the mechanisms of misfolding and aggregation, scientists must find alternative techniques due to the shortcomings of the current approaches. A study showed that the short peptide CGNNQQNY can be immobilised for force spectroscopy experiments by including a cysteine, which is a potent force spectroscopy method for analysing interactions between individual molecules. This peptide, which is a component of the Sup35 yeast prion protein, was discovered to be amyloidogenic, causing the protein as a whole to aggregate [155]. When the dimer formation occurs, it serves as a triggering step for the whole self-assembly process. The process of protein misfolding and aggregation will be better understood with increased use of single-molecule techniques. We believe that single-molecule approaches will improve our comprehension of processes including misfolding, aggregation, and the makeup of hazardous species. Such information is essential for managing these processes and creating potent therapeutics that can fend off linked neurodegenerative illnesses [156].

Carbon nanoparticles (CNPs): In a study investigated the impending effects of CNPs on the accretion of Prion proteins by combining a Thioflavin T fluorescence test with MD simulations [157]. The results validate that CNPs, such as graphene and carbon nanotubes, can inhibit the fibril formation. MD simulations of the PrP127-147 tetramer in nthe presence or absence of CNPs establish the chemical mechanism of the CNPs preventing prion aggregation, which can be related to the weaker interpeptide contacts because of the strong CNPs -peptide interfaces. The N-terminus of PrP127-147, peptides and CNPs interact with one another where. Graphene's inhibitory influence is more substantial than that of carbon nanotubes as it has a larger contact surface area. According to the interaction energy study, the interfaces amongst peptide residues and CNPs are mostly facilitated by aromatic amino acids mostly due to PrP127-147 peptide's aromatic ring stacking interaction with the CNPs' hexatomic ring. Overall, the results showed how CNPs affect Prion protein aggregation and the accompanying chemical process. This knowledge will be helpful for future applications of graphene and single-walled carbon nanotube in biological medicine and will help us better understand the interactions between nanoparticles and proteins connected to amyloid.

Dendrimers for Prion disease treatment: Dendrimers are synthetic, repeatedly branched polymers with numerous biological uses. They typically have a central monomer from which symmetrically organised monomer branches out, producing a spherical three-dimensional shape. The generation of dendrimers is controlled by the number of branch points, while the ionic charge is mostly determined by the surface functional groups. The anti-prion function of dendrimers was accidentally discovered, when the poly(amidoamine) (PAMAM) dendrimer in a transfection reagent was observed to eliminate prions from ScN2a cells. Dendrimers have a huge therapeutic promise because several different dendrimer types target amyloid proteins and maltotriose-modified PPI dendrimers have recently been shown to be able to cross the BBB [158]. Dendrimer can prevent Prion protein cellular form from being transformed to Prion protein scrapie form inside of cells. [159] stated that the research should contribute to identifying additional medications that could increase this impact by demonstrating that mPPIg5 prevents Prion protein cellular from being converted to Prion protein scrapie form. Future studies in animal models of prion disease is required to examine the potential of dendrimers in the treatment of prion and other NDD [159].

1.8.7   Treatment of Huntington disease with nanomaterials

Anti-HTT or ASO treatment: Since the discovery of genetic mutation researchers have started making use of it by creating HD model animals to study the gene and drug therapies in order to prevent or mitigate the disease [160]. Administering the HD mouse and the infrahuman models with antisense oligonucleotides (ASO) and anti-HTT small interfering RNA (siRNA) has shown a successful decrease of the disease progression. Based on these pre-clinical studies a clinical trial was set aiming at evaluating the safety, properties and tolerability of anti-HTT ASO (Ionis-HTTRx) in early stage of HD patients. Although the clinical trials were successful there is a need to inject the ASO 6 times per year through intrathecal route. The lumbar puncture deliver is fine for once in a while delivery but here the treatment is chronic and as a adverse side effect it might lead to local infection, radiculopathy, haemorrhage, post-LP headaches and arachnoiditis [161]. This obstacle enabled researchers to design nanoparticles NP as carrier for transporting and delivering the siRNA or ASO through intranasal route, in particularly targeting the cerebral cortex and the striatum in the brain. Using Mangafodir as a cross linking agent Chitosan nanoparticle was used to deliver the siRNA to the brain. Delivering the naked siRNA was prone to damage and the efficacy was low but when it was loaded in chitosan NP and delivered the efficacy was much better and the siNA was protected from damage [162]. The electrostatic interactions between positively charged moieties of amino groups and negatively charged phosphate moieties of the siRNA will exhibit electrostatic interactions which enables chitosan polymerization to form a NP (Yoshida et al., 1966). At the same time there is a weak interaction between the siRNA and chitosan which will facilitate easier delivery to the brain tissues [163]. Chitosan, being biodegradable, is digested by lysozymes produced by animals making it non-toxic (in mammals, with LD50 of 16 g/kg in rats [164,165]. Sedimentation, agglomeration or crystal growth are one such stability issues face while synthesising chitosan NP. By increasing viscosity and decreasing particle size the sedimentation issue was overcome by the researchers. In this particular study researchers recorded 50%decrease in the mutant htt protein whole compared to intrathecal space (CSF)administration which reported 42% decrease in the same mutant htt protein [166] [Fig: 8]

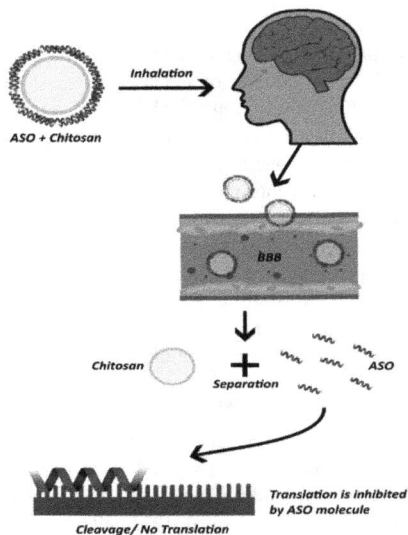

Fig.8- Chitosan nano particles based therapy

Selenium as nanoparticles: Studies have reported at Selenium (Se) homeostasis is disturbed in HD individuals [167] In strain HA759 C. elegans produced Htt – Q150) were excessively expressed inASH neuron leading to ASH degeneration and death. Treating HA759 C. elegans with Nano- se improved the survival rate up to 44% and after 3 days of consecutive treatment the expression of nlp- 3 and qui – 1 was upregulated. This indicated the recovery of physiological functions of ASH neurons which was analysed through response to external stimulus. The aggregation of insoluble HTT protein will lead to disruption of normal function the neurons. The Nano- se treatment of 0.02, 0.2 and 2μM concentrations decreased the aggregation by 18%, 22% and 30% in comparison with the control [168]. Not only the aggregates the reactive oxygen species which also lead to neurodegeneration decreased significantly. Down regulation of histone deacetylase (HDAC) such as analogue of HDA1 i.e. HDAC1, analogue ofHDA4 i.e. HDAC4 and SIRT1 results in inhibition of pathology in HD models [169, 168, 170].

Cholesterol nanoparticles: In HD individuals the cholesterol homeostasis is impaired in the pre and early stages of the disease [171,172]. The dysregulation of cholesterol is seen in the astrocytes which is probably due to mutant HTT acting on sterol regulatory element – binding proteins (SREBPs) and its target genes [173]. This leads to minimised transcription resulting in less cholesterol production for the uptake of the neuron. Consecutive delivery of g7-NPs-Chol (cholesterol NPs modified with glycopeptides which can pass through BBB much efficiently than unmodified one) has shown to rescue the synaptic communication, protected from cognitive decline and partially improved global activity in HD model mice. In this study scientists have used Polylactide-Co-Glycolide (PLGA), a polymeric NP that is biocompatible and biodegradable since it completely degrades into monomers such as glycolic acid and lactic acid which can be easily utilized by Krebs cycle. PLGA is also approved byFDA as a drug delivery system through parenteral administration [174,175] G7-NPs-Chol being an effective drug delivery system lacks immunological studies behind it. Furthermore, biochemical studies are necessary to be carried out to analyse the impact of g7-NPsdegradation and the molecules released by it to proceed with preclinical trials and translational developments of these NPs. Another limitation that g7-NPs posses in the current study is that drug loading is only 1% which does not allow the delivery of elevated amounts of cholesterol [166].

1.8.8   Treatment of amyotrophic lateral sclerosis with nanomaterials

Iron Oxide NPs for diagnosing (IONPs): The drugs available in the market do not cross the BBB (blood brain barrier) in order to achieve that NPs are an effective strategy. The current strategy focuses to integrate imaging, targeting and therapy in one system. IONPs) were used as drug carriers as well as MRI contrast agents (MRI CAs). One of the advantages of nanotechnology is that it increases the sensitivity of biomarker for detection [176]. MRI CAs such as Magnetite (Fe_3O_3) and Maghemite (γ- Fe_2O_3) are a pure form of iron oxides can be used as magnetic nanomaterials since their biocompatibility is good [177]. Iron oxides are considered safe, biologically tolerated and nontoxic which can be incorporated into natural process of metabolism in human body when injected [177]. The biomedical applications include Hyperthermia, protein separation and cell tracking. Along with the advantages, the disadvantages such as elimination by the spleen liver and kidney is seen. Accumulation in the tissues of liver and spleen are also studied [178]. To delay with these toxic effects iron homeostasis is maintained by combining the excess ions with ferritin. This will balance out the normal amount of iron required or present in the body. One of the studies has revealed that along with accumulation protein aggregation and oxidative stress are also one such toxic effect. The size, concentration and surface charges along with the what kind of coating and functional groups characterise the IONPs toxicity levels [179]. Researchers were able to trace and visualise the implanted stem cell MRI scan in the CNS of rodents with the help of IONPs. And administration of Super paramagnetic iron oxide nanoparticles (SPION) helped in tracing Amniotic fluid cells in a neurodegenerative environment in a Wobbler mouse (motor neuron degeneration model). These studies indicate that SPION can used as a tracer to monitor stem cells therapy in ALS [180, 181]

Neuroinflammation is one of the symptoms of ALS. Researchers visualised microglialactivation through Positron Emission Tomography (PET) in the brain of SOD1 G93A mice and living ALS indiviuals [182,183, 184]. Microglia gets over activated and releases pro-inflammatory cytokines. These cytokines accelerate neurone degeneration and its death. Hence microgliaoveractivation and its mediated inflammation has been the drug target. This inflammation also leads to loss of synapses, synaptic dysfunction and neuron damage. To deal with this reserchers have started working with anti-inflamatory drugs coated with NPs based on thei size, structure, polarity, function and target specificity. Reports also say that CeNC/IONC/MSN-T807 [185], chitosan-coated nanoemulsions [186], serum albumin [187], mSPAM [188] have showed anti-inflammatory properties in order to deal with the neuroinflammation. Oxidative stress is associated with ALS in human as well as SPD1 G93A mouse.Researchers have used Cerium oxide nanoparticles (CeNPs) to neutralise the reactivenitrogen and oxygen species that causes oxidative stress. When CeNPs were tested inSOD1 G93A mice it reduced the disease severity and increased the survival rate [189].

Riluzole coated nps: Riluzole which is been used as drug to treat ALS has been coated with Solid Lipid Nanoparticle (SLP). Due to their lipophilic character, they can cross the BBB easily through endocytosis and target the brain directly. Compared to the raw from of Riluzole the NP coated Riluzole were able to transport more medicine [189].In other studies, Riluzole was coated with Compritol® 888 ATO lipid matrix and with the help of Phophatidylcholine and Taurocholate Sodium Salt as surfactant and cosurfactant respectively. After several physiological settings of the NPs loaded with the drug system was characterised based on particle dimensions, z-potential and drug release profile. Once characterized it was injected into rats through intra-peritoneal route to study the preferred accumulation of the drug into the brain over the dispersion of water. It was observed and reported that the accumulation of the drug in the organs like spleen, kidneys, lung and heart were comparatively low and it was also easily transported to CNS [190]. Exosomes, being a natural NPs has an advantage of non-immunogenic property and a wide range of payload capacity. Mesenchymal Stem Cells of Human (MSCs) were used to create the exosomes which then were treated with anti-inflammatory, neuroprotective RNA and protein components loaded Interferon-gamma (IFN). The prepared complex was administered intravenously to an autoimmuneencephalomyelitis (EAE). This study demonstrated strong penetration of the drug through BBB as well as regaining motor ability increased encouragingly [191].

MiRNA deregulation: Deregulation of miRNA is one of the critical conditions that marks the onset of considerable number of neurodegenerative diseases. Considering the greater impact of miRNAs on homeostasis and their aberrant expression, it has become a drug target to the researchers [192,193,194]. Thereby researchers are targeting on these aberrant miRNAs to design new therapies[195].Compared to other NPs in the market graphene a 2D structures hold an advantage of having a better ability to penetrate the cell membranes and a larger surface-to-volume ratio for nucleic acid and limited cytotoxicity. Due to their pi-pi bonding the study says it will be able to trap nucleic acids easily [196, 197]. Not only that but also the Sp2 and Sp3 hybridization provides greater heat tolerance, greater mechanical strength along with optical and mechanical properties [198]. Hence [199] says that graphene derivatives such as graphene oxide or reduced graphene oxide should be considered and explored.

Curcumin is challenging to employ in vivo due to its low absorption, high metabolism, and quick excretion; however, these drawbacks can be circumvented using a variety of strategies. To repair mitochondrial damage in a brain cell line with TDP-43 mutations, analogue dimethoxy curcumin increased the transmembrane potential, increased electron transfer chain complex I activity, and up-regulated UCP2. Due to increased heme oxygenase-1 production, monocarbonyl dimethoxycurcumin C, an improved curcumin analogue, prevented aggregation and decreased oxidative stress [197]. Nanoparticle administration is another strategy to increase the bioavailability of curcumin. Curcumin-loaded inulin-d-alfa-tocopherol succinate micelles successfully transported curcumin into mesenchymal stromal cells, showing the possibility for ALS treatment [200].

1.9   Challenges and future prospects

Nanoparticle based treatments are still in progress and not many studies have been done. Nanoparticles have been proved excellent biocompatibility, stability and their use as drug delivery system is being used and studied more. The use of nanoparticles for Alzheimer's is still on level 1 and there are plenty of areas to be studied and analysed for further treatment purpose. The answer on how to remove nanoparticles from the Central Nervous system is still a question.

Senolytics, enhanced autophagy, anti-inflammation, and stem cell therapy studies using AD models have demonstrated its huge promise as DMT, including reducing both pathology and symptom in AD. Although they are currently in the research and development phases, they might soon become useful clinical treatments. Enhancing the biocompatibility involves managing the medication release in the disease-affected area and keeping an eye on the action of the pharmaceuticals and nanocarriers. The results of this study provide an essential basis for developing siRNA delivery systems that can be applied to AD treatments. In order to corroborate and completely appreciate the mechanisms of internalisation and trafficking, additional study will be conducted on the delivery of the nanoconjugates by exposing primary neuron and astrocyte co-cultures. The investigations would also be beneficial for a more precise calculation of endosomal escape rates and the numerous parameters involved in achieving superior results.

Further studies and research conducted should focus on:
- What composition of drug conjugate can be taken?
- How much of the nanoparticle encapsulated with the drug can be supported by the body?
- Ways of excreting the nanomaterial out of the body
- Enhancing the ability of nanomaterial to cross the BBB and reach the affected site of the brain.

Once these factors are figured out, we can be sure of treating any NDD with the help of nanomaterials [71].

1.10   Ongoing work in our lab

Our lab focusses on green synthesis of *Bacopa monnieri* silver bionanoparticles and were characterized using standard techniques. Further mitigation studies in *SH-SY5Y* cell lines against acrylamide induced neurotoxicity was carried out MTT assay results revealed a dose-dependent protection against cell death. It was observed that the production of mitochondrial reactive oxygen species and protecting the cells from Acrylamide induced reduction in GSH levels (Sanjana et al., unpublished data).

Acknowledgements

Authors thank the management of Kristu Jayanti College, autonomous for their unwavering support.

Conflict of interest

There is no conflict of interest to declare.

References

[1] K.A. Jellinger, Basic mechanisms of neurodegeneration: a critical update, J. Cell Mol. Med. 14(2010) 457-87. https://doi.org/10.1111/j.1582-4934.2010.01010

[2] Information on https://medlineplus.gov/degenerativenervediseases.html.

[3] A.S. Schachter, K.L. Davis, Alzheimer's disease, Dialogues, Clin. Neurosci.2 (2000)91-100. https://doi.org/10.31887/DCNS.2000.2.2/asschachter.

[4] M.W. Bondi, E.C. Edmonds, D.P. Salmon, Alzheimer's Disease: Past, Present, and Future' J. Int. Neuropsychol. Soc. 23 (2017) 818-831. https://doi.org/10.1017/S135561771700100X

[5] Z. Breijyeh, R. Karaman, Comprehensive Review on Alzheimer's Disease: Causes and Treatment, Molecules. 25(2020)5789. https://doi.org/10.3390/molecules25245789

[6] Information on Alzheimer's Association .https://www.alz.org/alzheimers-dementia/what-is-alzheimers.

[7] A. Burns, S. Iliffe' Alzheimer's Disease, B.M.J. 338(2009)467-477. b158. https://doi.org/10.1136/bmj.b158

[8] M.V.F Silva, C.d.M.G. Loures, L.C.V Alves, L.C de Souza, K.B.G. Borges, M.D.G. Carvalho, Alzheimer's disease: risk factors and potentially protective measures, J. Biomed. Sci. **26**, 33 (2019). https://doi.org/10.1186/s12929-019-0524-y

[9] J. Jankovic, E.K. Tan, Parkinson's disease: etiopathogenesis and treatment, J. Neurol. Neurosur. & Psy. 91(2020)795-808. http://dx.doi.org/10.1136/jnnp-2019-322338

[10]   Information on https://www.medicalnewstoday.com/articles/323396.Parkinson's disease: Early signs, causes, and risk factors.

[11]   Information on https://www.news-medical.net/condition/Parkinsons-Disease

[12]   T.F. Outeiro, D.J. Koss, D. Erskine, L. Walker, M. Kurzawa-Akanbi, D. Burn, P Donaghy, C. Morris, J. Taylor, A. Thomas, J. Attems, I McKeith , Dementia with Lewy bodies: an update and outlook., Mol. Neurodeg. **14**(2019). https://doi.org/10.1186/s13024-019-0306-8

[13]   S.D. Capouch, M. R. Farlow, J .R. Brosch, A Review of Dementia with Lewy Bodies' Impact, Diagnostic Criteria and Treatment, Neurol Ther. 7(2018) 249–263. https://doi.org/10.1007/s40120-018-0104-1

[14]   Information on https://www.nia.nih.gov/health/what-lewy-body-dementia-causes-symptoms-and-treatments. What Is Lewy Body Dementia? Causes, Symptoms, and Treatments.

[15]   Information on Lewy body dementia: Information for patients, families, and professionals. https://www.nia.nih.gov/health/lewy-body-dementia

[16]   I.G. McKeith, B.F. Boeve, D.W. Dickson, G Halliday, J.P. Taylor, D. Weintraub, K. Kosaka, Diagnosis and management of dementia with Lewy bodies: fourth consensus report of the DLB Consortium. Neurology, 89(2017), 88-100. https://doi.org/10.1212/WNL.0000000000004058

[17]   Information on https://www.theaftd.org/what-is-ftd/disease-overview/Disease Overview.

[18]   Information on Alzheimer's Association: https://www.alz.org/alzheimers-dementia/what-is-dementia/types-of-dementia/frontotemporal-dementia

[19]   Information on National Institute on Aging: https://www.nia.nih.gov/health/frontotemporal-disorders-information-people-early-stage

[20]   T. O. John, A. Thomas, Vascular dementia. Non-Alzheimer's dementia, Lancet, 386(2015)1698-1706.https://doi.org/10.1016/S0140-6736(15)00463-8

[21]   V.D. Flier, W. Skoog, I. Schneider, J.A. Schneider,L. Pantoni L,V. Mok, C.L.H. Chen, P. Scheltens, Vascular cognitive impairment. Nat. Rev. Dis. Pri.15(2018)18003. https://doi.org/10.1038/nrdp.2018.3

[22]   E.R. McGrath, S. Alexa, Beiser, A. Donnell, J.H. Jayandra, P.P. Matthew P. Claudia, L. Satizabal, S. Sudha, Determining vascular risk factors for dementia and demntia risk prediction across mid to later life, Neurology,99 (2022). https://doi.org/https://doi.org/10.1212/WNL.0000000000200521

[23]   Information in https://www.ninds.nih.gov/Disorders/Patient-Caregiver-Education/Fact-Sheets/Spinocerebellar-Ataxia-Fact-Sheet.

[24]   R. Sullivan, W.Y. Yau, E. O'Connor, H. Houlden, Spinocerebellar ataxia: an update, J. Neurol. 266(2019)533-544. https://doi.org/10.1007/s00415-018-9076-4

[25]   J. Collinge, J, Prion diseases of humans and animals: their causes and molecular basis. Ann. Rev. Neurosci, 24(2001)519-550.DOI: 10.1146/annurev.neuro.24.1.519 https://doi.org/10.1146/annurev.neuro.24.1.519

[26]   L. Westergard, H.M. Christensen, D.A. Harris, The cellular prion protein (PrP(C)): its physiological function and role in disease, Biochim Biophys Acta. 1772(2007:629-44. https://doi.org/10.1016/j.bbadis.2007.02.011

[27]   M. Imran, S. Mahmood, An overview of human prion diseases. Virol. J. **8**, 559 (2011). https://doi.org/10.1186/1743-422X-8-559

[28]   G.K. Shinomol, M.M.S. Bharath, Muralidhara, Neuromodulatory Propensity of *Bacopa monnieri* Leaf Extract Against 3-Nitropropionic Acid-Induced Oxidative Stress: *In Vitro* and *In Vivo*, Neurotox Res.22(2012)102-14. https://doi.org/10.1007/s12640-011-9303-6

[29]   G.K. Shinomol, H. Ravikumar, Muralidhara, Prophylaxis with *Centella asiatica* confers protection to prepubertal mice against 3-nitropropionic-acid-induced oxidative stress in brain, Phytother Res.24(2010)885-92. https://doi.org/10.1002/ptr.3042

[30]   R.A. Roos, Huntington's disease: a clinical review, Orphanet. J. Rare. Dis. 5(2010). https://doi.org/10.1186/1750-1172-5-40

[31]   P. Masrori, P. Van Damme, Amyotrophic lateral sclerosis: a clinical review, Eur. J. neurol. 27(2020)1918-1929. https://doi.org/10.1111/ene.14393

[32]   C.M. Bellettato, M. Scarpa, Possible strategies to cross the blood–brain barrier. Ital. J. Pediatr. 44 (2018). https://doi.org/10.1186/s13052-018-0563-0

[33]   B. Obermeier, A. Verma, R.M. Ransohoff, Chapter 3 - The blood–brain barrier, S.J. Pittock, A.Vincent (Eds.), Handbook of Clinical Neurology, Elsevier, 133, 2016, 39-59. ISSN 0072-9752, ISBN 9780444634320, https://doi.org/10.1016/B978-0-444-63432-0.00003-7

[34]   W.M. Pardridge, Drug transport across the blood-brain barrier,J. Cereb. Blood Flow Metab. 32(2012)1959-1972. https://doi.org/10.1038/jcbfm.2012.126

[35]   S. Jafari, I.S. Baum, O.G. Udalov, Y. Lee, O. Rodriguez, S.T. Fricke, M. Jafari, M. Amin, R. Probst, X. Tang, C. Chen, D.J. Ariando, A. Hevaganinge, L.O. Mair, C. Albanese, I.N. Weinberg, Opening the Blood Brain Barrier with an Electropermanent Magnet System. Pharmaceutics. 20(2022)1503. https://doi.org/10.3390/pharmaceutics14071503

[36]   X. Dong, Current Strategies for Brain Drug Delivery, Theranostics, 8(2018)1481-1493. https://doi.org/10.7150/thno.21254

[37]   M.D. Sweeney, A.P. Sagare, B.V. Zlokovic, Blood-brain barrier breakdown in Alzheimer disease and other neurodegenerative disorders, Nat. Rev. Neurol. 14(2008)133-150. https://doi.org/10.1038/nrneurol.2017.188

[38]   D.A. Kuhn, D. Vanhecke, B. Michen, F. Blank, P. Gehr, A. Petri-Fink, Rothen-Rutishauser, B. Beilstein, J. Nanotechnol. 5(2014)1625–1636. https://doi.org/10.3762/bjnano.5.174

[39]   S. Salatin, K. A Yari, Overviews on the cellular uptake mechanism of polysaccharide colloidal nanoparticles, J. Cell Mol. Med. 21(2017)1668-1686. https://doi.org/10.1111/jcmm.13110

[40]   Y. Zhou ,J. Li, F. Lu, J. Deng, J. Zhang, P. Fang, X. Peng, S.F. Zhou. A study on the hemocompatibility of dendronized chitosan derivatives in red blood cells. Drug Des Devel Ther. 9(2015)2635-45. https://doi.org/10.2147/DDDT.S77105

[41]   H. Lu, S. Zhang, J. Wang, Q. Chen, A Review on Polymer and Lipid-Based Nanocarriers and Its Application to Nano-Pharmaceutical and Food-Based Systems, Front. Nutr. 8(2021)783831. https://doi.org/10.3389/fnut.2021.783831

[42]   D. Chenthamara, S. Subramaniam, S.G. Ramakrishnan, S. Krishnaswamy, M.M. Essa, F.H. Lin, M.W. Qoronfleh, Therapeutic efficacy of nanoparticles and routes of administration, Biomater. Res. 23(2019)20. https://doi.org/10.1186/s40824-019-0166-x

[43]   Y. Chen, L. Liu, Modern methods for delivery of drugs across the blood-brain barrier, Adv. Drug Deliv. Rev. 64(2012) 640-665. https://doi.org/10.1016/j.addr.2011.11.010

[44]   L.P. Sovan, J. Utpal, P.K. Manna, G.P. Mohanta, R. Manavalan, Nanoparticle: An overview of preparation and characterization., J. Appl. Pharma. Sci. 01 (2011) 228-234. https://www.japsonline.com/admin/php/uploads/159_pdf.pdf

[45]   Information in NIH Stem Cell Information Home Page. In Stem Cell Information. Bethesda, MD: National Institutes of Health, U.S. Department of Health and Human Services, 2016 https://stemcells.nih.gov/info/basics/stc-basics

[46]   W. Tang, Challenges and advances in stem cell therapy, Biosci. Trends, 13(2019)286. https://doi.org/10.5582/bst.2019.01241

[47]   L. Yang, L.S.T.D. Chueng, Y. Li, P. Misaal, R. Christopher, D. Gangotri, L Wang, L. Cai, K.B. Lee, A biodegradable hybrid inorganic nanoscaffold for advanced stem cell therapy, Nat. Commun. 9(2018) 3147. https://doi.org/10.1038/s41467-018-05599-2

[48]   C. Ying, Z. Shiwei, L. Lingling, Z. Jun-jie, Nanomaterials-based sensitive electrochemi luminescence biosensing, Nan. Tod. 12(2017) 98-115, ISSN 1748-0132,https://doi.org/10.1016/j.nantod.2016.12.013

[49]   P. Su, Y. Hongmei, Y. Xiaoyu, Y. Xiaohong W. Yan, L. Qinyi, J. Liliang, Y. Yudan, Application of Nanomaterials in Stem Cell Regenerative Medicine of Orthopedic Surgery, J. Nanomat. 2017(2017), Article ID 1985942, https://doi.org/10.1155/2017/1985942

[50]   M. Wei, S. Li, W. Le, Nanomaterials modulate stem cell differentiation: biological interaction and underlying mechanisms, J Nanobiotechnol. 15(2017). https://doi.org/10.1186/s12951-017-0310-5

[51]   S.B. Tiwari, MM. Amiji, A review of nanocarrier-based CNS delivery systems. Curr Drug Deliv. (2006)219-232. https://doi.org/10.2174/156720106776359230

[52]   P. Mittal, S. Anjali, V. Ravinder, M.A. Farag, M.A. Altalbawy, G.E.B. Alfaidi, A. Wahida, G.K. Rupesh, M. Sahab Uddin, M.S. Rahman, Dendrimers: A New Race of Pharmaceutical Nanocarriers, BioMed Res.Int.2021, Article ID 8844030. https://doi.org/10.1155/2021/8844030

[53]   S. Batra, S. Sharma, N.K. Mehra, Carbon Nanotubes for Drug Delivery Applications. In: J. Abraham, S. Thomas, N. Kalarikkal. (Eds.), Handbook of Carbon Nanotubes. Springer publishers 2021, https://doi.org/10.1007/978-3-319-70614-6_39-1

[54]   L. Dykman, N. Khlebtsov, N, Gold nanoparticles in biomedical applications: recent advances and perspectives, Chem. Soc. Rev. 41(2012) 2256-2282. https://doi.org/10.1039/C1CS15166E

[55]   R. Dinali, A. Ebrahiminezhad, M. Manley-Harris, Y. Ghasemi, A. Berenjian, Iron oxide nanoparticles in modern microbiology and biotechnology, Crit. Rev. Microbiol. 43(2017)493-507. https://doi.org/10.1080/1040841X.2016.1267708

[56]   R.K. Kesrevani, A.K. Sharma, 2 - Nanoarchitectured Biomaterials: Present Status and Future Prospects in Drug Delivery in: Alina Maria Holban, Alexandru Mihai Grumezescu (Eds.), For Smart Delivery and Drug Targeting, William Andrew Publishing,2016,PP.35-66,ISBN 9780323473477, https://doi.org/10.1016/B978-0-323-47347-7.00002-1

[57]   R.P. Mrunali, B.P. Rashmin, D.T. Shivam, 29 - Nanoemulsion in drug delivery, in: Inamuddin, M.A. Abdullah, M. Ali, Woodhead Publishing Series in Biomaterials, Applications of Nanocomposite Materials in Drug Delivery,Woodhead Publishing,2018,pp 667-700,ISBN 9780128137413,https://doi.org/10.1016/B978-0-12-813741-3.00030-3

[58]   E.E. Ngowi, Y. Wang, L. Qian, Y.A. Helmy, B. Anyomi, T. Li, M. Zheng, E. Jiang , S. Duan, J. Wei, D. Wu, X. Ji, The Application of Nanotechnology for the Diagnosis and Treatment of Brain Diseases and Disorders. Front.s in Bioeng. and Biotech. 9 (2021). https://doi.org/10.3389/fbioe.2021.629832

[59]   C. Spuch, O. Saida, C. Navarro, Advances in the treatment of neurodegenerative disorders employing nanoparticles, Rec. Pat. on Drug Deliver. Formu. 6(2012) 2–18. https://doi.org/10.2174/187221112799219125

[60]   W.H. De Jong, P.J. Borm, Drug delivery and nanoparticles: applications and hazards, Int. J. Nanomed. 3(2008):133-49. https://doi.org/10.2147/ijn.s596

[61]   L. Cui, X. Ren, M. Sun , H. Liu, L. Xia. Carbon Dots: Synthesis, Properties and Applications, Nanomaterials, 11(2021)12:3419. https://doi.org/10.3390/nano11123419

[62]   D. Yadav, K. Sandeep, D. Pandey, R.K. Dutta, Liposomes for Drug Delivery. J Biotechnol. Biomater.7(2017)276. https://doi.org/10.4172/2155-952X.1000276

[63]   T. Fukuta, N. Oku, K. Kogure, Application and Utility of Liposomal Neuroprotective Agents and Biomimetic Nanoparticles for the Treatment of Ischemic Stroke. Pharmaceutics, 14(2)361. https://doi.org/10.3390/pharmaceutics14020361

[64]   S.R.K. Pandian, K.K. Vijayakumar, S. Murugesan, S. Kunjiappan, Liposomes: An emerging carrier for targeting Alzheimer's and Parkinson's diseases. Heliyon, 8(2022):e09575. https://doi.org/10.1016/j.heliyon.2022.e09575

[65]   G. Sancini, M. Gregori, E. Salvati, I. Cambianica, F.R., F. Ornaghi, M. Canovi, C. Fracasso, A. Cagnotto, M. Colombo, C. Zona, M. Gobbi, M. Salmona, B. La Ferla, F. Nicotra, and M. Masserini, Functionalization with TAT-Peptide Enhances Blood-Brain Barrier Crossing In vitro of Nanoliposomes Carrying a Curcumin-Derivative to Bind Amyloid-b Peptide. J Nanomed Nanotechol 4(2013)171. https://doi.org/10.4172/2157-7439.1000171

[66]   I.F. Uchegbu, S.P. Vyas. Non-ionic surfactant based vesicles (niosomes) in drug delivery, Int. J. Pharma. 172 (1998) 33-70, doi.org/10.1016/S0378-5173(98)00169-0

[67]   M.I. Teixeira, C.M. Lopes, M.H. Amaral, P.C. Costa, Current insights on lipid nanocarrier-assisted drug delivery in the treatment of neurodegenerative diseases. Eur. J.

Pharm. Biopharm. 149(2020)192-217. https://doi.org/https://doi.org/10.1016/S0378-5173(98)00169-0

[68]   I. Cacciatore, M. Ciulla, E. Fornasari, L. Marinelli, A. Di Stefano, Solid lipid nanoparticles as a drug delivery system for the treatment of neurodegenerative diseases, Exp. Opin. Drug Deliv. 13(2016)1121-31. https://doi.org/10.1080/17425247.2016.1178237

[69]   S. Dhivya, A.N. Rajalakshmi, Curcumin Nano drug delivery systems: A Review on its type and therapeutic application, PharmaTutor. 5(2018), 30-39. https://doi.org/10.29161/PT.v5.i12.2017.30

[70]   Y.W. Lin, C.H. Fang, C.Y. Yang, Y.J. Liang, F.H. Lin, Investigating a Curcumin-Loaded PLGA-PEG-PLGA Thermo-Sensitive Hydrogel for the Prevention of Alzheimer's Disease, Antioxidants (Basel), 11(2022)727. https://doi.org/10.3390/antiox11040727

[71]   D. Tuncel, H.V. Demir, Conjugated polymer nanoparticles, Nanoscale, 2(2010)484-94. https://doi.org/10.1039/b9nr00374f

[72]   Y. Zhang, Z. Zou, S. Liu, S. Miao, H. Liu, Nanogels as Novel Nanocarrier Systems for Efficient Delivery of CNS Therapeutics 10 (2022) https://doi.org/10.3389/fbioe.2022.954470

[73]   N. Habibi, A. Mauser, Y. Ko, J. Lahann, Protein Nanoparticles: Uniting the Power of Proteins with Engineering Design Approaches, Advanced Science, 9 (2022) 8 . https://doi.org/10.1002/advs.202104012

[74]   A.V. Kabanov , H.E Gendelman, Nanomedicine in the diagnosis and therapy of neurodegenerative disorders, Prog. Polym. Sci.32(2007)1054-1082. https://doi.org/10.1016/j.progpolymsci.2007.05.014

[75]   T. Kim, T. Heyon, Application of inorganic nanoparticles as therapeutic agents, Nanotechnology, 25(201301200125. https://doi.org/10.1088/0957-4484/25/1/012001

[76]   M.A. Cotta, Quantum dots and their applications. what lies ahead : Application, ACS Appl. Nano Mater. 3(2020)4920–4924. https://doi.org/10.1021/acsanm.0c01386

[77]   J. Jampilek, K. Zaruba, M. Oravec, M. Kunes, P. Babula, P. Ulbrich, I. Brezaniova, R. Opatrilova, J. Triska, P. Suchy, Preparation of silica nanoparticles loaded with nootropics and their in vivo permeation through blood-brain barrier, Biomed Res Int. 2015(2015)812673. https://doi.org/10.1155/2015/812673

[78]   J. Qian, X. Li, M. Wei , X. Gao, Z. Xu, S .He , Bio-molecule-conjugated fluorescent organically modified silica nanoparticles as optical probes for cancer cell imaging, Opt Expr.16(2008)19568-19578.doi: 10.1364/oe.16.019568

[79]   R. Khalifehzadeh, H. Arami, Biodegradable calcium phosphate nanoparticles for cancer therapy, Adv Colloid Interface Sci. 279(2020)102157. https://doi.org/10.1016/j.cis.2020.102157

[80]   C. Qiu , Y. Wu ,Q. Guo,Q. Shi, J. Zhang, Y.Q. Meng, F. Xia, J. Wang, Preparation and application of calcium phosphate nanocarriers in drug delivery, Mater Today Bio. 22(2022)17:100501. https://doi.org/10.1016/j.mtbio.2022.100501

[81]   S. Svenson, D.A. Tomalia, Dendrimers in biomedical applications--reflections on the field, Adv. Drug Deliv. Rev. 14(2005)2106-2129. https://doi.org/10.1016/j.addr.2005.09.018

[82]   S. Sharma, S. Gupta, Dendrimers as Nanocarriers for Drug Delivery Applications. J. Nanosci. and Nanotech.19(2019) 4590-4617. https://doi.org/10.1166/jnn.2019.17005

[83]   P. Nirale, A. Paul , K.S. Yadav, Nanoemulsions for targeting the neurodegenerative diseases: Alzheimer (39) Parkinson (39) and Prion(39.), Life Sci. 15(2020)245:117394. https://doi.org/10.1016/j.lfs.2020.117394

[84]   E. Park, L. Li, Y. He, C. Abbasi, A.Z. Ahmed, T. Foltz, W.D. Flaherty, R. Zain, M. Bonin, R.P. Rauth, A.M. Fraser, P.E. Henderson, J. T. Wu, X.Y., Brain-Penetrating and Disease Site-Targeting Manganese Dioxide-Polymer-Lipid Hybrid Nanoparticles Remodel Microenvironment of Alzheimer: Disease by Regulating Multiple Pathological Pathways. (2023)2207238. https://doi.org/10.1002/advs.202207238

[85]   G. Birolini, M. Valenza, I. Ottonelli, A. Passoni, M. Favagrossa, T.J. Duskey, M. Bombaci, M.A. Vandelli, L. Colombo, R. Bagnati, C. Caccia, V. Leoni, F. Taroni, F. Forni, B. Ruozi, M. Salmona, G. Tosi, E. Cattaneo, Insights into kinetics, release, and behavioral effects of brain-targeted hybrid nanoparticles for cholesterol delivery in Huntington, 330(2021) 587-598. https://doi.org/10.1016/j.jconrel.2020.12.051

[86]   M. Dehvari, A. Ghahghaei,The effect of green synthesis silver nanoparticles (AgNPs) from *Pulicaria undulata* on the amyloid formation in α-lactalbumin and the chaperon action of α-casein, J.Biol. Macromol. 108(2018)1128-1139,doi.org/10.1016/j.ijbiomac.2017.12.040

[87]   S. Jadoun, R. Arif, N.K. Jangid, R.K. Meena, Green synthesis of nanoparticles using plant extracts: a review, Environ. Chem. Lett. 19(2020)355-374.https://doi.org/10.1007/s10311-020-01074-x

[88]   P.B. Gonçalves, A.C.R. Sodero, Y. Cordeiro, Green Tea Epigallocatechin-3-gallate (EGCG) Targeting Protein Misfolding in Drug Discovery for Neurodegenerative Diseases. Biomolecules. 11(2021)767. https://doi.org/10.3390/biom11050767

[89]   Md. M. Rahman, Md. R. Islam, S. Akash, Md. H.-O.-Rashid, T.K. Ray, Md. S. Rahaman, M. Islam, F. Anika, Md. K. Hosain, F.I. Aovi, H.A. Hemeg, A. Rauf, P. Wilairatana. Saidur Rahaman, Mahfuzul Islam, Fazilatunnesa Anika, Md. Kawser Hosain, Farjana Islam Aovi, Hassan A. Hemeg, Abdur Rauf, Polrat Wilairatana, Recent advancements of nanoparticles application in cancer and neurodegenerative disorders: At a glance. Biomed. Pharmacother. 153 (2022), https://doi.org/10.1016/j.biopha.2022.113305

[90]   S.M. Asil, J. Ahlawat, G.G. Barrosoc, M. Narayan, Nanomaterial based drug delivery systems for the treatment of neurodegenerative diseases, Biomat. Sci. 15(2020) 15, https://doi.org/https://doi.org/10.1039/D0BM00809E

[91]   N.S. Mohd Sairazi, K.N.S. Sirajudee, Natural Products and Their Bioactive Compounds: Neuroprotective Potentials against Neurodegenerative Diseases. Evid. Based Complement. Alternat. Med. 14 (2020)6565396. https://doi.org/10.1155/2020/6565396

[92]   A. Ahmad, A. Husain, M. Mujeeb, S.A. Khan, A.K. Najmi, N.A. Siddique, Z.A. Damanhouri, F. Anwar, A review on therapeutic potential of *Nigella sativa*: A miracle herb,

Asian Pac. J. Trop. Biomed. 3(2013)337-352. https://doi.org/10.1016/S2221-1691(13)60075-1

[93]   M. Hajialyani, M. Hosein Farzaei, J. Echeverría, S. Nabavi, E. Uriarte, E. Sobarzo-Sánchez, Hesperidin as a Neuroprotective Agent: A Review of Animal and Clinical Evidence, Molecules, 24(2019), 648. https://doi.org/10.3390/molecules24030648

[94]   S. Parham, A.Z. Kharazi, H.R. Bakhsheshi-Rad, H. Nur, A.F. Ismail, S. Sharif, S. Rama Krishna, F. Berto. Antioxidant, Antimicrobial and Antiviral Properties of Herbal Materials. Antioxidants, 9(2020):1309. https://doi.org/10.3390/antiox9121309

[95]   N. Suganthy, V. Sri Ramkumar, A. Pugazhendhi, G. Benelli, G. Archunan. Biogenic synthesis of gold nanoparticles from *Terminalia arjuna* bark extract: assessment of safety aspects and neuroprotective potential via antioxidant, anticholinesterase, and antiamyloidogenic effects. Environ. Sci. Pollut. Res. Int. 25(2018)10418-10433. https://doi.org/10.1007/s11356-017-9789-4

[96]   T. Bhattacharya, G.A.B. Soares, H. Chopra, M.M. Rahman, Z. Hasan, S.S. Swain, S. Cavalu, Applications of Phyto-Nanotechnology for the Treatment of Neurodegenerative Disorders, Materials,15(2022),804. https://doi.org/10.3390/ma15030804

[97]   J. K. Vasir, V. Labhasetwar , Chapter 56-Biodegradable Nanoparticles, in Gene Transfer: Delivery and Expression of DNA and RNA, Friedmann,Rossi(Eds.), https://www.sigmaaldrich.com/IN/en/technicaldocuments/protocol/genomics/gene-expression-and-silencing/biodegradable-nanoparticles.

[98]   J. Panyam,V. Labhasetwar, Biodegradable nanoparticles for drug and gene delivery to cells and tissue, Adv. Drug Deliv. Rev. 55(2003)329-347. https://doi.org/10.1016/s0169-409x(02)00228-4

[99]   A. Kumari, S.K.Yadav, S.C Yadav, Biodegradable polymeric nanoparticles based drug delivery systems, Collo.Surf B Biointer.75(010)1-18. https://doi.org/10.1016/j.colsurfb.2009.09.001

[100]   R.K. Jain, T. Stylianopoulos , Delivering nanomedicine to solid tumors, Nat. Rev. Clin. Oncol.7(2010)653-664, https://doi.org/10.1038/nrclinonc.2010.139

[101]   M.A. Shim, A. Jyoti, G.B. Gileydis, N. Mahesh, Nanomaterial based drug delivery systems for the treatment of neurodegenerative diseases, Biomat. Sci.15(2020). https://doi.org/10.1039/D0BM00809E

[102]   B. Olesja, S. Mart, Neurotrophic Factors in Parkinson's Disease: Clinical Trials, Open Challenges and Nanoparticle-Mediated Delivery to the Brain, Common Pathways Linking Neurodegenerative Diseases – The Role of Inflammation, Fronts. Cell. Neuro. 5(2021). https://doi.org/10.3389/fncel.2021.682597

[103]   Y.B. Hanie, G.K. Maryam, P. Marzieh, Curcumin-loaded nanoparticles: a novel therapeutic strategy in treatment of central nervous system disorders, Intl. J.f nanomed.14(2019)4449-4460. https://doi.org/10.2147/IJN.S208332

[104]   G.R.P. Rúben, A.J.Coutinho,M.Pinheiro,A.R.Neves, Nanoparticles for Targeted Brain Drug Delivery: What Do We Know?, Int J Mol Sci. 22(2021)11654. https://doi.org/10.3390/ijms222111654

[105]   L. Zhang, F.X. Gu, J.M. Chan, A.Z. Wang, R.S. Langer, O.C. Farokhzad, Nanoparticles in medicine: therapeutic applications and developments, Clin. Pharmacol. & Therapeu.83(2008), 761-769. https://doi.org/10.1038/sj.clpt.6100400

[106]   Y. Yao, Y. Zhou, L. Liu, Y. Xu, Q. Chen, Y. Wang, S. Wu, Y. Deng, J. Zhang, A. Shao, Nanoparticle-Based Drug Delivery in Cancer Therapy and Its Role in Overcoming Drug Resistance, Fronts. in Mol. Biosci.7(2020)193. https://doi.org/10.3389/fmolb.2020.00193

[107]   K. Johan, J. Vaughan Hannah, G. J.Jordan, Biodegradable Polymeric Nanoparticles for Therapeutic Cancer Treatments, Ann. Rev. of Chem. Biomol. Eng. 9(2018)105-127. https://doi.org/10.1146/annurev-chembioeng-060817-084055

[108]   S. Gavas, S. Quazi, T.M. Karpiński, Nanoparticles for Cancer Therapy: Current Progress and Challenges, Nano. Res. Lett. 16(2021)173. https://doi.org/10.1186/s11671-021-03628-6

[109]   A. Kaushik, Biomedical Nanotechnology Related Grand Challenges and Perspectives, Front. Nanotech.1(2019) https://www.frontiersin.org/articles/10.3389/fnano.2019.00001

[110]   J.R. Heath, Nanotechnologies for biomedical science and translational medicine, *PNAS*. *112*(2015) 14436–14443. https://doi.org/10.1073/pnas.1515202112

[111]   J.J. Chu, W.B. Ji, J.H. Zhuang, B.F. Gong, X.H. Chen, W.B. Cheng, W.D. Liang, G.-R. Li, J. Gao, Y. Yin, Nanoparticles-based anti-aging treatment of Alzheimer's disease, Drug Del. 29(2022), 2100-2116, https://doi.org/10.1080/10717544.2022.2094501

[112]   Y.C. Kuo, Y.I. Lou, R. Rajesh, Dual functional liposomes carrying antioxidants against tau hyperphosphorylation and apoptosis of neurons, J. of Drug Targ. 28(2020). https://doi.org/10.1080/1061186X.2020.1761819

[113]   S.K. Tiwari, S. Agarwal, B. Seth, A. Yadav, S. Nair, P. Bhatnagar, M. Karmakar, M. Kumari, L.K.S. Chauhan, D.K. Patel, V. Srivastava, D. Singh, S.K. Gupta, A. Tripathi, R.K. Chaturvedi, K.C. Gupta, Curcumin-loaded nanoparticles potently induce adult neurogenesis and reverse cognitive deficits in Alzheimer's disease model via canonical Wnt/β-catenin pathway, ACS Nano, 8(2014), 76–103. https://doi.org/10.1021/nn405077y

[114]   A.R. Neves, J.F. Queiroz, S. Reis, Brain-targeted delivery of resveratrol using solid lipid nanoparticles functionalized with apolipoprotein E, J. of Nanobiotech.14 (2016) 27. https://doi.org/10.1186/s12951-016-0177-x

[115]   S. Palmal, A.R. Maity, B.K. Singh, S. Basu, N.R. Jana, Inhibition of amyloid fibril growth and dissolution of amyloid fibrils by curcumin-gold nanoparticles, *Chemistry (Weinheim an Der Bergstrasse, Germany)*, *20* (2014) 6184–6191. https://doi.org/10.1002/chem.201400079

[116]   M. Bilal, M. Barani, F. Sabir, A. Rahdar, G.Z. Kyzas, Nanomaterials for the treatment and diagnosis of Alzheimer's disease: An overview, NanoImpact, 20(2020) 100251. https://doi.org/10.1016/j.impact.2020.100251

[117] W. Li, Q. Guo, H. Zhao, L. Zhang, J. Li, J. Gao, W. Qian, B. Li, H. Chen, H. Wang, J. Dai, Y. Guo, Novel dual-control poly(N-isopropylacrylamide-co-chlorophyllin) nanogels for improving drug release, *Nanomedicine, 7* (2012) 383–392. https://doi.org/10.2217/nnm.11.100

[118] S. Shah, N. Rangaraj, K. Laxmikeshav, S. Sampathi, Nanogels as drug carriers— Introduction, chemical aspects, release mechanisms and potential applications, *Intl. J. Pharma.581(2020)* 119268. https://doi.org/10.1016/j.ijpharm.2020.119268

[119] N. Lopez-Barbosa, J.G. Garcia, J. Cifuentes, L.M. Castro, F. Vargas, C. Ostos, G.P. Cardona-Gomez, A.M. Hernandez, J.C. Cruz, Multifunctional magnetite nanoparticles to enable delivery of siRNA for the potential treatment of Alzheimer's, *Drug Del.*27(2020),864–875. https://doi.org/10.1080/10717544.2020.1775724

[120] S.I. Laura, M.C. Christina, K.S. Georgina, P.R.J. Angus, Nanoescapology: Progress toward understanding the endosomal escape of polymeric nanoparticles. *WIREs Nanomed. and Nanobiotech. 9*(2017) e1452. https://doi.org/10.1002/wnan.1452

[121] T. Hamaguchi, K. Ono, M. Yamada, REVIEW: Curcumin and Alzheimer's Disease, *CNS Neuroscie .& Therapeu.16*(2010) 285–297. https://doi.org/10.1111/j.1755-5949.2010.00147.x

[122] K.G. Goozee, T.M. Shah, H.R. Sohrabi, S.R. Rainey-Smith Brown, B. Brown, G. Verdile, R.N. Martins, Examining the potential clinical value of curcumin in the prevention and diagnosis of Alzheimer's disease, *The Bri. J. Nut.115*(2016) 449–465. https://doi.org/10.1017/S0007114515004687

[123] M. Tang, C. Taghibiglou (2017). The Mechanisms of Action of Curcumin in Alzheimer's Disease, *J. of Alz. Dis.: JAD, 58*(2017) 1003–1016. https://doi.org/10.3233/JAD-170188

[124] S. Fan, Y. Zheng, X. Liu, W. Fang, X. Chen, W. Liao, X. Jing, M. Lei, E. Tao, Q. Ma, X. Zhang, R. Guo, J. Liu, Curcumin-loaded PLGA-PEG nanoparticles conjugated with B6 peptide for potential use in Alzheimer's disease, *Drug Delivery, 25*(2018) 1091–1102. https://doi.org/10.1080/10717544.2018.1461955

[125] S. Pillay, V. Pillay, Y.E. Choonara, D. Naidoo, Riaz.A. Khan, Lisa C. du Toit, V.M.K. Ndesendo, G. Modi, M.P. Danckwerts, S.E. Iyuke, Design, biometric simulation and optimization of anano-enabled scaffold device for enhanced delivery of dopamine to the brain, Intl. J.of Pharma. 382(2009) 277–290. https://doi.org/10.1016/j.ijpharm.2009.08.021

[126] A. Trapani, E. De Giglio, D. Cafagna, N. Denora, G. Agrimi, T. Cassano, S. Gaetani, V. Cuomo, G. Trapani, Characterization and evaluation of chitosan nanoparticles for dopamine brain delivery, Intl. J. of Pharma. 419(2011) 296–307. https://doi.org/10.1016/j.ijpharm.2011.07.036

[127] S. Mao, W. Sun, T. Kissel, Chitosan-based formulations for delivery of DNA and siRNA, Advanced Drug Delivery Reviews, 62(2010) 12–27. https://doi.org/10.1016/j.addr.2009.08.004

[128] K. Jagaran, M. Singh, Lipid Nanoparticles: Promising Treatment Approach for Parkinson's Disease, Intl. J. of Mol. Sci. 23(2022), Article 16. https://doi.org/10.3390/ijms23169361.

[129] B.N. Aldosari, I.M. Alfagih, A.S. Almurshedi, Lipid Nanoparticles as Delivery Systems for RNA-Based Vaccines, Pharmaceutics, 13(2021) 206. https://doi.org/10.3390/pharmaceutics13020206

[130] S. Chakraborty, G.S. Dhakshinamurthy, S.K. Misra, Tailoring of physicochemical properties of nanocarriers for effective anti-cancer applications, J. Biomed. Ma. Res. Part A, 105(2017) 2906–2928. https://doi.org/10.1002/jbm.a.36141

[131] N. Dudhipala, T. Gorre, Neuroprotective Effect of Ropinirole Lipid Nanoparticles Enriched Hydrogel for Parkinson's Disease: In Vitro, Ex Vivo, Pharmacokinetic and Pharmacodynamic Evaluation, Pharmaceutics, 12(2020)448. https://doi.org/10.3390/pharmaceutics12050448

[132] A. Waris, A. Ali, A.U. Khan, M. Asim, D. Zamel, K. Fatima, A. Raziq, M.A. Khan, N. Akbar, A. Baset, M.A.S. Abourehab, Applications of Various Types of Nanomaterials for the Treatment of Neurological Disorders, Nanomaterials (Basel, Switzerland), 12 (2022)2140. https://doi.org/10.3390/nano12132140

[133] E. Alimohammadi, A. Nikzad, M. Khedri, M. Rezaian, A.M. Jahromi, N. Rezaei, R. Maleki, Potential treatment of Parkinson's disease using new-generation carbon nanotubes: Abiomolecular in silico study, Nanomedicine, 16 (2021)189–204. https://doi.org/10.2217/nnm-2020-0372

[134] B.A. Rzigalinski, C.S. Carfagna, M. Ehrich, Cerium oxide nanoparticles in neuroprotection and considerations for efficacy and safety, Wil.Interdiscip. Rev. Nanomed Nanobiotechnol. 9(2017). https://doi.org/10.1002/wnan.1444

[135] B.S. Inbaraj, B.-H. Chen. An overview on recent in vivo biological application of cerium oxide nanoparticles, Asi. J. Pharmaceu. Sci.15 (2020)558-575, ISSN 1818-0876, https://doi.org/10.1016/j.ajps.2019.10.005

[136] R. Li, T. Liang, L. Xu, N. Zheng, K. Zhang, X. Duan, Puerarin attenuates neuronal degeneration in the substantia nigra of 6-OHDA-lesioned rats through regulating BDNF expression and activating the Nrf2/ARE signaling pathway., Brain Res.1523(2013) 1–9. https://doi.org/10.1016/j.brainres.2013.05.046

[137] G. Zhu, X. Wang, S. Wu, Q. Li. Involvement of activation of PI3K/Akt pathway in the protective effects of puerarin against MPP+-induced human neuroblastoma SH-SY5Y cell death, Neurochem. Intl.60 (2012)400–408. https://doi.org/10.1016/j.neuint.2012.01.003

[138] T. Chen, W. Liu, S. Xiong, D. Li, S. Fang, Z. Wu, Q. Wang, X. Chen,Nanoparticles Mediating the Sustained Puerarin Release Facilitate Improved Brain Delivery to Treat Parkinson's Disease, ACS Appl. Mater. Interfaces 11(2019) 45276–45289. https://doi.org/10.1021/acsami.9b16047

[139] P. Edison, C.C. Rowe, J.O. Rinne, S. Ng, I. Ahmed, N. Kemppainen, V.L. Villemagne, G. O'Keefe, K. Någren, K.R. Chaudhury, C.L. Masters, D.J. Brooks, Amyloid load in

Parkinson's.disease dementia and Lewy body dementia measured with [11C]PIB positron emission tomography, J. Neurol. Neurosur. and Psy. 79 (2008) 1331–1338. https://doi.org/10.1136/jnnp.2007.127878

[140] S. Masoudi Asil, J. Ahlawat, G. Guillama Barroso, M. Narayan, Nanomaterial based drug delivery systems for the treatment of neurodegenerative diseases, Biomat. Sci. 8 (2020)4109–4128. https://doi.org/10.1039/D0BM00809E

[141] P. Turcano, C.D. Stang, J.H Bower, J.E. Ahlskog, B.F. Boeve, M.M. Mielke, R. Savica, Levodopa-induced dyskinesia in dementia with Lewy bodies and Parkinson disease with dementia, Neurology: Clin. Prac.10(2020), 156–161. https://doi.org/10.1212/CPJ.0000000000000703

[142] G. Leyva-Gómez, H. Cortés, J.J. Magaña, N. Leyva-García, D. Quintanar-Guerrero, B. Florán, Nanoparticle technology for treatment of Parkinson's disease: The role of surfacephenomena in reaching the brain, Drug Discov. 20(2015) 824–837. https://doi.org/10.1016/j.drudis.2015.02.009

[143] T. Nie, He, Z., Zhu, J., K. Chen, G.P. Howard, J. Pacheco-Torres, I. Minn, P. Zhao, Z.M. Bhujwalla, H.Q. Mao, L. Liu, Y. Chen, Non-invasive delivery of levodopa-loaded nanoparticlesto the brain via lymphatic vasculature to enhance treatment of Parkinson's disease, Nano. Res. 14(2021) 2749–2761. https://doi.org/10.1007/s12274-020-3280-0

[144] D. Kim, J.M. Yoo, H. Hwang, J. Lee, S.H. Lee, S.P. Yun, M.J. Park, M. Lee, S. Choi, S.H. Kwon, S. Lee, S.H. Kwon, S. Kim, Y.J. Park, M. Kinoshita, Y.H Lee, S. Shin, S.R. Paik, S.J. Lee, S. Lee, B.H. Hong, H.S. Ko, Graphene quantum dots prevent α-synucleinopathy in Parkinson's disease, Nat. Nanotech. 13(2018) 812–818. https://doi.org/10.1038/s41565-018-0179-y

[145] N.K. Bhatia, A. Srivastava, N. Katyal, N. Jain, M.A.I Khan, B. Kundu, S. Deep, Curcumin binds to the pre-fibrillar aggregates of Cu/Zn superoxide dismutase (SOD1) and alters its amyloidogenic pathway resulting in reduced cytotoxicity, Biochim.Et Biophy. Acta, 1854(2015),426–436. https://doi.org/10.1016/j.bbapap.2015.01.014

[146] J.S. Rane, P. Bhaumik, D. Panda, (2017). Curcumin Inhibits Tau Aggregation and Disintegrates Preformed Tau Filaments in vitro. J. Alz. Dis.: JAD, 60(2017) 999–1014. https://doi.org/10.3233/JAD-170351

[147] J. den Haan, T.H.J. Morrema, A.J. Rozemuller, F.H. Bouwman, & amp; J.J.M Hoozemans, Different curcumin forms selectively bind fibrillar amyloid beta in post mortem Alzheimer's diseasebrains: Implications for in-vivo diagnostics. Acta Neuropathol. Commun. 6(2018), 75.https://doi.org/10.1186/s40478-018-0577-2

[148] M.M. Khan, A. Ahmad, T. Ishrat, M.B. Khan, M.N. Hoda , G. Khuwaja ,S.S. Raza, A. Khan, H. Javed, K. Vaibhav, F. Islam, Resveratrol attenuates 6-hydroxydopamine-induced oxidative damage and dopamine depletion in rat model of Parkinson's disease, Brain Res. 1328(2010) 139-51. https://doi.org/10.1016/j.brainres.2010.02.031

[149] X. Cui, Q. Lin, Y. Lian, Plant-Derived Antioxidants Protect the Nervous System From Aging by Inhibiting Oxidative Stress, Front. Aging Neurosci. Neuroinfl. and Neuropathy, 12(2020). https://doi.org/10.3389/fnagi.2020.00209

[150] A. Yadav, A. Sunkaria, N. Singhal, R. Sandhir, Resveratrol loaded solid lipid nanoparticles attenuate mitochondrial oxidative stress in vascular dementia by activating Nrf2/HO-1 pathway, Neurochem Int. 112(2018)239-254. https://doi.org/10.1016/j.neuint.2017.08.001

[151] M. Jaiswal, R. Dudhe, P.K. Sharma, Nanoemulsion: An advanced mode of drug delivery system, 3 Biotech, 5(2015) 123–127. https://doi.org/10.1007/s13205-014-0214-0

[152] J.W. Russell, L. Yang, Y. Guangze, Z. Chun-Xia Zhao, Nanoemulsions for drug delivery, Particuology,64(2022) 85-97. https://doi.org/10.1016/j.partic.2021.05.009

[153] M. Mizrahi, Y. Friedman-Levi, L. Larush, K. Frid, O. Binyamin, D. Dori, N. Fainstein, H. Ovadia, T. Ben-Hur, S. Magdassi, R. Gabizon, Pomegranate seed oil nanoemulsions for the prevention and treatment of neurodegenerative diseases: the case of genetic CJD, Nanomedicine. 10(2014)1353-63. https://doi.org/10.1016/j.nano.2014.03.015

[154] Y. Friedman-Levi, Z. Meiner, T. Canello, K. Frid, G.G. Kovacs, H. Budka, D. Avrahami, R. Gabizon, Fatal Prion Disease in a Mouse Model of Genetic E200K Creutzfeldt-Jakob Disease, PLOS Pathogens, 7 (2011) e1002350. https://doi.org/10.1371/journal.ppat.1002350

[155] M. Balbirnie, R. Grothe, D.S, Eisenberg, An amyloid-forming peptide from the yeast prion Sup35 reveals a dehydrated beta-sheet structure for amyloid, PNAS.98 (2001) 2375–2380. https://doi.org/10.1073/pnas.041617698

[156] A.V. Krasnoslobodtsev, A.M. Portillo, T. Deckert-Gaudig, V. Deckert, Y.L. Lyubchenko, Nanoimaging for prion related diseases, Prion, 4(2010) 265–274. https://doi.org/10.4161/pri.4.4.13125

[157] S. Zhou, Y. Zhu, X. Yao, H. Liu, Carbon Nanoparticles Inhibit the Aggregation of Prion Protein as Revealed by Experiments and Atomistic Simulations, J Chem Inf Model .59 (2019)1909–1918. https://doi.org/10.1021/acs.jcim.8b00725

[158] A. Janaszewska, B. Ziemba, K. Ciepluch, D. Appelhans, B. Voit, B. Klajnert, M. Bryszewska, The biodistribution of maltotriose modified poly(propylene imine) (PPI) dendrimers conjugated with fluorescein—Proofs of crossing blood–brain–barrier, *New J. Chem.36* (2012)350–353. https://doi.org/10.1039/C1NJ20444K

[159] J.M McCarthy, D. Appelhans,J. Tatzelt, M.S. Rogers, Nanomedicine for prion disease treatment. *Prion*, 7(2013), 198–202. https://doi.org/10.4161/pri.24431

[160] S.Ramaswamy, J.L. McBride, J.H. Kordower, Animal Models of Huntington's Disease, ILAR Journal, 48(2007)356–373. https://doi.org/10.1093/ilar.48.4.356

[161] S.J Tabrizi, B.R. Leavitt, G.B. Landwehrmeyer, E.J. Wild, C. Saft, R.A. Barker, N.F. Blair, D. Craufurd, J. Priller, H. Rickards, A. Rosser, H.B. Kordasiewicz, C. Czech, E.E. Swayze, D.A. Norris, T. Baumann, I. Gerlach, S.A. Schobel, E. Paz, … Phase 1–2a IONIS-HTTRx Study Site Teams, Targeting Huntingtin Expression in Patients with Huntington's Disease, NEJM. 380(2019)2307–2316. https://doi.org/10.1056/NEJMoa1900907

[162] M.C. Didiot, L.M. Hall, A.H. Coles, R.A. Haraszti, B.M. Godinho, K. Chase, E. Sapp, S. Ly, J.F. Alterman, M.R. Hassler, D. Echeverria, L. Raj, D.V. Morrissey, M. Di Figlia, A. Neil , K. Anastasia, Exosome-mediated Delivery of Hydrophobically Modified siRNA for

Huntingtin mRNA Silencing, Mol Ther.24(2016)1836-1847.
https://doi.org/10.1038/mt.2016.126

[163] D.D. Ojeda-Hernández, A.A. Canales-Aguirre, J. Matias-Guiu, U. Gomez-Pinedo, J.C.
       Mateos-Díaz, Potential of Chitosan and Its Derivatives for Biomedical Applications in the
       Central Nervous System, Front. Bioeng. Biotechnol. 8 (2020)389.
       https://doi.org/10.3389/fbioe.2020.00389

[164] G.M Escott, D.J. Adams, Chitinase activity in human serum and leukocytes, Infec.
       Immunity, 63(1995)4770–4773. https://doi.org/10.1128/iai.63.12.4770-4773.1995

[165] T. Chandy, C.P. Sharma, Chitosan—As a biomaterial, Biomaterials, Artificial Cells, and
       Artificial Organs, 18(1990) 1–24. https://doi.org/10.3109/10731199009117286

[166] V. Sava, O. Fihurka, A. Khvorova, J. Sanchez-Ramos, Enriched chitosan nanoparticles
       loaded with siRNA are effective in lowering Huntington's disease gene expression following
       intranasal administration, Nanomedicine nanomed-nanotechnol 24(2020) 102119.
       https://doi.org/10.1016/j.nano.2019.102119

[167] A. Shahar, K.V Patel, R.D Semba, S. Bandinelli, D.R. Shahar, L. Ferrucci, & J.M
       Guralnik, Plasma selenium is positively related to performance in neurological task
       sassessing coordination and motor speed, Movt. Disor. 25(2010) 1909–1915.
       https://doi.org/10.1002/mds.23218

[168] W. Cong, R. Bai, Y-F. Li, L. Wang, C. Chen, Selenium Nanoparticles as an Efficient
       Nanomedicine for the Therapy of Huntington's Disease, ACS Appl. Mat. Interfaces,
       11(2019) 34725–34735. https://doi.org/10.1021/acsami.9b12319

[169] M. Mielcarek, C. Landles, A. Weiss, A. Bradaia, T. Seredenina, L. Inuabasi, G.F.
       Osborne, K. Wadel, C. Touller, R. Butler, J. Robertson, S.A. Franklin, D.L. Smith, L. Park,
       P.A. Marks, E.E. Wanker, E.N. Olson, R. Luthi-Carter, H. van der Putten, B. Vahri, G.P.
       Bates, HDAC4 Reduction: A Novel Therapeutic Strategy to Target Cytoplasmic Huntingtin
       and Ameliorate Neurodegeneration, PLoS Biol 11 (2013)e1001717.
       https://doi.org/10.1371/journal.pbio.1001717

[170] M.R. Smith, A. Syed, T. Lukacsovich, J. Purcell, B.A Barbaro, S.A, Worthge, S.R. , Wei,
       G. Pollio, L. Magnoni, C. Scali, L. Massai, D. Franceschini, M. Camarri, M. Gianfriddo, E.
       ,Diodato, R. Thomas, O. Gokce, S.J. Tabrizi, A. Caricasole, J.L. Marsh, Apotent and
       selective sirtuin 1 inhibitor alleviates pathology in multiple animal and cell models of
       huntington's disease, Hum. Mol. Gen. 23(2014). https://doi.org/10.1093/hmg/ddu010

[171] V. Leoni, C. Mariotti, S.J. Tabrizi, M. Valenza, E.J. Wild, S.M. Henley, N.Z. Hobbs,
       M.L. Mandelli, M. Grisoli, I. Björkhem, E. Cattaneo, S. Di Donato, Plasma 24S-
       hydroxycholesterol and caudate MRI in pre-manifest and early Huntington's disease, *131*
       (2008) 2851–2859. https://doi.org/10.1093/brain/awn212

[172] V. Leoni, J.D. Long, J.A. Mills, S. Di Donato, J.S. Paulsen JS, Plasma 24S-
       hydroxycholesterol correlation with markers of Huntington disease progression. *Neurobiol
       Dis*55(2013)37–43. https://doi.org/10.1016/j.nbd.2013.03.013

[173] A.B. Ziegler,T. Christoph, T. Federico, H. Astrid , S. Peter, T. Gaia , Cell-Autonomous Control of Neuronal Dendrite Expansion via the Fatty Acid Synthesis Regulator SREBP, Cell Reports 21 (2017) 3346–3353. https://doi.org/10.1016/j.celrep.2017.11.069

[174] M.S. Shive, J.M. Anderson, Biodegradation and biocompatibility of PLA and PLGA microspheres, Adv. Drug Deliv. Rev. 28(1997)5-24. https://doi.org/10.1016/s0169-409x(97)00048-3

[175] F. Danhier, E. Ansorena, J.M. Silva, R. Coco, A. Le Breton, V. Préat, PLGA-based nanoparticles: an overview of biomedical applications, J. Contr. Rel.,161(2012)505-22. https://doi.org/10.1016/j.jconrel.2012.01.043

[176] J. Gupta, M.T. Fatima, Z. Islam, R.H. Khan, V.N. Uversky, P. Salahuddin, Nanoparticle formulations in the diagnosis and therapy of Alzheimer's disease, Intl.J. Biol. Macromol. 130 (2019) 515–526. https://doi.org/10.1016/j.ijbiomac.2019.02.156

[177] D. Ling, T. Hyeon, Chemical design of biocompatible iron oxide nanoparticles for medical applications, Small (Weinheim an Der Bergstrasse, Germany), 9(2013), https://doi.org/10.1002/smll.201202111

[178] R. Adami, D. Bottai, Curcumin and neurological diseases, Nutr. Neurosci. 25 (2022)441–461. https://doi.org/10.1080/1028415X.2020.1760531

[179] Z. Yarjanli, K. Ghaedi, A. Esmaeili, S. Rahgozar, A. Zarrabi, Iron oxide nanoparticles may damage to the neural tissue through iron accumulation, oxidative stress, and protein aggregation, BMC Neurosci. 18 (2017). https://doi.org/10.1186/s12868-017-0369-9

[180] P. Bigini, V. Diana, S. Barbera, E. Fumagalli, E. Micotti, L. Sitia, A. Paladini, C. Bisighini, L. de Grada, L. Coloca, L. Colombo, P. Manca, P. Bossolasco, F. Malvestiti, F. Fiordaliso, G. Forloni, M. Morbidelli, M. Salmona, D. Giardino, L. Cova. Longitudinal tracking of human fetal cells labeled with super paramagnetic iron oxide nanoparticles in the brain of mice with motor neuron disease, PLoS One, 7 (2012) https://doi.org/10.1371/journal.pone.0032326

[181] L. Canzi, V. Castellaneta, S. Navone, S. Nava, M. Dossena, I. Zucca, T. Mennini, P. Bigini, E. A. Parati. Human skeletal muscle stem cell antiinflammatory activity ameliorates clinical outcome in amyotrophic lateral sclerosis models. Mol. Med. 18 (2012). https://doi.org/10.2119/molmed.2011.00123

[182] S. Gargiulo, S. Anzilotti, A.R.D. Coda, M. Gramanzini, A. Greco,M. Panico, A. Vinciguerra, A. Zannetti, C. Vicidomini, F. Dollé, G. Pignataro, M. Quarantelli, L. Annunziato, A. Brunetti, M. Salvatore, S. Pappatà, Imaging of brain TSPO expression in amouse model of amyotrophic lateral sclerosis with 18F-DPA-714 and micro-PET/CT, Eur. J. Nu. Med.2016. 1–12. https://doi.org/10.1007/s00259-016-3311-y

[183] B.E. Clarke, R. Patani, The microglial component of amyotrophic lateral sclerosis. Brain. 143(2020):3526-3539. https://doi.org/10.1093/brain/awaa309

[184] F.E. Turkheimer, G. Rizzo, P.S. Bloomfield, Howes, P. Zanotti, A. Bertoldo, M. Veronese, M. (2015). The methodology of TSPO imaging with positron emission tomography, Biochem. Soc. Transac. 43(2015). https://doi.org/10.1042/BST20150058

[185] Q. Chen, Y. Du, K. Zhang, Z. Liang, J. Li, H. Yu, R. Ren, J. Feng, Z. Jin, F. Li, J. Sun, M. Zhou, Q. He, X. Sun, H. Zhang, M. Tian, & amp; D. Ling, (2018). Tau-Targeted Multifunctional Nanocomposite for Combinational Therapy of Alzheimer's Disease. ACS Nano, 12(2), Article 2. https://doi.org/10.1021/acsnano.7b07625

[186] F.N.S. Fachel, M. Dal Prá, J.H. Azambuja, M. Endres, V.L. Bassani, L. Koester, A.T. Henriques, A.G. Barschak, H.F. Teixeira, E. Braganhol, Glioprotective Effect of Chitosan-Coated Rosmarinic Acid Nanoemulsions Against Lipopolysaccharide-Induced Inflammation and Oxidative Stress in Rat Astrocyte Primary Cultures, Cell. Mol. Neurobiol. 40(2020) https://doi.org/10.1007/s10571-019-00727-y

[187] S. Liu, G. Zhen, R. Chi Li, & amp; S. Dore, Acute bioenergetic intervention orpharmacological preconditioning protects neuron against ischemic injury. J. Exp. Stroke Transl. Med. 06(2013). https://doi.org/10.4172/1939-067X.1000140

[188] S.K. Rajendrakumar, V. Revuri, M. Samidurai, A.Mohapatra, J.H. Lee, P. Ganesan, J. Jo, Y.K. Lee, & amp; I.-K. Park, Peroxidase-Mimicking Nanoassembly Mitigates Lipopolysaccharide-Induced Endotoxemia and Cognitive Damage in the Brain by Impeding Inflammatory Signaling in Macrophages, Nano Lett. 18(2018) https://doi.org/10.1021/acs.nanolett.8b02785

[189] M.L. Bondì, E.F. Craparo, G. Giammona, & amp; F. Drago, Brain-targeted solid lipidnanoparticles containing riluzole: Preparation, characterization and biodistribution, Nanomedicine (London, England), 5(2010). https://doi.org/10.2217/nnm.09.67

[190] Z. Mazibuko,Y.E. Choonara, P. Kumar, L.C. Du Toit, G. Modi, D. Naidoo, V. Pillay, A review of the potential role of nano-enabled drug delivery technologies in amyotrophic lateral sclerosis: Lessons learned from other neurodegenerative disorders, J. Pharma.Sci. 104(2015). https://doi.org/10.1002/jps.24322

[191] G.Y. Wang, S.L. Rayner, R. Chung, B.Y. Shi, X.J. Liang, Advances in nanotechnology-based strategies for the treatments of amyotrophic lateral sclerosis, Mat. Today Bio, 6(2020) 100055. https://doi.org/10.1016/j.mtbio.2020.100055

[192] M. Benigni, C. Ricci, A.R. Jones, F. Giannini, A.Al-Chalabi, S. Battistini, Identification of miRNAs as Potential Biomarkers in Cerebrospinal Fluid from Amyotrophic Lateral Sclerosis Patients, Neuro Mol. Med. 18(2016). https://doi.org/10.1007/s12017-016-8396-8

[193] Pegoraro, V., Merico, A., & amp; Angelini, C. (2019). MyomiRNAs Dysregulation in ALS Rehabilitation. Brain Sciences, 9(1), Article 1. https://doi.org/10.3390/brainsci9010008

[194] A. Freischmidt, K. Müller, L. Zondler, P. Weydt, A.E. Volk, A.L. Božič, M. Walter, M. Bonin, B. Mayer, C.A.F. von Arnim, M. Otto, C. Dieterich, K. Holzmann, P.M. Andersen, A.C. Ludolph, K.M. Danzer, J.H. Weishaupt, Serum microRNAs in patients with genetic amyotrophic lateral sclerosis and pre-manifest mutation carriers, Brain, 137(2014). https://doi.org/10.1093/brain/awu249

[195] B. Niccolini, V. Palmieri, M. De Spirito, M. Papi, Opportunities Offered by Graphene Nanoparticles for MicroRNAs Delivery for Amyotrophic Lateral Sclerosis Treatment, Materials (Basel, Switzerland), 15(2021). https://doi.org/10.3390/ma15010126

[196] L. Cui, Z. Chen, Z. Zhu, X. Lin, X. Chen, C.J. Yang, Stabilization of ssRNA on Graphene Oxide Surface: An Effective Way to Design Highly Robust RNA Probes, Analy. Chem. 85(2013). https://doi.org/10.1021/ac303179z

[197] A.J. Patil, J.L. Vickery, T.B. Scott, S. Mann, Aqueous Stabilization and Self-Assembly of Graphene Sheets into Layered Bio-Nanocomposites using DNA, Adv. Mat. 21(2009). https://doi.org/10.1002/adma.200803633

[198] A.K. Geim, K.S. Novoselov, The rise of graphene. In Nanoscience and Technology 2009 (pp. 11–19). Co-Published with Macmillan Publishers Ltd, UK. https://doi.org/10.1142/9789814287005_0002

[199] V. Novak, B. Rogelj, V. Župunski, Therapeutic Potential of Polyphenols in AmyotrophicLateral Sclerosis and Frontotemporal Dementia. Antioxidants, 10(2021). https://doi.org/10.3390/antiox10081328

[200] G. Tripodo, T. Chlapanidas, S. Perteghella, B. Vigani, D. Mandracchia, A. Trapani, M. Galuzzi, M.C. Tosca, B. Antonioli, P. Gaetani, M. Marazzi, M.L. Torre, Mesenchymal stromalcells loading curcumin-INVITE-micelles: A drug delivery system for neurodegenerative diseases.Colloids and Surfaces, Biointerfaces, 125(2015)300–308. https://doi.org/10.1016/j.colsurfb.2014.11.034

Nanoparticles in Healthcare: Applications in Therapy, Diagnosis and Drug Delivery   Materials Research Forum LLC
11Materials Research Foundations 160 (2024) 247-295   https://doi.org/10.21741/9781644902974-10

Chapter 10

Promising Role of Theranostic Nanoparticles in Lung Diseases

M. Sonia Angeline[1,*,] Vinita Ernest[2], Nitin John[1]

[1]Department of Life Sciences, Kristu Jayanti College, Autonomous, Bangalore-560077, Karnataka, India

[2]Dept of Pulmonary Medicine, Christian Medical College Vellore, Ranipet Campus, Ranipet - 632517, Tamil Nadu, India

*sonia.m@kristujayanti.com

Abstract

The role of nanoparticles is inevitable in the medical sector in the present era of technology. Though the field of medicine deals with the diagnosis and treatment of various diseases with the development of pharmaceuticals, nanoparticles are nevertheless   gaining much attention with the discovery and usage in recent years. Nanoparticles due to their smaller size, efficacy and other potentials such as bio-compatible nature and functionalization have been increasingly used as theranostic agents (diagnosis plus therapy) in various diseases. This chapter reviews the potential and significant role of theranostic nanoparticles in lung diseases.

Keywords

Theranostic Nanoparticle, Lung Diseases, Chronic Bronchitis, Tuberculosis

Contents

1.   Introduction

As a next generation treatment, theranostic (Therapy and Diagnostic) nanoparticles are of great interest in the recent decades where the development of an innovative field of precision/personalized medicine has emerged for specific targeted diagnosis and therapy using nanoparticles simultaneously for the treatment of various diseases. A basic and simple definition of theranostic nanoparticle is that a nanoparticle containing both therapeutic and diagnostic agent on the same nanomaterial.

Nanoparticles provide numerous advantages and benefits over the conventional drugs due to its low toxicity level, size, charge and chemical and bio-functional capabilities. In diagnosis, the nanoparticle acts as an imaging agent based on its large surface to volume ratio whereas in therapeutics, the same nanoparticle is functionalized using a carrier which effectively delivers the drug at the specified site enabling it to control the drug release from the matrix [1].

Though various diseases use theranostic nanoparticles, lung diseases are of much interest for clinicians as well as researchers since the lung is an attractive target for nanoparticle-mediated drug delivery for many reasons: a) non-invasive, b) high systemic bio-availability, c) large surface area for absorption, d) limited proteolytic activity. These factors contribute greater potential for local delivery of drugs for the treatment of lung diseases [2]. It is also beneficial that these nanodrugs are able to deposit at the disease site directly thus avoiding rapid metabolism.

Theranostic nanoparticles have been administered in different ways so far such as oral and intravenous and the most recent way of administering is locoregional which possess a greater scope for higher bio-availability at the administered site. Theranostic nanoparticles commonly contain hydrophobic organic drugs, drug-polymer conjugates, polymeric/magnetic nanoparticles, solid lipid nanoparticles, dendrimers, liposomes, micelles, gold nanoparticles, carbon nanomaterials, proteins, and peptides as therapeutic agents. Along with therapeutic agents, diagnostic agents are also used in theranostic nanoparticles. For optical imaging, fluorescent dyes or quantum dots are commonly employed. Superparamagnetic metals such as iron oxides are utilized for magnetic resonance imaging (MRI), while radionuclides are used for nuclear imaging. Computed tomography (CT) utilizes heavy elements like iodine.

Quantum dots are particularly significant among the diagnostic agents used in theranostic nanoparticles. They offer advantages over organic dyes as they are highly photo-stable, emit intense signals, possess larger absorption coefficients, and exhibit high brightness levels [3].

Theranostic nanoparticles offer effective and controlled delivery of medicines in minute amounts with greater sensitivity to diagnosis and performs accurate targeting with the specific receptor molecule. Theranostic nanoparticles are a promising approach for the diagnosis and treatment of lung diseases. These nanoparticles can be designed to target specific cells and tissues in the lungs, and they can simultaneously deliver therapeutic agents and imaging probes to monitor the disease progression.

One example of theranostic nanoparticles for lung diseases is the use of carbon nanotubes (CNTs). CNTs can be functionalized by targeting ligands such as antibodies or peptides, which can enable them to specifically bind to lung cancer cells. CNTs can also carry therapeutic agents such as chemotherapy drugs or gene therapy vectors, which can be released directly into the cancer cells for treatment. Additionally, CNTs can be labelled with imaging probes such as fluorescent dyes or magnetic nanoparticles, which can allow for real-time monitoring of the therapeutic response [4].

Another example of theranostic nanoparticles for lung diseases is the use of mesoporous silica nanoparticles (MSNs). MSNs can be functionalized by targeting ligands such as peptides or aptamers, which can enable them to specifically bind to lung cancer cells. MSNs can also carry therapeutic agents such as chemotherapy drugs or siRNA molecules, which can be released directly into the cancer cells for treatment. Additionally, MSNs can be labelled with imaging probes such as quantum dots or iron oxide nanoparticles, which can allow for real-time monitoring of the therapeutic response. Overall, theranostic nanoparticles offer a promising approach for the diagnosis and treatment of lung diseases Figure 1, and ongoing research is exploring new ways to design and optimize these nanoparticles for improved efficacy and safety.

This chapter focuses primarily on the lung diseases affecting:

- The airways – Asthma, Chronic Obstructive Pulmonary Disease (COPD), Chronic Bronchitis, Cystic fibrosis (CF)

- The air sacs (alveoli) – Pneumonia, Tuberculosis, Lung cancer and Acute respiratory distress syndrome (ARDS)

The lungs are one of the major organs of the body that functions consistently, ensuring the exchange of gases. This multilobed organ predominantly consists of the bronchi, alveolar tubes, alveoli, and pulmonary blood vessels. Its special sponge-like tissues play a crucial role in fulfilling the oxygen demand of the body [2]. Inhalation of the ambient air for oxygen risks exposure to other environmental factors like smoke, pollen grains, dust, chemicals, etc. Exposure to these toxic substances can lead to several disorders [5]. In general, it encompasses a variety of conditions that affect the function and structure of the lungs and is predicted to be responsible for 20% of all deaths worldwide by 2030 [6].

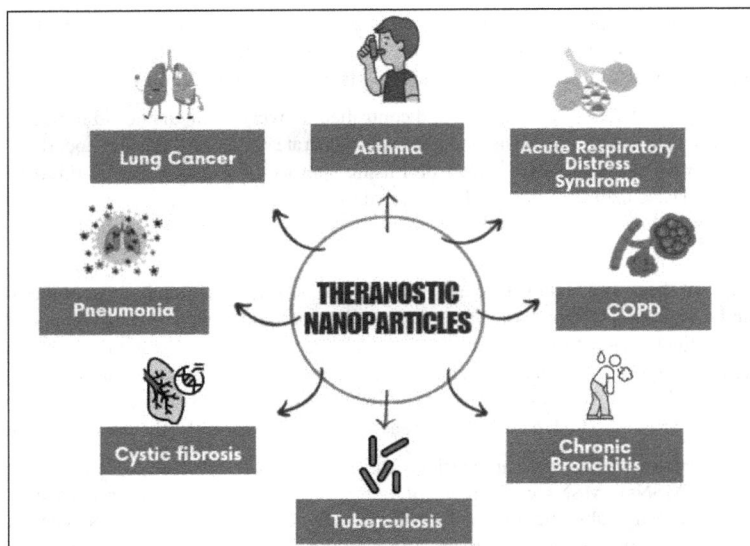

Figure 1: Effect of theranostic nanoparticles on lung diseases

Lung disease has a wide range of causative agents, some of which are medical or hereditary in origin, others of which are environmental. Some of the most common causes are depicted in Figure 2. The leading cause of lung disease is mainly attributed to smoking. Inhalation of cigarette smoke contains toxic chemicals causing damaging effects to the lung tissue and airways. Smoking is known to reduce the intake of oxygen into the bloodstream due to inflamed air sacs and it also decreases the number and effectiveness of tiny hair-like structures called cilia that line the airways and helps to remove mucus and other particles from the lungs [7]. Another significant contributor to pulmonary disease is the degree of pollution in the air around us. Ranked as the 9th most significant risk factor for mortality, air pollution is known to cause 3.2 million deaths annually [8]. Particulate matter (PM), ozone (O_3), sulfur dioxide (SO_2), nitrogen oxides (NOx), carbon monoxide (CO), and lead are the six common air pollutants. Exposure to these pollutants can influence the development of immune function and lung mechanics [9]. Children are more susceptible to these toxicants as compared to adults due to their smaller airways and immature detoxification [8]. The sepsis of a virus, bacteria, or fungus in the lungs can cause lung diseases. Upper respiratory infections are mainly of viral etiology, inducing epithelial damage resulting in edema and hemorrhage [10]. Lower respiratory infections, on the other hand, are viral and bacterial that cause bronchitis and community-acquired pneumonia respectively. An impaired mucociliary function and inflammation are observed when affected by pathogens [10]. Numerous familial genes are another well-known factor contributing to this disease. Some of these genes are inherited in an autosomal recessive manner, in disorders such as Alpha-1 Antitrypsin Deficiency, Cystic Fibrosis, and Hereditary Pulmonary Alveolar Proteinosis, whereas familial interstitial pulmonary fibrosis is an autosomal dominant disorder that can trigger difficulty in breathing and respiratory failure. It is crucial to understand that not all lung conditions are inherited and that external elements like environmental and occupational exposures can also cause lung conditions to emerge [11]. Exposure to specific substances (such as coal mines & silica dust, fumes, chemicals, metals) at work may induce or exacerbate respiratory conditions called Occupational Lung Diseases (OLDs). The most frequently diagnosed OLD is occupational asthma [12]. An aging lung generally involves molecular and physiological alterations that result in varied lung functions and decreased ability of the lung tissue to repair and regenerate. With a massive growth in the aging population, it is important to examine the enhanced risk of vulnerability to infections and the advancement of chronic disorders such as chronic obstructive pulmonary disease [13].

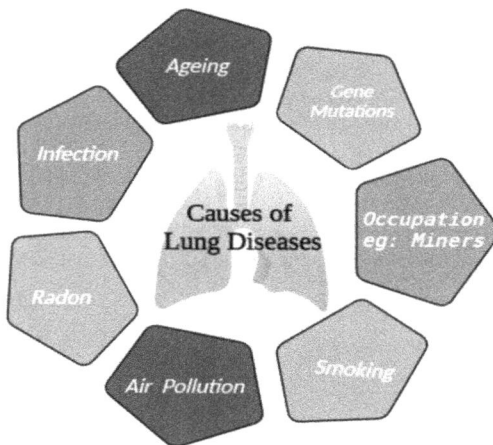

Figure 2: Causes of Lung Diseases

There exist several types of lung diseases based on which part of the pulmonary tract is affected, impacting individuals of various age groups and demographics. The most prevalent of them worldwide include asthma, Chronic Obstructive Pulmonary Disease (COPD), respiratory infections, tuberculosis, and lung cancer [14]. Bronchial asthma refers to a recurring condition that ensued due to inflammation and contraction of the airways. With symptoms such as difficulty in breathing, wheezing, and coughing, this disorder was estimated to have a prevalence of around 262 million people, in the year 2019 [15]. COPD is a chronic respiratory disorder exemplified by restricted airflow, but unlike asthma is only partially reversible. Being the third leading cause of death globally, it is estimated that around 391.9 million people suffer from COPD [16] Symptoms of COPD are akin to that of asthma but tend to exacerbate over time, negatively affecting the quality of life [2]. Lung infections are characterized by the infiltration of pathogenic agents into the pulmonary system via inhalation of airborne particles or direct contact with contaminated surfaces. While pneumonia refers to a viral, bacterial, or fungal infection that advances to inflamed air sacs, tuberculosis is caused by the infection of the bacteria *Mycobacterium tuberculosis,* majorly affecting the lungs. The severity of these infections can vary from mild to life-threatening, as symptoms portrayed include chest pain, fever, and restricted breathing [17]. Of all cancers diagnosed, 12.4% are attributed to lung cancers or bronchogenic carcinoma and are known to be responsible for 1.8 million deaths worldwide. This cancer originates from the bronchial mucosal epithelium and alveoli and is primarily of two types non-small cell lung cancer and small cell lung cancer. Smoking is known to be the leading cause of lung cancer. Exposure to cigarette smoke and other carcinogens results in abnormal cell growth in the lung epithelium, gene mutations influencing protein formation, and hence stimulating carcinogenesis. Symptoms such as coughing up blood, weight loss, and fatigue are usually evident during the latter stages of the disorder [18].

The nature and intensity of the disorder will determine the accessible therapies for lung conditions Table 1. Although a wide range of extensive research on the pathology of several infectious and non-infectious pulmonary diseases. While bronchodilators (albuterol & ipratropium) and long acting β2 agonists (LABAs) have proved to be beneficial in curing the faint symptoms of recurring lung diseases, unfortunately, they have failed to be completely useful in treating chronic asthma and COPD. Additionally, extreme variants of this disorder are caused due to irreversible tissue injury which cannot be addressed by the currently offered therapies. Hence, infectious respiratory diseases are a significant reason for worry. As a result of natural selection, genetic mutations, and resistance to drugs due to their overuse, more efficient therapy to confront this concern [19].

Table 1: Lung diseases and its therapeutic applications

Therapeutic Applications	Disease Assessed	Components	Advantages	Disadvantages	Reference
Oral Drug Delivery	Cystic Fibrosis, COPD, Asthma	Phosphodiesterase (PDE) inhibitors, theophylline, chronic macrolides, leukotriene modifiers	Enhanced delivery of the drug to areas inaccessible to traditional inhalation agents Cost-efficient and easy administration Distinct mode of action	Intake of hydrophilic drugs is prevented due to the epithelial barrier Ineffective treatment of the lesion Toxic effects may arise due to the disintegration of the drug during the gastrointestinal transport	[20, 21] [2]

Intravenous Administration	Bacterial infections, Pneumonia	Cyclophosphamide, aminophylline, IV antibiotics	Able to penetrate the mucosal barrier Can control the rate of administration and concentration Drugs affected by the gastrointestinal tract are better administered intravenously	Inconvenience to the patient Higher medical costs and possible side effects The tendency of the body to partially filter or breakdown active therapeutic agents	[22, 2]
Inhalation Therapy	Lung cancer, infections, asthma, COPD	Albuterol, levalbuterol, theophylline, formoterol	Elevated drug absorption and accumulation in the lungs Limiting or preventing drug accumulation in other healthy tissues and the bloodstream.	Modern drugs are frequently very potent. This raises the chance of unfavourable side effects on healthy tissues while also improving various respiratory disorders. It could be accompanied by the loss of active substances, which could make it challenging to achieve an exact dose of the active component. Although these inhaled antibiotic drugs can prevent and reduce the damaging pulmonary tissue, complete elimination is not possible.	[23, 24, 2]

With the advancement of science and technology in the past few decades, nanotechnology has offered an upgrade in the workable platform by enhancing detection and drug delivery not only for traditional lung diseases such as cancer, asthma, COPD, and cystic fibrosis but also for a wide range of emerging infectious diseases. In comparison to conventional drug delivery systems, the use of engineered nanoparticles has extended blood circulation, elevated drug loading capacity, reduced cytotoxicity, and minimal immunogenicity [6]. In the following chapters, we will be discussing the favourable role of theranostic nanoparticles in different types of lung diseases.

2. Lung diseases affecting the airways

The incidence of airway diseases has surged in recent decades, despite notable therapeutic advancements. These maladies encompass a range of medical conditions that afflict the air passages [25]. This section aims to expound upon the four primary classifications of airway diseases, detailing their etiology, symptoms, treatment options, and the potential advantages of nanoparticle utilization.

2.1   Asthma

Asthma is a disorder characterized by chronic inflammation and restriction of bronchial tubes. Affecting 334 million people globally, this inflammation is associated with hyperresponsiveness activated by specific agents like viruses and allergens. The disorder is observed to be about 12%-16% more prevalent in individuals from developed countries than from less developed countries. Studies have shown that children are more likely to be diagnosed with asthma, although 66% of all cases were noted to be below 18 years of age, and about 50% of them ceased to produce symptoms in early adulthood [26]. Severe cases of asthma generally have co-existing symptoms and hence mandate a multidimensional therapy, which entails a chain of assessments regarding the risk factors involved [27].

Several different risk factors are attributed to this disorder which includes a wide range of categories such as allergens (pollen), irritants (smoke/chemicals), exercise, climate change, certain medications, etc. [28]. Although the causative agents are more or less similar, there are slight variations in the different forms of asthma. The most common type of asthma is when the phenotypes are observed in children between the ages of 5-14 years. Childhood asthma can be life-threatening, triggered by various agents such as second-hand tobacco smoke, genetic factors, infections, pollutants, stress, etc. On the other hand, adult-onset asthma is known to develop at any stage of adulthood. Factors responsible for asthma in adults include obesity, pregnancy, stress, hormonal factors, allergens, etc. [29]. The severity of childhood and adult-onset asthma varies as the former group experiences intermittent symptoms, whereas the latter group faces more persistent and poorly managed symptoms. In adults, lung functions and structure are observed to decline at a faster rate than in children [30]. Occupational asthma is caused by exposure to an allergen or pollutant at work. It is observed that approximately 1 in 6 adult-onset conditions begin at the workplace [31]. Allergens inducing asthma, present in the milieu at specific periods of the year are known as Seasonal asthma. Individuals facing this condition experience asthma throughout the year, however, only suffer the symptoms when triggered by the cold breeze of the winter or pollen in spring [32].

The four fundamental symptoms detected in individuals suffering from asthma are Wheezing, Shortness of breath, Coughing, and Chest tightness [33]. While an individual appears symptom-free for a prolonged duration but is disrupted by suddenly deteriorated symptoms called asthma attacks. However, others may also experience symptoms on a regular basis. Asthma attacks occur as the outcome of muscle tightening (called bronchospasm) when triggered. The attack initiates inflammation of the airways and excessive mucus production, which triggers the fundamental symptoms previously mentioned [34]. Symptoms of childhood asthma usually begin to appear by 5 years of age. Nonetheless, a symptom frequently observed in childhood asthma is the three wheezing phenotypes. The transient wheezing phenotype is noticed during 3-5 years of age, with a range of factors such as declined lung function, smoking during pregnancy, exposure to other children, etc. Non-atopic wheezing phenotype is associated with those children who suffer from wheezing until adolescence triggered by viral infection exposed during the initial 3 years of life. IgE-mediated (atopic) wheezing/ classic asthma is characterised by constant wheezing as a result of early allergic sensitization and airway hyperresponsiveness [28].

Once aware of the symptoms, it's crucial to address them instantly as the severity of the disorder can aggravate quickly. A delay in instant treatment can lead to difficulty in breathing. Asthma management aims to control, prevent, and reduce asthma-related symptoms. Control-based asthma care has been shown to enhance the condition at a community level, which bases therapeutic adjustments on the prevalence of symptoms [35]. As part of asthma self-management, every patient, despite the level of severity should have a customized 'Asthma Action Plan', which entails the signs that signify how to manage the deteriorating control and aggravation. Individuals who received self-management training were admitted 36% less frequently than those without [36]. Once in control of the disorder is determined by the extremities of the symptoms, the medication is to be reduced with the goal to attain the lowest effective concentration of the drug.

Although there is no known solution for asthma at the moment, there are a number of efficient treatments that can help control and suppress the condition's symptoms and enhance the quality of life. The treatment administered at the initial stages of asthma is the best way to treat this disorder. Presently available therapy is broadly classified into two groups of drugs, anti-inflammatory drugs, and bronchodilators [37]. The anti-inflammatory drugs reduce the inflammation that restricts the airways. These drugs include leukotriene modifiers, mast cell stabilizers, and corticosteroids (administered intravenously, orally, inhalation). For instance, montelukast and zileuton are leukotriene modifiers that help inhibit asthma by hindering the formation of leukotrienes that are known to trigger bronchoconstriction. Likewise, cromolyn and nedocromil are mast cell stabilizers, provided to prevent asthma in children and individuals suffering from exercise-induced asthma. This anti-inflammatory drug controls the inflammatory chemicals produced by mast cells thereby decreasing the likelihood of the airways shrinking. Corticosteroids are the strongest type of anti-inflammatory medication, they have played a significant role in the management of asthma for many years. While the inhaled form of the drug is used to avoid an attack and enhance lung function, the injected form is used in high concentration to alleviate serious attacks, the oral medication is administered after the attack as a long-term solution. Prolonged usage of corticosteroids can progressively lower the probability of asthma, by making the lungs less susceptible to a variety of aggravating triggers. Although oral usage of the drug over an extended period of time has several adverse side effects like obesity, easy bruising, insomnia, cataracts, etc. [38].

Similarly, to ease up and dilate the lungs bronchodilators are also administered. This class of drugs is known to produce several side effects including rapid heartbeats and muscle tremors ensuing from its reaction with all beta-adrenergic receptors throughout the body. Hence beta-2-adrenergic receptors present in the lungs are targeted to reduce the possible side effects. Bronchodilators comprise beta-adrenergic drugs (for both short-term symptom alleviation and long-term management), anticholinergics, and methylxanthines. Based on how long the impacts of the drug last, beta-adrenergic drugs are further categorized. Short-acting beta-adrenergic drugs (SABA) are generally used to aid individuals during asthma attacks and exercise-induced asthma. The impacts of many SABA drugs (albuterol/levalbuterol), particularly those inhaled, last from only two to six hours. The need for additional use, more than recommended, should seek medical attention as it signals bronchoconstriction. On the other hand, Long-acting beta-adrenergic (LABA) drugs are known to last for 12 hours and hence patients are given two doses a day. They are utilized more frequently to steer clear of asthma attacks than to cure them. It was observed individuals who take LABA drugs (salmeterol, formoterol, and olodaterol) alone had an increased likelihood of dying, and therefore these medications are always administered coupled with oral corticosteroids [39]. Additionally, Ultra-long-acting beta-adrenergic drugs are known to be administered to patients as their effects are known to last for 24 hours, along with inhaled corticosteroids (ICS), similar to LABA, when administered alone can be lethal [40]. The most common form of delivering beta-adrenergic drugs is with the aid of metered-dose inhalers. The cartridges containing gas are converted to a fine spray of measured concentration due to the pressure. It increases the amount of drug and time taken to reach the lungs, however, there is a possibility that the drug may not be able to reach the severely restricted airways. Yet another medication widening the bronchi is theophylline, a type of methylxanthine. Despite its main goal of inhibiting asthma, theophylline can be fatal to an individual. Excess traces of the drug in the blood can be life-threatening causing abnormal heartbeat, with additional side effects such as insomnia, agitation, and seizures. Therefore, this medication is less commonly used than before [38].

Initial therapy suggests that for several individuals suffering from mild asthma, a combination of inhaled corticosteroid with formoterol be prescribed at low concentrations. LABA is favoured in this scenario as it is considered quick to act in comparison to the other beta-agonists. Alternatively, inhaled corticosteroids along with SABA have also been proved to enhance lung function and be cost-effective [41]. Whereas patients with repeated symptoms are administered a low dose of inhaled corticosteroid as an initial treatment. In some cases, increased doses of ICS-LABA do not prevent asthma, in such situations, additional therapies are considered. Currently, biological drugs have been gaining recognition for the treatment of severe asthma. Evaluation of the patient's phenotype and endotype is a necessary stage before treatment with biological agents. Type 2 (T2) High asthma falls under the category of types of severe asthma with respect to the level of eosinophils expressed. T2 high asthma comprises of both allergic and non-allergic eosinophilic asthma [42]. While Immunoglobulin E (IgE) plays a vital role in allergic asthma, T2 cytokine plays an important role in non-allergic asthma. Omalizumab, an anti-IgE medication, blocks IgE from interacting with its receptor in mast cells and basophils. It is known to alleviate asthma symptoms by 25%–50%. Mepolizumab is yet another example; known to bind to IL-5 and prevent it from interacting with its target. It has been discovered to reduce asthma by 50%. A drug called dupilumab binds to the IL-4 receptor, inhibiting the IL-4 and IL-3 signaling system. There has been a 50–70% decrease in asthma exacerbations [43]. Newer biological drugs under study like Tezepelumab and Astegolimab are monoclonal antibodies known to reduce symptoms and enhance airway functions [44]. Another therapy for patients with allergic asthma is immunotherapy. This involves the administration of extrinsic aeroallergen to which a patient is triggered either subcutaneously or sublingually (SLIT) with the goal of lessening the IgE-associated allergy linked to asthma. For example, in Australia therapies registered beneficial against asthma comprise dust mites and grass pollen SLIT, however, this therapy ought to be employed only under supervision as it has the potential to exacerbate the symptoms [45].

Existing therapeutic medications have their own drawbacks, particularly corticosteroids, which can have several adverse side effects and raise mortality. Numerous studies have shown that using nano-delivery devices to transport drugs can increase their efficacy. Along with delivery, with the aid of nanosensors, a device was created to help differentiate and detect asthma patients from healthy individuals. The electronic nose/ E-nose was built to identify volatile organic compounds (VOC) from exhaled gas. This non-invasive device with its excellent precision makes it possible to differentiate the four inflammatory asthma phenotypes based on the sputum analysis [46, 47]. Drugs coupled with NPs are known to have a better therapeutic impact by boosting efficacy, accurate delivery system, and increased concentration of drug in the lung [48].

Inhalation therapy is known to be crucial for the treatment of asthma. With the growth in the use of dry powder inhalers, researchers formulated nano-embedded microparticle (NEM) powders that possess nucleic acids, enhance drug delivery, and increase lung permeability making the most beneficial for inhalation therapy [49]. A study investigated the influence of NPs when coupled with steroid-based drugs such as Beclomethasone dipropionate and dexamethasone. A surge in drug accumulation and a fall in the side effects were observed in the former, while dexamethasone NPs assisted in the transportation and in avoiding allergic inflammation caused by eosinophils [50, 51]. Similarly, another study worked on Salbutamol a frequently prescribed bronchodilator when modified into NPs has proved to be more favourable as the total lung deposition escalated 2.3 times. This increase in concentration at the target site is known to be instrumental in the long-term recovery of bronchospasm [52].

Furthermore, nanosized non-ionic surfactant-based vesicles (niosomes) were developed as drug-delivering agents with the additional property of controlled release over a period. The study concluded that a spherical niosome entrapped 66.19% of salbutamol, which was released at a controlled rate for 8 hours [53]. Drug targeting T helper type 2 (Th2) cytokines are also administered as therapy for asthmatic patients. A crucial molecule called bavachinin obtained from *Psoralea corylifolia* has been found to be beneficial in asthma treatment. A study conducted in 2018, aimed to develop bavachinin-loaded PEG5000-PLGA NPs, as the molecule alone was least potent due to its minimal water solubility. The results portrayed anti-asthma effects in murine models [54]. Likewise, Andrographolide (AG) is a diterpenoid with an anti-asthmatic role, unlike corticosteroids with various side effects. To solve issues such as low bioavailability and short life,

AG enclosed NPs were developed and orally administered to murine models. The study observed an increase and decrease in bioavailability and inflammatory agents respectively [55].

Nanoimmunotherapy has been upgraded from its conventional method, as better therapeutic outcomes have been observed. Studies have investigated the association between the sizes of PLGA molecules and the incidence of cytosine-phosphate-guanine (CpG) motifs against an antigen. Results showed a significant reduction in inflammation when provided with smaller PLGA molecules. Moreover, CpG-enclosed PLGA particles induced Th1-type immune response and hence plays a vital role in the inhibition of house dust mite-induced allergy [56]. In the majority of allergic disorders, chemokine receptor systems are regarded as a crucial component of the immune response. It has been established that unique chemokine profiles regulate the clinical manifestations of asthma. The primary component in the advancement of asthma therapy is chemokine signal transduction via CC chemokine receptor 3 (CCR3). A group of researchers formulated a novel NP CCR3 inhibitor (R321) peptide to prevent signal transduction. It was observed that R321 successfully restricted eosinophil accumulation in the airways and blood and also inhibited airway hyperresponsiveness in murine models [57].

Gene therapy coupled with nanotechnology is known to be highly beneficial against asthma. For instance, in a study, CIN therapy was used on mice with ovalbumin-induced allergic asthma. Patients were found to produce low levels of IFN-γ, a Th-1 cytokine that downregulates Th2-associated inflammation and hyperresponsiveness. While several studies worked on adenovirus-mediated IFN-γ gene transfer, an obstacle encountered was the frequency of the gene delivered. Hence, they developed chitosan/IFN-γ pDNA nanoparticles (CIN), which was advantageous as it demonstrated elevated IFN-γ with a decline in Th2-cytokines and other effects of asthma [58].

2.2   Chronic Obstructive Pulmonary Disease (COPD)

Chronic Obstructive Pulmonary Disease (COPD) is a condition driven by restricted lung airflow caused due to exposure to several triggers. This inflammatory disease is known to affect pulmonary vasculature and lung parenchyma. Protease-antiprotease imbalance and oxidative stress are thought to play a vital role in this process. Emphysema is the generic term given to one of the structural alterations associated with COPD, where the alveoli are damaged causing obstructive physiology. While elastin is broken down by proteases, there is a lack of elastic recoil, which causes a compressed airway while exhaling [59]. A sporadic form of Emphysema called alpha-1 antitrypsin deficiency (AATD) is characterised by a loss in antiproteases, leaving the lung parenchyma at risk for damage by the enzymes. The mutant protein misfolding accumulated in the liver is known as the primary cause of this disorder [60]. A global study conducted in 2019, stated that 212.3 million cases of COPD were reported, where COPD was deemed responsible for 3.3 million deaths. Countries such as Denmark, Myanmar, and Belgium had the highest incidence of COPD, with Nepal regarded as having the highest death rate per 100,000 [61].

COPD exacerbations can be triggered by several factors. A substantial amount of research suggests that smoking status (most common cause), ambient particulate matter pollution, occupational exposure to particulate matter, gases, household air pollution from solid fuels, AATD, ambient ozone pollution, and low and high temperatures are the risk factors that lead to COPD [59]. A study in 2019 indicated that 47.8% of global patients with disability-adjusted life years were credited to exposure to 87 environmental, occupational, and behavioural agents. The study compared the exposure value of these risk factors over 204 countries between the years 1990 and 2019. Death rates due to COPD have decreased with a decline in critical hazards associated with SDI, including inadequate water, sanitation, air pollution, vitamin A, and zinc shortages. The study also depicted a slight fall in exposure to tobacco and a more than 1% rise in the three risk factors such as high BMI, particulate matter pollution, and high FPG in one year [62].

Breathing difficulties are one of the main symptoms of COPD, particularly post a physical strain, as an injury to the lung makes it difficult for the entry and exit of air. Chronic coughing and wheezing caused due to inflammation and excessive mucus production are frequently observed in COPD patients. Another typical COPD symptom leading to uneasiness is chest tightness. Fatigue and unintended weight loss are also possible symptoms encountered by patients with COPD due to the additional energy required to breathe, moreover, a decreased appetite is also observed as a result of the shortness of breath [63]. Other observations include increased anterior-posterior chest wall diameter which is a condition where the chest appears to be inflated and also central cyanosis describes a blue discolouration of the skin, mouth, nails, etc due to low levels of arterial oxygenation [64].

Except in the case of alpha-1 antitrypsin deficiency no standardized biomarker is used to identify COPD during its early stages. However, there are several biomarkers under investigation that are associated with the pathogenesis of COPD [65]. For instance, elevated levels of C-reactive protein and fibrinogen are known to increase the risk of COPD. Regulatory factors for these markers such as IL-1, IL-6, and TNF-alpha are heavily involved in the pathogenesis of COPD. Enzymes like superoxide dismutase 3 and epoxide hydrolase-1 are known to hinder elevated oxidative stress triggered due to smoking, making them a suitable marker for COPD detection [65]. Another area of investigation as markers is matrix metalloproteinases (MMPs) are possible markers (MMP-8, MMP9, MMP12) for COPD. Evidence suggests that sputum MMP-8 levels are greater in symptomatic smokers than in asymptomatic smokers, hence limiting detection at the latter stages of COPD. Likewise, escalated levels of cathepsin S and cystatin C in symptomatic smokers were observed against asymptomatic smokers [66]. A novel approach aimed to obtain ECG Derived Respiration (EDR) to classify COPD patients as EDR can be effortlessly derived from the ECG. The respiratory pattern in people with COPD differs from that of healthy individuals due to variations in the underlying mechanisms. The study reported that EDR can be developed into a COPD detection device as morphological variations were observed in COPD patients in both respiration and EDR waveform [67]. A recent study attempted to identify COPD patients utilizing artificial intelligence methods. This was presented by a unique framework that uses an ultra-wideband (UWB) radar-based temporal and spectral properties to assemble machine learning models. This new respiration data was derived from a non-invasive technique from 1.5 metres distance with the aid of UWB radar. The results are encouraging, with a high COPD detection accuracy score of a 100% [68].

Since there is now no known medication that may stop or reverse the course of COPD, it is imperative that the disease be discovered as soon as possible, while lung damage may still be treatable. The effectiveness of therapies in early illness will eventually decide the usefulness of current efforts to uncover COPD. Initial efforts need to be taken to aid reduce the several symptoms observed [69]. Several studies have shown that early smoking cessation has the ability to slow down or even halt the deterioration of lung function, illustrating the importance of early detection of the disease. A randomized clinical trial investigated 3,926 smokers suffering from mildly restricted airways. The test sample was randomly categorized into smoking cessation and non-intervention groups. Individuals who ceased smoking showed an improvement in FEV1 in just one year. The study suggests that smokers with airway obstruction have a positive therapeutic effect after quitting [70]. Few therapies are developed to target specific peptides associated with COPD. Roflumilast aids in preventing the enzyme phosphodiesterase-4 (PDE-4) and also hinders inflammation. Studies have observed that Roflumilast has demonstrated significant progress in the treatment of COPD [71]. Anti-inflammatory medications, like N-acetylcysteine (NAC) have primarily been employed to minimize the frequency and length of exacerbations, that are associated with loss of lung function. A HIACE study involving Chinese COPD patients was administered with high-dose NAC. The results obtained indicated an improvement effect on the small airways [72]. Biological agents that cause infections are known to be influencing factors for COPD exacerbations. Studies have evaluated the successful effects of influenza and pneumococcal vaccinations being able to reduce exacerbations, risk of death, and even preserved lung function [73]. Several inhaled bronchodilator products have been studied in this mode of therapy for COPD. According to the Lung Health Study, the FEV1 increased in two of the study groups. Although this rise in FEV1 was observed in one year period and remained the same for the next 4 years. Additionally, the progress observed with the use of the product reversed when discontinued indicating no long-term benefit of the therapy [74]. The effectiveness of inhaled Budesonide to treat COPD smokers was studied by Pawels et al, where individuals selected included people with early-stage COPD between the age of 30-65. The study depicted that the inhalation of Budesonide was associated with a modest one time increase in FEV1, although it has no impact on those experiencing a long-term reduction in FEV1 [75]. The IMPACT Study (Informing the Pathway of COPD Treatment) has revealed information on the advantages of single inhaler triple treatment (ICS/LABA/LAMA) in comparison to ICS/LABA. The study reported that this therapy aided in the decline in exacerbations and ameliorated lung function. Although oxygen supplementation is known to significantly assist pulmonary hypertension, only a few studies have shown that long-term oxygen therapy can reduce the mortality rate in COPD patients [76].

According to a study conducted by the WHO, COPD has become the third leading cause of mortality, accounting for almost 4 million deaths yearly, and still increasing. The main obstacle in treating COPD is the absence of effective medications [77]. Those that are administered are not specific to the disorder and prolonged exposure to it might potentially have a major side effect on the patient. Additionally, with traditional respiratory drugs, limitations such as the challenges faced to supply and deposit the drug effectively persist. They also fail to eliminate the dense bacterial biofilm formed due to the presence of bacterial growth in the excessive mucus formed as a result of COPD. For example, Amikacin (AM) is a popular antibacterial medication prescribed against COPD coupled with gram-negative bacilli lung infections. Studies conducted stated that when combined, AM-NPs raise the amount of drug content and reduce the usually observed side effects [78]. To overcome the mucus barrier issue, a group of scientists worked on an ion crosslinking technique with chitosan, and black phosphorus quantum dots (QDs) and positively accomplished the development of a nano drug facilitated by black phosphorus QDs (BPQDs-AM NPs) that is PEG surface modified. The outcomes demonstrated that the positively charged chitosan and hydrophilic PEG might aid in the nanocarrier's ability to breach the mucus barrier and attach to the epithelial cells. The BPQDs are then known to quickly oxidize and damage to produce phosphate ions, which cleave the nanosphere to assist drug release. Concurrently in order to prevent the formation of a biofilm, the antibacterial property of chitosan was boosted by its protonation ensued by the oxidative degradation of BPQDs. Animal studies indicated that the use of BPQDs-AM NPs can greatly minimize the toxicity, the adverse effects, and relax the airway obstruction caused due to COPD [79].

Identifying ways to suppress ROS levels and glucocorticoid resistance (by decreasing HDAC2) will aid in curbing the development of COPD. Researchers constructed a core-shell lipid-polymer NP (LPNs), consisting of a polylactic acid shell that is covered with an antioxidant Mn porphyrin dimer (MnPD) and a core that comprises a cationic lipid propane shell and an HDAC2-encoding pDNA. The expression levels of HDAC2 and ROS was found to increase and decrease respectively as a result of LPN therapy. Furthermore, LPN therapy also decreases the expression of IL-8 in COPD patients [80].

The treatment of COPD focusing on miRNAs has received more attention in recent years. miR-146a in particular when formulated including a cationic lipid DOTAP as a carrier is known to reduce the expression levels of IL-1 receptor associated kinase- 1(IRAK1) gene in individuals suffering from COPD. Additionally, miR-146a NP was found to effectively bring the level of IRAK1 down up to 40% [81].

Despite the multiple studies based on PLGA nanosystem therapy for COPD, some studies report that PLGA NPs have limitations. For instance, the surface of NPs is known to be negatively charged due to the use of emulsifier polyvinyl alcohol. The negative charge resulted in the quick removal of the NPs by the airway defence system. To overcome this challenge researchers coated the surface of PLGA-NPs with PEG, to escape the defence system, expand the retention time, target specific cell types in the body, as well as to penetrate the mucus barrier [82].

A combination of drugs has also been studied as a treatment for COPD. Researchers formulated liposomes that consisted of formoterol (beta 2 selective receptor agonist) and beclomethasone (glucocorticoid steroid) that aimed to improve lung function and retain drug effect for extended periods respectively. Hence proved crucial to treat COPD [83]

2.3 Chronic bronchitis

Chronic Bronchitis (CB) is a form of COPD, described as a cough lasting longer than three months appearing over a period of 2 years. The path of the disorder includes excessive production and secretion of mucus by the goblet cell [84]. The hallmark of this disorder includes the release of inflammatory agents like cytokines and IL-8, and a decrease in angiotensin-converting enzymes and neutral endopeptidase as a response to toxic, infectious triggers. This is followed by airflow restriction and a decline in lung function. The clogged airways induce a sense of irritation and cough. A study conducted within a group of COPD patients reported that CB patients had worsened respiratory, exercise ability, lung function, and recurrent exacerbations than non-CB COPD patients [85]. The prevalence of CB globally has been reported to alter between 3%-7% in healthy individuals. Although, from the adults already diagnosed with COPD about 74% of these patients have been reported with CB [86]. Interestingly, in 2009 the National Centre for Health Sciences stated that 67.8% of adults suffering from CB were women. Similarly, a 10-year study among 21,130 Dutch adults mentioned that chronic mucus production was 10.7% and 8.7% in women and men respectively. While the reasons for this incline in the incidence of CB in women are unknown it is suggested to be due to the hormonal influence or difference in symptom reporting between sexes [87].

Similar to as observed in COPD the major cause of CB is smoking. A study mentioned that the combined incidence of CB among present smokers is 42%. It should however be emphasized that CB has been reported in about 4-22% of non-smokers. Hence indicating the presence of other causative agents such as biomass fuels, chemicals, dust, etc. [87]. According to a national household study conducted among individuals in South Africa, exposure to smoky home fuels and occupational exposure are risk factors for CB in women and men, respectively [88]. The occurrence of gastroesophageal reflux is another possible risk factor for CB, potentially due to the pulmonary aspiration of refluxed gastric components. The consequence of aggravating the esophageal mucosa could cause acid-induced damage, infections, and neural-facilitated bronchoconstriction [89]. Another risk factor observed contributing to CB is Livestock farming. 4735 Norwegian farmers picked for a study had depicted an elevated risk of CB and declined lung performance in comparison to crop farmers due to the raise in levels of allergens, endotoxins, peptidoglycans, dust, etc. Additionally, a meta-analysis stated the link between exposure to agricultural pesticides and the incidence of CB [90, 91]. A comprehensive investigation performed by the Committee on the Medical Effects of Air Pollutants (COMEAP) in the UK described a relationship between the prevalence of CB as a result of long-term exposure to air pollution [92]. While infections are generally responsible for acute bronchitis, repeated exposure can lead to chronic bronchitis. Influenza type A/B and Staphylococcus/Streptococcus are some of the major known causative viruses and bacteria. Individuals subjected to environmental pollutants like dust or chemicals including ammonia and sulfur dioxide have an increased risk of CB [84].

Symptoms in CB patients generally are absent or mild and are known to get aggravated as the severity progresses. Although once considered to represent a range of disorders under the term COPD, it is understood to be a distinct entity that can exist with or without an airflow restriction CB [93]. Individuals with CB-type symptoms along with airflow obstruction indicate COPD or CB, whereas the absence of airflow restriction is diagnosed as. Other clinical features, similar to that of COPD include frequent coughing along with mucus, wheezing, shortness of breath, tightness of the chest, and in extreme cases weight loss, weakness, and swelling of feet [94]. A study conducted among Korean subjects aimed to differentiate CB from non-CB patients within a group of individuals suffering from COPD. They observed that with regard to the symptoms and exercise capacity non-CB patients scored better than CB patients. Psychological scores depicted that the non-CB group was comparatively less depressed and anxious than the CB group. CB patients had reduced lung function tests and more exacerbations in severe cases. The investigation concluded by stating that the CB group had the worst symptom, quality of life, increase mortality, and aggravated exacerbations [95].

The primary goal of treatment for CB is to alleviate the symptoms (mucoactive agents and beta-agonists), prevent the decline in lung function (smoking cessation), reduce exacerbations (phosphodiesterase-4 inhibitors), inhibit excessive mucus production, and regulate inflammation [96]. Both pharmacological and nonpharmacological approaches can be used to accomplish this goal. A few examples of pharmacological interventions include:

- Bronchodilators: Aid in clearing airway lumen and improving ciliary function. Methylxanthines and SABAs are generally provided to stimulate mucus clearance. Clinical investigations have observed the beneficial effects of theophylline against CB by enhancing lung function, however, no improvement in cough and sputum was reported [97]. LABAs are known to have favourable effects by decreasing hyperinflation and elevating peak expiratory flow. For instance, formoterol is known to substantially enhances mucociliary clearance in CB patients as compared to the placebo group [87].

- Glucocorticoids: These drugs reduce the generation of mucus and inflammation, thereby causing a decline in exacerbations. Beclomethasone, budesonide, and triamcinolone are some examples of inhaled corticosteroids that are valuable as therapy for CB because of their anti-inflammatory properties. Although advantageous, long-term exposure can cause osteoporosis, diabetes, and hypertension [98].

- Antibiotics: Macrolide therapy has proven to have anti-bacterial, and immune-modulatory effects and therefore may play a crucial role in treating CB [99]. 1142 COPD patients, highly susceptible to exacerbations were treated with Azithromycin (250 mg/day) in a MACRO study conducted in 2011. The study concluded that azithromycin was indeed the most beneficial to inhibit exacerbations [100].

- Phosphodiesterase-4 inhibitors: Promotes smooth muscle relaxation of the airways and decreases inflammation by hindering the hydrolysis of cAMP, which releases inflammatory agents upon degradation. cAMP is known to regulate the activity of the cystic fibrosis transmembrane receptor (CFTR) and ciliary beat frequency, which are the two important factors in efficient mucociliary clearance [87]. Roflumilast has been proven to enhance pre- and post-bronchodilator FEV1 and dyspnea against CB linked with COPD [101]. A study conducted included 1945 patients with CB and airflow obstruction at entry who were administered roflumilast (500 µg/day) for 52 weeks as part of receiving Appropriate Combination Therapy (REACT) [102]. Likewise, a study enlisted 2254 individuals for the Roflumilast effect on Exacerbations in Patients on Dual (LABA/ICS) Therapy (RESPOND) [103] and observed that the exacerbation frequency of this study was similar to that obtained from REACT. From these studies, GOLD 2018 suggested the administration of roflumilast for CB patients without ICs/LABA. However, several side effects such as diarrhea, nausea, weight loss, insomnia, etc. were observed during the initial stages [104].

- Bronchial rheoplasty: A novel treatment that used non-thermal pulsed electric fields with the goal of removing the excess mucus producing cells restricting the survival of healthy epithelial cells. These energy fields are targeted to cause cell death by disturbing cellular homeostasis. The study concluded via histological studies, the statistically substantial decline in the number of epithelial goblet cells, and furthermore estimated the feasibility and safety of the therapy in CB patients [105].

- Mucoactive agents: The primary aim of these drugs is to prevent the overproduction of mucus, elevate ciliary transport, and decrease mucus tenacity. The agents are classified into four based on their mechanism of action, including mucoregulators, mucolytics, and mucokinetics, and expectorants [106]. The most familiar drug investigated is N-acetylcysteine (NAC), which has antioxidant and anti-inflammatory properties. A meta-analysis involving 4155 patients suffering from COPD were prescribed NAC had lesser exacerbations related to CB, and fascinatingly, this was more obvious in patients without airflow restriction [107].

Non-Pharmacological measures similar to COPD include smoking cessation. Enhanced mucociliary function and reduced goblet cell hyperplasia is observed in several CB patients post-smoking cessation. However, due to the paucity of data regarding smoking cessation and its effect on CB, further studies are required [108]. Treatment for CB often also includes pulmonary rehabilitation, which includes education, modifying one's lifestyle, engaging in regular physical exercise, and avoiding exposure to toxins at work or in one's surroundings at home [109].

Gold NPs due to their effortlessly adjustable optical and photodynamic features for disease treatment have been widely explored in drug delivery studies. In a study performed by Geiser et al., inhaled gold NPs (AuNPs) were formulated as carriers for CB associated COPD treatment. Scnn1b-transgenic (Tg) mice models were utilized as they brought about CB and mucus hypersecretion in-vivo. Post-administration of AuNP aerosols, the particles that formed small agglomerates less than 100 nm in diameter were located at the surface of the lung epithelial cells, whereas those larger than 100 nm were found in macrophages. Both Wt and Scnn1b-Tg animals had inhaled AuNPs accumulated in the alveolar epithelium, nonetheless, the Tg mice demonstrated slower macrophagic uptake and increased epithelial internalisation of the AuNPs. These findings concluded that inhaled AuNPs are reliable delivering agents to the lung epithelium in CB patients [110].

2.4   Cystic fibrosis (CF)

CF is a hereditary disorder triggered by a defective cystic fibrosis transmembrane conductance regulator (CFTR) gene located on chromosome 7, which causes mucus accumulation due to a decrease in salt and water transportation. The CFTR gene is primarily responsible for the transportation of salt across various human tissues [111]. In 1949, Lowe et al. proposed that CF is an autosomal recessive pattern of inheritance, and the high salt concentrations in the sweat were suggested as an anomaly in the electrolyte transport from the sweat gland. However, further investigation revealed that CF was an autosomal recessive trait of the exocrine gland [112]. It is crucial to recall that CF patients have healthy lungs at and after birth, however, post-exposure to infection and the accompanying inflammatory response, the disease is found to develop as a cascade of events. CF has been estimated to affect more than 70,000 individuals globally and is known to frequently affect non-hispanic white people [113].

There are approximately 2000 distinct mutagens, all of which are classified into five different classes. The classes of dysfunction are known to cause mutations of the CFTR gene. Class 1 dysfunction is caused by nonsense, frameshift, or splice-site mutations that hinder the translation of the CFTR protein, resulting in its total absence. Class 2 dysfunction is the abnormal post-translational processing of the CFTR protein. This process is known to be vital for the accurate intracellular transit of proteins and hence this dysfunction causes the improper location of the protein in the cell. Class 3 dysfunction is characterized by reduced protein activity, which is observed as a completely developed, non-functional protein channel in the cell membrane. Class 4 dysfunction involves a correctly produced and localised CFTR protein, however, the rate of chloride ion flow and duration of channel activation is reduced. Class 5 dysfunction is identified by a reduced concentration of CFTR channels in the cell membrane as a result of the quick degradation of cellular function [114].

Typical respiratory symptoms such as persistent cough, hyperinflation of the lungs, and restricted airways are observed in CF patients. Repeated infections linked to inflammatory cell aggregation and expulsion of cell content are found to damage bronchial walls as the condition worsens, resulting in loss of bronchial cartilaginous support and muscle tone. Further progress of the disorder can induce exacerbations of cough, dyspnea, raise in sputum production, weight loss, and often leads to permanent loss of lung function over time, etc. [115]. Patients suffering from CF over long periods have generally airways infected with either *Staphylococcus aureus* or *Pseudomonas aeruginosa*. Additionally, other microbes that are found to colonise the airways of CF patients are *Stenotrophomonas, Achromobacter, xylosoxidans*, nontuberculous mycobacteria, *Aspergillus fumigatus,* etc. Prolonged exposure to this airway colonisation instigates an inflammatory response, worsening CF symptoms [116]. Furthermore, CF patients have thick mucus in their pancreas, which prevents the release of digestive enzymes. Diabetes associated CF can also occur as a result of pancreatic injury. Studies have shown that most men with CF are infertile, whereas CF women may struggle to become pregnant. Abnormal functioning sweat glands are responsible for the elevated risk of dehydration in CF patients [114].

CF is a systemic disease that, if not well managed, can have significant negative effects on the quality of life. Hence, the treatment goal with regard to lung function should be to maximise function by tackling respiratory infections and mucus clearance [117]. *P. aeruginosa* frequently causes infectious etiologies, hence antibiotic treatment should address this pathogen's whole spectrum. Sputum is, however, cultured, and the pathogens present should have a sensitivity profile determined. Oral antibiotics may be effective for mild exacerbations, but intravenous therapy is necessary for more severe exacerbations. Inhaled bronchodilators such as albuterol and ipratropium bromide should be administered to assist ventilation and oxygenation. In order to ease the blockage, anti-inflammatory medications such as glucocorticoids are also utilised to help relax the airways. Studies also recommend the use of hypertonic saline despite the severity. This sterile saline solution in distinct concentrations (3%, 3.5%, 7%) basically attracts the water content thereby causing thinned secretions, enabling a CF patient to expectorate [111].

A novel class of drugs called CFTR modulator therapy was designed against a specific class of dysfunction that caused the alteration of the CFTR gene. For instance, Ivacaftor was prescribed when a mutation at G551D was the major aberration (class 3 dysfunction). It functions by attaching to the damaged CFTR protein at the cell surface and opening the chloride channel, reversing the protein dysfunction. This was the first drug to directly target protein channels, as opposed to just treating the symptoms. Another medication designed was the Lumacaftor, a chaperone molecule that aids in transporting the damaged CFTR protein from the organelles to the cell surface. When administered alone, this medication was found to have minimal clinical value. However, the 92-week PROGRESS research revealed slight benefits when coupled lumacaftor/ivacaftor was prescribed to individuals who tolerate the medication in both pulmonary function tests and body mass index (BMI). Likewise, another medication similar to lumacaftor is Tezacaftor which functions as a CFTR chaperon enhancing the intracellular processing of CFTR. The clinical efficacy of the drug was observed only when combined with ivacaftor [118].

For CF patients, airway clearance treatments (ACT) are an essential part of their treatment. To maintain healthy airways, and avoid inflammation and infection, patients must perform ACT to assist clear the secretions. The CF foundation in 2009, compared the effectiveness of all treatments that exist, and it was reported that Act was similar to the other forms of therapy with regard to its virtue [119].

Despite significant improvements in CF medication therapy, the disease process is still progressing, and without surgical intervention, the lungs would eventually collapse early from the disease load. The preferred course of end-stage treatment for advanced CF is lung transplantation [111]. The International Society of Heart and Lung Transplantation issued a list of situations to be used to determine transplant referral, which takes into account the FEV1 falling to 30% of predicted values, rapidly declining FEV1 despite optimal therapy, a 5-year predicted survival of less than 50%, etc. For almost all lung transplants for CF, both lungs must be replaced. This is due to the fact that a native, sick lung would serve as a source of infectious secretions that would endanger the transplanted lung and possibly cause respiratory failure [120].

From the several existing therapies, gene therapy offers a wide range of potential applications considering that CF is caused due to the mutation of the CFTR gene. The treatment includes the use of viral or non-viral vectors to transfer the CFTR gene to the target site. However, it is important to note that non-viral vectors (such as NPs) offer more benefits than viral vectors, including lower production difficulty and expense, increased shelf life, reduced immune regulatory response, improved drug tolerance, etc [121]. A research investigation employed plasmid DNA containing the CFTR gene enclosed in PEG NPs. They observed that this approach was safe and had an increased gene transfer efficiency [122]. A comparison study performed to assess several non-viral vectors reported that cationic liposome preparation GL67A was the most useful vector. The initial phases of the study depicted that aerosolised PGM169 plasmid DNA containing the CFTR gene when delivered with the aid of liposome NPs was beneficial and had positive therapeutic effects. The next phase of the study attempted to observe the clinical effects of PGM129/Gl67A which resulted in the improvement of lung function. Furthermore, when prescribed once a month for a year, the results displayed a rise in FEV1 and enhanced lung function [123].

Recently, a promising approach in treating cystic fibrosis (CF) has emerged through the use of nanoparticles (NPs) for RNA-based therapy. Robinson et al. conducted a study where they utilized lipid-based NPs (LNP) to encapsulate chemically modified CFTR mRNA (cmCFTR), with the aim of curing CF and restoring chloride secretion. Their findings demonstrated successful expression of the desired transcripts using chemically modified mRNA (cmRNA) [124]. When CFTR knockout mice were administered LNP-cmCFTR in the nasal cavity for a 14-day period, chloride secretion was effectively revived. The study also revealed peak CFTR activity three days post-transfection. The mechanism and effectiveness of LNP-cmRNA were comparable to those of ivacaftor, a known CFTR modulator.

In another study, Haque et al. developed chitosan-coated PLGA NPs and administered intratracheal and intravenous doses of chemically modified human CFTR (hCFTR) mRNA to CFTR-deficient mice [125]. This approach resulted in significant recovery of key lung function measures, including a notable increase in forced expiratory volume in one second (FEV1) and a decrease in chloride secretion.

Furthermore, Tagalakis et al. formulated a receptor-targeted nanocomplex (RTN) for CF treatment, which combined specific peptides, liposomes, and siRNA. Their study demonstrated that NP-mediated siRNA transfection could reverse mucociliary abnormalities in vivo [126].

Overall, these studies highlight the potential of utilizing NPs for RNA-based treatments in CF, showing promising results in restoring chloride secretion, recovering lung function measures, and reversing mucociliary abnormalities.

A novel non-viral vector based on cell-penetrating peptides was constructed to specifically deliver genes using glycosaminoglycan binding enhanced transduction (GET). The GET was altered by researchers with PEG to stabilize and improve gene transfer. The study demonstrated that the PEG-GET complex could quickly diffuse into CF patient sputum and navigate the mucus network. Since PEGylated particles have proven to be advantageous in terms of distribution, safety, and efficiency, the study concluded that the PEG-GET complex for enhanced transfection is an apt treatment for CF patients [127].

Similar to other lung diseases, CF patients also experience a poor outcome from medications due to the presence of the mucus barrier. To overcome this obstacle, a few investigators developed a phage transport assay to filter T7 phage-displayed cysteine restricted heptapeptide libraries in CF mucus. This aided in isolating the net-neutral charged peptides that can penetrate the mucus exvivo, with ~600-fold better penetration, and also enhance the uptake of the drug to the epithelial cells, as compared to PEGylated particles [128]. Due to their decreased conformational flexibility and increased stability, cysteine-constrained peptides have a greater receptor-specific affinity. This biomolecule-based technique has the potential to overcome several delivery impediments and improve the potency of treatments for conditions like CF [2].

As mentioned earlier, lumacaftor/ivacaftor combination therapy is known to be a beneficial therapy for CF patients. In the context that, when inhaled the therapeutic effects of these drugs are enhanced, a nanostructured lipid carrier (NLC) was formulated for direct inhalation. The study concluded that NPs reversed the abnormal expression of CFTR protein and increased drug loading capacity [129].

Recent research has formulated two types of liposome antibiotics which include Liposome Amikacin (AM) and Liposome Ciprofloxacin. While liposome AM was authorised for clinical use in 2018, liposome ciprofloxacin has completed phase 2. Experimental studies have proven liposome ciprofloxacin to be more favourable than traditional ciprofloxacin, as encapsulated ciprofloxacin PLGA NPs were found to have good anti-protective antigen strain efficacy and significantly improved mucus penetrating ability [130]. Additional advantages include these NPs had increased antibacterial activity and larger capacity for drug loading. Furthermore, Tobramycin was also coupled in PEG-coated PLGA NPs which was observed to alleviate CF [131].

3. Lung diseases affecting air sacs (alveoli)

Alveolar diseases are categorized based on the contents of the air sacs (alveoli). When materials obstruct the alveoli's normal physiological function (ventilation), alveolar disorders are present. Alveoli can be overinflated, stretched, deflated, or even completely destroyed. This section aims to explicate upon the four primary classifications of air sac diseases, detailing their etiology, symptoms, treatment options, and the potential advantages of nanoparticle utilization.

3.1   Pneumonia

Pneumonia is a lung infection that can range in severity from moderate to severe. It occurs when an infection causes the air sacs (alveoli) in the lungs to fill with fluid or pus, making it difficult to get enough oxygen into the bloodstream. Pneumonia can affect anyone, but certain groups are at a higher risk, including children under the age of two and individuals over the age of 65. These age groups may have less robust immune systems, making them more susceptible to the infection.

Pneumonia can affect one or both lungs and can be caused by bacteria, viruses, or fungi. If the infection is caused by bacteria or a virus, it can be transmitted from one person to another. Common causes of pneumonia include flu viruses, cold viruses, the RSV virus (a leading cause of pneumonia in infants aged one and under), Streptococcus pneumoniae bacteria, and Mycoplasma pneumonia.

In some cases, pneumonia can occur in hospitalized patients and is referred to as "ventilator-associated pneumonia." This type of pneumonia is associated with the use of ventilators in medical settings [132]. If pneumonia occurs while in the hospital and not ventilator associated, it is called "hospital-acquired" pneumonia. However, the vast majority of people have "community-acquired pneumonia," which means they did not catch it in a hospital. Smoking cigarettes and consuming too much alcohol might significantly increase your chances of getting pneumonia. Pneumonia can strike at any age. This is a common respiratory infection that makes breathing difficult. This is due to the fact that it causes the alveoli, or air sacs, in the lungs to fill with fluid. Pneumonia is especially dangerous for young children, persons with chronic conditions like asthma, and the elderly.

Some of the risk factors for this condition include viral respiratory infections, chronic lung disease, cerebral palsy, heart disease, recent surgery, and cigarette smoking. Pneumonia symptoms differ from person to person based on age and overall health. Persistent cough, high fever, chills, yellow or greenish phlegm, bloody mucus, shortness of breath, severe and stabbing chest pain, headaches, heavy perspiration, nausea, vomiting, diarrhea, exhaustion, loss of appetite, and loss of energy are some of the most prevalent symptoms. Infants with this disease often appear dull and listless, but the elderly may also begin to show indications of bewilderment.

The aetiology of pneumonia might also have an impact on the symptoms. Bacterial pneumonia is characterised by a high fever that reaches 105 degrees Fahrenheit. Excessive sweating, as well as an elevated heart and pulse rate, are common symptoms of this form of fever. It can also cause the patient to get confused. As a result of the lack of oxygen, the patient's lips and nails may turn blue. The symptoms of viral pneumonia are similar to those of the regular flu. These symptoms include a mild fever, a dry cough, headaches, fatigue, and muscle pains. The patient should also expect some mucous. Pneumonia can lead to complications such as bacteremia, which occurs when bacteria enter the bloodstream [133]. Septic shock and organ failure might result from this. Breathing difficulties can be seen which require the use of a breathing machine until the lungs heal. Fluid accumulation between the tissue layers that border the lungs and chest cavity. This fluid can become contaminated as well. When a pocket of pus grows inside or around your lung, you have a lung abscess.

Pneumonia is easily treatable, but if not treated promptly, it can lead to a variety of consequences, including respiratory failure, sepsis, and lung abscesses. Such problems are particularly common in newborns and the elderly. Pneumonia treatment might take anywhere from 3 to 8 weeks, depending on the patient's overall condition. It is often treated with antibiotics taken orally. In some situations, antibiotics must be administered intravenously, which necessitates hospitalisation. In addition to medication, the patient should obtain enough rest and drink plenty of water.

Pneumonia is readily avoidable. Getting vaccinated against bacterial pneumonia and flu is one approach to reduce your risk of contracting this disease. This is particularly important for children and individuals over the age of 65 [134]. Washing your hands before eating and cooking will help keep pneumonia at bay. A balanced diet and lifestyle can also assist to enhance immunity and reduce the incidence of pneumonia. Tobacco use is significantly more dangerous since it affects the lungs and lowers the lungs' ability to fight infections. For bacterial pneumonia, antibiotics may be recommended. After one to three days of antibiotic treatment, most people begin to feel better [135]. However, the antibiotics must be taken as directed by the doctor. Antiviral medication is occasionally recommended for viral pneumonia. These medications, however, are not effective against every infection that causes pneumonia. Fungal pneumonia is treated with antifungal medications. Pneumonia is a common but dangerous infectious disease and ranks as the sixth leading cause of death. When foreign pathogens such as viruses, fungi, and bacteria enter the lungs, they trigger an inflammatory response, leading to the accumulation of fluid in the bronchioles and alveoli. While pharmacotherapies have shown effectiveness in treating infections, there is ongoing development of advanced approaches to tackle challenging cases involving viral infections, lung inflammation, and acute lung injury (ALI).

Inflammation-regulating nanoparticles (NPs) have emerged as a promising solution for alleviating pneumonia and ALI. These NPs can reduce immune cell activation and inhibit the production of inflammatory chemicals in the lungs. By modulating the inflammatory cells, cytokines, chemokines, and microenvironments associated with pneumonia, they can enhance therapeutic outcomes [136]. Poly (lactic-co-glycolic acid) (PLGA) nanoparticles have been utilized for the treatment of pneumonia, offering advantages such as non-toxicity and biodegradability. Polyethylene glycol (PEG) is another commonly used substance in lung drug delivery systems. FDA-approved materials for drug delivery system carriers include PLGA and chitosan [137]. Additionally, liposomes, micelles, and inorganic nanoparticles like gold nanoparticles, iron nanoparticles, and quantum dots have been explored as carriers for antibiotics in addition to polymer nanoparticles. These advancements in nanotechnology-based drug delivery systems hold promise for improving the treatment of pneumonia and related lung conditions [138].

Theranostic nanoparticles have the potential to provide a non-invasive and effective method of diagnosing and treating pneumonia. One potential application of theranostic nanoparticles for pneumonia is in the delivery of antibiotics directly to the site of infection. By incorporating antibiotics into the nanoparticles, they can be delivered specifically to the infected lung tissue, reducing the risk of systemic side effects and improving treatment efficacy [139]. In addition, the diagnostic capabilities of the nanoparticles can be used to monitor the effectiveness of treatment and provide real-time feedback on the progression of the infection. Another potential application for theranostic nanoparticles in pneumonia is early detection and diagnosis. They can identify the presence of lung inflammation and other early symptoms of pneumonia by integrating imaging molecules into the nanoparticles. This early discovery can allow for rapid treatment and the avoidance of more serious problems. Theranostic nanoparticles show promise as a targeted and personalised strategy to diagnosis and treatment of pneumonia [140]. However, additional study is required to adequately assess their safety and efficacy in this situation.

3.2   Tuberculosis (TB)

Although medical research and medicines have come a long way, tuberculosis (TB) continues to be the leading cause of death and social catastrophe for millions of people worldwide. Throughout recorded history and human prehistory, it has affected humanity. The bacteria Mycobacterium tuberculosis is the primary cause of the lethal infectious illness tuberculosis [141]. In 1882, Robert Koch received the Nobel Prize for his outstanding discovery. Mycobacterium tuberculosis (Mtb) is an intracellular pathogen and an acid-fast bacillus that has evolved a number of defence mechanisms against macrophage eradication [142, 143]. It is regarded as the most successful pathogen since it may live for years in a host without causing illness [144]. One of the leading causes of death and disease worldwide is tuberculosis.

The World Health Organisation estimates that roughly one-third of the world's population is infected with M. tuberculosis (Mtb), which results in more than 9 million new cases and 2 million deaths each year. The remaining infected individuals do not exhibit any symptoms. After AIDS among infectious diseases, tuberculosis is the infectious disease that causes the most fatalities each /year. As a result, WHO deemed tuberculosis a global health emergency in 1993 [145].

Mycobacterium mostly affects the lungs, but it can also harm the kidney, lymphatic system, central nervous system, circulatory system, genitourinary system, joints, and bones.

Tuberculosis (TB) is a preventable and treatable disease. It is estimated that one-quarter of the world's population has been infected with TB bacteria. However, only around 5-10% of those infected will develop symptoms and progress to TB illness. It's important to note that individuals who are infected but not yet ill are unable to spread the disease [146].

The primary treatment for tuberculosis involves the use of antibiotics. Prompt and appropriate treatment is crucial, as tuberculosis can be lethal if left untreated. The disease can spread when an infected person coughs, sneezes, or sings, releasing microscopic droplets containing the TB bacteria into the air. When another person breathes in these droplets, the bacteria can enter their lungs, leading to infection. Tuberculosis spreads more easily in places where people congregate or live in crowded settings [147]. The risk of contracting tuberculosis is higher in individuals with conditions such as HIV/AIDS and other immune system disorders compared to those with a healthy immune system. It is important to note that some forms of tuberculosis bacteria have developed resistance to certain antibiotics, making treatment more challenging. These drug-resistant strains require alternative treatment strategies to effectively combat the infection [148].

After being inhaled, TB bacteria can enter the lungs and begin to develop there. From the lungs, the bacteria can travel through the bloodstream to other organs such as the kidneys, spine, and brain. TB infection in the throat or lungs can be contagious, meaning the bacteria can spread to other individuals. However, TB in other body parts, such as the kidney or spine, is typically not contagious. The first stage of TB infection is called the primary infection. During this stage, immune system cells locate and capture the pathogens [142]. In many cases, the immune system is able to completely eliminate the bacteria. However, some bacteria may survive and multiply. The primary infection is usually asymptomatic, although some individuals may experience flu-like symptoms such as low-grade fever, tiredness, and cough.

When TB bacteria are inhaled and an infection occurs, the body's immune system is typically able to control the bacteria and prevent them from spreading. In cases of latent TB infection, the bacteria remain in the body without causing illness. Individuals with latent TB infection usually have a positive TB skin test or blood test. Without treatment for latent TB infection, there is a risk of developing TB disease. However, those with latent TB infection do not feel unwell and cannot spread TB bacteria to others. Many people with latent TB infection never develop symptoms of active TB disease and carry inactive TB bacteria for the rest of their lives without experiencing any illness. If the immune system fails to control the growth of TB bacteria, they become active and start multiplying within the body. This leads to TB disease, where individuals become ill. People with TB disease have the potential to transmit the bacteria to those they come into frequent contact with [147].

It is worth noting that many individuals with latent TB infection never develop symptoms of TB illness. However, some patients may become ill with TB disease shortly after being infected, typically within a few weeks, before their immune system can effectively halt the multiplication of TB germs [142]. Others may develop the disease later on if their immune system becomes compromised due to other factors. The risk of developing TB disease is significantly higher for individuals with weakened immune systems, especially those who are HIV-positive, compared to individuals with healthy immune systems. Babies and young children in several nations receive the Bacille Calmette-Guérin (BCG) vaccine to prevent tuberculosis (TB). The vaccine shields against TB outside the lungs but not within. In general, isoniazid, rifampin, pyrazinamide, and ethambutol are used in combination therapy for several months to cure tuberculosis. These medications are used orally and are incredibly powerful against Mtb. The Mtb strain transforms into the more advanced form of TB known as (multi drug resistant) MDR-TB when it becomes resistant to the most potent first-line medications, isoniazid and rifampin [149]

Aminoglycosides such as amikacin and kanamycin, polypeptides like capreomycin, viomycin, and enviomycin, fluoroquinolones including ciprofloxacin, levofloxacin, and moxifloxacin, as well as thioamides like ethionamide, prothionamide, and cycloserine are among the medications used for the treatment of multidrug-resistant tuberculosis (MDR-TB). These medications are considered second-line drugs and are typically reserved for cases where the first-line medications are ineffective or the bacteria have developed resistance to them [150]. It is important to note that these second-line medications are generally more potent, but they also come with increased toxicity, higher costs, and may require a longer duration of therapy compared to first-line medications. Rifabutin, linezolid, thioridazine, arginine, vitamin D, and macrolides such as clarithromycin and thioacetazone are some other medications that have been used in the treatment of TB. However, their efficacy may not be as well-established as that of other TB medications, or their effectiveness may still be under investigation [150]. The choice of medications for MDR-TB treatment is complex and requires careful consideration based on individual patient factors and drug susceptibility testing.

For the treatment of tuberculosis, new cutting-edge technologies including the creation of carrier-based drug delivery systems are currently examined. To decrease the dosage and length of treatment, biodegradable polymers, liposomes, and microspheres have been developed [151]. Compared to routinely used medications, the pharmaceuticals were released gradually with a high concentration and minimal toxicity. Despite the effectiveness of the current anti-TB medications, urgent mechanisms needs to be established to ensure their administration. In this regard, nanotechnology is one of the most promising avenues for the creation of more incisive and efficient drug delivery systems for the treatment of tuberculosis, as well as the effective method for the creation and distribution of next-generation TB vaccines.

Nanotechnology-based drug delivery systems hold potential in enhancing the tolerability of toxic chemotherapy, providing controlled and prolonged drug release, and improving bioavailability. Particle sizes ranging from 50 to 200 nm are preferred for effective drug localization following inhalation delivery. The ability to achieve delayed, sustained, and controlled release from biodegradable nanoparticles is a notable advantage of nanobead delivery systems [152]. In the context of developing polymer-based antibiotic therapies against Mycobacterium tuberculosis, various animal models have been tested. However, none of these models have fully met expectations or accurately replicated all the characteristics of human TB.

Nanoparticles can be taken up through different mechanisms, including transcytosis through M-cells, intracellular uptake and transport via intestinal mucosal epithelial cells, and uptake by Peyer's patches. The drugs released from nanoparticles exhibit stability and slow release, enabling their administration through oral routes. Rifampin, isoniazid, and pyrazinamide are three effective antitubercular medications commonly used in tuberculosis treatment. Microemulsion is a thermodynamically stable liquid solution consisting of water, oil, and an amphiphile (surfactant and co-surfactant) [153]. Microemulsions have gained significant attention in recent years for the development of innovative drug delivery systems due to their thermodynamic stability, high diffusion and absorption rates, ease of preparation, and high solubility. Nanotechnology-based drug delivery systems, including nanoparticles and microemulsions, offer promising avenues for improving the efficacy and targeted delivery of medications for tuberculosis treatment [154]. They help to increase medication bioavailability, resistance to enzymatic hydrolysis, and toxicity. Microemulsion droplets are extremely tiny and thermodynamically stable. Since the droplet diameter in stable microemulsions is typically in the range of 10-100 nm, these systems are often referred to as nanoemulsions [155]. Microemulsions are widely used in colloidal drug delivery systems for medication targeting and controlled release.

3.3   Lung cancer

The largest cause of cancer-related fatalities worldwide is lung cancer. Lung cancer is also the most prevalent type of second primary malignancy, which refers to a person's being diagnosed with a new cancer that is unrelated to an earlier cancer diagnosis [156]. Small cell lung cancer (SCLC) and non-small cell lung cancer (NSCLC) are the two main subtypes of lung cancer. Adenocarcinoma, squamous cell carcinoma, and large cell carcinoma are the three subtypes of NSCLC, which affects about 85% of people with lung cancer.

Symptoms of lung cancer can vary from person to person, and some individuals may experience lung-related issues, while others may have symptoms related to metastasis (spread of cancer to other parts of the body). In some cases, people with lung cancer may only exhibit general signs of illness. It's important to note that many lung cancer patients may not show symptoms until the disease has advanced. Persistent or worsening cough, chest pain, difficulty breathing, wheezing, coughing up blood, extreme exhaustion, and weight loss are the symptoms of lung cancer. Lung cancer can present with additional alterations, such as repeated episodes of pneumonia and swollen or enlarged lymph nodes near the lungs. These symptoms can occasionally accompany lung cancer and should be evaluated by a healthcare professional [157].

The staging of lung cancer is primarily based on factors such as the size of the initial tumor, its invasiveness into surrounding tissues, and whether it has spread to lymph nodes or other organs. The development of lung cancer is influenced by a combination of environmental factors, particularly tobacco smoke, and genetic predisposition. Rare germ line mutations in genes such as p53, retinoblastoma, and epidermal growth factor receptor (EGFR) significantly increase the risk of developing cancer. Reduced effectiveness of DNA repair mechanisms may also play a role in the development of lung cancer [158].

Chemicals present in tobacco smoke have been strongly implicated in the development of lung cancer. Recent research highlights the involvement of tyrosine kinases in the pathogenesis of lung cancer. Abnormal activation of tyrosine kinases, downstream signaling pathways, overexpression, and autocrine/paracrine stimulation can contribute to the development and progression of lung cancer. Some specific tyrosine kinases, such as EGFR, PIK3CA, MET, and others, are frequently found to be activated in non-small cell lung cancer (NSCLC) and can serve as therapeutic targets. It is important to note that lung cancer is a complex disease influenced by multiple factors, and a comprehensive evaluation by healthcare professionals is necessary for accurate diagnosis, staging, and treatment planning [159, 160].

Diagnosis of lung cancer typically involves a combination of blood tests, imaging techniques (such as X-rays, CT scans, and PET scans), and biopsies of fluid or tissue samples to confirm the presence of cancer cells. Treatment options for lung cancer include surgery, radiofrequency ablation (RFA), radiation therapy, chemotherapy, targeted medication therapy, and immunotherapy. The choice of treatment depends on various factors, including the type and stage of lung cancer, as well as the overall health of the patient. Surgery is a common option for non-small cell lung cancer (NSCLC) that has not spread extensively and small cell lung cancer (SCLC) that is localized to a single tumor. It involves the removal of the tumor and nearby lymph nodes [161]. Radiofrequency ablation (RFA) is a procedure used to treat NSCLC tumors that are located near the edges of the lungs. It utilizes high-energy radio waves to heat and destroy cancer cells. Radiation therapy uses high-intensity beams of radiation to kill cancer cells. It can be administered as a standalone treatment or in combination with surgery or chemotherapy. In palliative care, radiation therapy may be used to reduce tumor size and alleviate symptoms.

Chemotherapy, which involves the use of drugs to inhibit the growth of cancer cells, is often administered as a combination of different medications. It can be used before, after, or alongside other treatments, such as immunotherapy. Chemotherapy is typically delivered intravenously. Targeted medication therapy is employed in certain cases where specific mutations in the lung cancer cells promote the growth of the disease. These targeted drugs aim to block or inhibit the activity of these specific mutations [162]. Angiogenesis inhibitors are a class of medications that can prevent the growth of new blood vessels that tumors need to spread. By inhibiting angiogenesis, these drugs aim to restrict the tumor's blood supply and hinder its growth.

Immunotherapy is a treatment approach that harnesses the body's immune system to fight cancer. It works by exposing cancer cells to the immune system, enabling it to recognize and attack the cancer cells more effectively. It's important for lung cancer patients to consult with their healthcare team to determine the most appropriate treatment plan based on their specific diagnosis and individual circumstances. Some lung cancer therapies are used to treat symptoms like pain and breathing difficulties. These include treatments to drain fluid from around your lungs and prevent it from returning as well as therapies to lessen or eliminate tumours that are obstructing airways. One of the quickly expanding topics in biomedical science is nanotechnology, which has been used to treat and diagnose a variety of biological issues. Recent years have seen a significant increase in the use of nanotechnology in the treatment of a wide range of illnesses, including cancer, diabetes, bacterial infections, cardiovascular diseases, etc. [163, 164]. Scientists and researchers have concentrated on developing nanoscale therapeutic agents, with the delivery system including liposomal nanoparticles, polymeric nanoparticles, metal nanoparticles, and bio-nano particles, due to several limitations in the conventional therapeutic strategies for lung cancers.

Due to their small size, which allows them to specifically aggregate in tumour cells due to an enhanced permeability and retention effect (EPR), these nanoparticles have shown to be an efficient theranostic tool for treating lung cancer [165]. Additionally, nanoparticles have a high drug loading capacity and are simple to functionalize due to their huge surface area to volume ratio [166]. In addition, due to their superior biocompatibility and ability to avoid kidney clearance, nanoparticles have an advantage over traditional medicinal approaches. Additionally, a number of nanoparticles shown multifunctional properties such as imaging, diagnostics, treatments, and sensing, which enables researchers to use these nanomaterials for multifunctional biomedical applications in lung cancer theranostics [167] Figure 3.

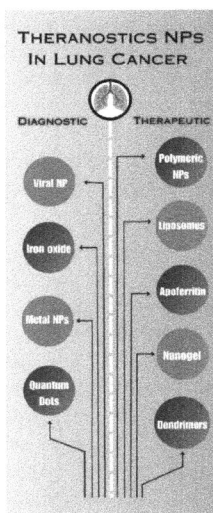

Figure3: Theranostic nanoparticles in lung cancer treatment.

Due to their excellent biocompatibility and biodegradability, bio-nanoparticles are frequently used in lung cancer theranostics. The complexity of the synthesis strategies, however, might occasionally drive up the cost and length of the production process. Therefore, more research is required to produce these bio-nanoparticles at industrial size rather than just on a lab scale. Numerous applications for the detection and therapy of malignant development are related to improvements in nanoparticle designing and formulations [168]. The current developments in lung cancer imaging and treatment using nanotheranostics is porphyrins. Due to their advantageous photophysical characteristics, porphyrins have proven particularly effective for photodynamic treatment (PDT) and cancer imaging. Since 1993, more than 120 nations have approved the use of hematoporphyrin derivative (HpD), porfimer sodium for the detection and photodynamic therapy of oesophagus, lung, superficial bladder, gastric, cervical, and endobronchial malignancies [169].

Nanomaterials outstanding adaptability is based on more than just their capacity to deliver different chemicals or genes in variable doses to predetermined locations [170]. Even though this area of research has received the majority of attention in recent years, nanoscale materials stand on their own as effective therapeutic agents. Theranostics, or personalized cancer treatment, involves the use of an individualized prescription based on evidence-based medicine, ensuring the right treatment at the right time for improved efficacy and patient outcomes. Nanotechnology-based delivery platforms are utilized in theranostics to deliver both therapeutic and imaging agents to specific locations in the body. Nanomaterials have demonstrated their usefulness in clinical applications and can overcome challenges related to solubility and stability through surface modification and formulation techniques. Nanostructures possess unique physical properties that enable them to carry a higher therapeutic payload and provide a larger surface area for interactions [171]. Functionalized nanoparticles can be designed by combining ligands, drugs, biomolecules, and imaging agents, allowing for customized drug administration and diagnostics. The targeted delivery of therapeutic payloads to cancer sites, either through passive or active targeting, using nanoparticle-based cancer cell targeting strategies can minimize nonspecific toxicity and enhance treatment efficacy.

However, there are several challenges that need to be addressed before lung cancer theranostics can be widely used in clinical practice. These challenges include scaling up production, ensuring cost-effective manufacturing processes, understanding drug pharmacokinetics in the context of nanoparticles, and optimizing imaging constructs. Additionally, issues related to nanotoxicity, regulatory standards, and other barriers must be overcome to fully leverage the potential of nanotechnology in lung cancer and other cancer treatments. While significant progress has been made in the field of nanotechnology, there is still untapped potential that needs to be explored to fully realize the benefits of theranostics in lung cancer and other cancer types [166]. Ongoing research and advancements in nanotechnology will contribute to addressing these challenges and expanding the applications of personalized cancer treatment approaches.

3.4   Acute respiratory distress syndrome (ARDS)

Acute Respiratory Distress Syndrome (ARDS) is a severe and life-threatening lung condition characterized by the accumulation of fluid in the air sacs of the lungs. It was first identified in 1967 and is associated with poor oxygenation and stiff lungs. The condition is marked by injury to the capillary endothelium and widespread damage to the alveoli. The excessive fluid in the lungs can impair the exchange of oxygen and carbon dioxide, leading to reduced oxygen levels in the bloodstream and potentially increased carbon dioxide levels. This can deprive organs of the necessary oxygen, eventually resulting in organ failure. ARDS most commonly affects critically ill patients in a hospital setting but can also be caused by severe trauma, pneumonia, or infections originating from other parts of the body.

Symptoms of ARDS typically manifest within a day or two after the initial illness or injury and can include sudden and severe shortness of breath, gasping for air, and difficulty breathing [172].

Pulmonary artery vasoconstriction and pulmonary hypertension can also develop in patients with ARDS. ARDS carries a significant mortality rate, and effective treatment options for this life-threatening condition are limited. It is considered a medical emergency with potentially fatal consequences. The primary cause of ARDS is damage to the small blood vessels in the lungs, resulting in the leakage of fluid into the air sacs. These air sacs, known as alveoli, are responsible for the exchange of oxygen and carbon dioxide during breathing. When they become filled with fluid, oxygen uptake is compromised.

Various factors can contribute to the development of ARDS, including inhalation of toxic substances like smoke or chemicals, severe infections such as pneumonia, severe blood infections, chest or head injuries, and overdosing on certain medications like sedatives or tricyclic antidepressants. These conditions can lead to lung damage, inflammation, and the subsequent development of ARDS. Early recognition and prompt treatment of the underlying cause, along with supportive care to maintain adequate oxygenation and ventilation, are crucial in managing ARDS. Interventions such as mechanical ventilation, administration of oxygen, and medications to reduce inflammation may be employed based on the severity of the condition and the individual patient's needs [173].

In rare situations, an ARDS diagnosis might cause or be linked to other health problems that must be treated. Multiple organ failure, pulmonary hypertension (a rise in blood pressure), blood clots forming during treatment, atelectasis (collapse of the lung's small air pockets), and other problems are possible. Researchers have worked to better understand and manage ARDS since the first instance in 1967. This has aided clinicians and scientists in learning about the signs and risk factors of ARDS, as well as determining the most effective diagnosis and treatment strategies. There is much more study to be done on ARDS, and scientists are attempting to develop effective methods for determining subphenotypes and endotypes in ARDS cases. This is done to establish which treatment method is optimal for each person [173].

The major goal of ARDS treatment is to ensure that the patient receives adequate oxygen to avoid organ failure. A doctor may use a mask to give oxygen. A mechanical ventilation equipment can also be used to drive air into the lungs while decreasing fluid in the air sacs. Another ARDS therapy option is fluid management. This can aid in maintaining an optimum fluid balance. Excess fluid in the body can cause fluid buildup in the lungs. However, a lack of fluid can strain the organs and the heart [174]. Side effects are frequently prescribed to people with ARDS. Among these are the following medications:

- analgesics to alleviate discomfort
- antibiotics for infection treatment
- blood thinners to prevent clots in the lungs or legs

Individuals recovering from ARDS may require pulmonary rehabilitation. This is a method of strengthening the respiratory system and expanding lung capacity. To aid in the rehabilitation from ARDS, these programmes may involve exercise training, lifestyle classes, and support groups. The prognosis and fatality rate of ARDS can vary depending on the cause of ARDS and the overall condition of the individual. Factors such as age, underlying health conditions, and the severity of the underlying injury or infection can impact the outcome. While many ARDS patients do recover completely within a few months, there are cases where individuals may experience chronic lung damage. This can lead to long-term respiratory complications and a reduced quality of life. Additionally, ARDS can have other adverse effects, including muscle weakness, fatigue, and mental health deterioration [174]. In terms of treatment, there are ongoing developments in medicines and stem cell therapies for ARDS. These advancements aim to improve outcomes and provide more effective interventions for patients. However, it's important to note that there is currently no specific medicinal therapy available for ARDS, and management primarily involves supportive care measures to address symptoms and provide respiratory support.

ARDS is a significant cause of respiratory failure in critically ill patients, often accompanied by complications such as sepsis, alveolar damage, increased lung permeability, noncardiogenic pulmonary edema, and low oxygen levels (hypoxemia). These factors contribute to the severity and complexity of ARDS, making it a challenging condition to manage [175]. ARDS currently lacks a specific medicinal therapy, and management primarily involves supportive care measures. These measures include lung-protective ventilation strategies, extracorporeal membrane oxygenation (ECMO) for severe cases, and conservative fluid management to prevent fluid overload. These approaches aim to support the patient's respiratory function and address the underlying causes and complications of ARDS. Diagnosing ARDS relies on clinical symptoms such as acute onset of respiratory distress, along with radiographic evidence of bilateral infiltrates on chest imaging. However, accurately assessing the extent of lung injury and determining the true clinical burden can be challenging, potentially leading to delayed identification and treatment initiation.

Despite multiple randomized controlled studies conducted for ARDS, the effectiveness of therapies has been limited, and successful outcomes are not consistently achieved. Pharmacotherapies may face obstacles in delivering drugs effectively to the damaged alveoli, achieving sufficient drug accumulation in the lungs, overcoming short circulation half-life, and crossing physiological barriers like mucus and alveolar fluid for systemic delivery [174]. These challenges highlight the need for further research and development in finding more targeted and effective therapies for ARDS. The field of nanomedicine and drug delivery systems holds promise in addressing some of these limitations by enhancing drug delivery to specific lung regions and improving drug effectiveness. However, more investigation and clinical trials are necessary to validate these approaches and ensure their safety and efficacy in ARDS management [176].

Indeed, the use of nano-based drug delivery systems has gained significant attention in the scientific community due to their potential to enhance the delivery of bioactive substances to pulmonary tissues. These nano-drug delivery systems offer several advantages that can potentially improve ARDS clinical intervention strategies. One advantage is the selective targeting of drugs to specific pulmonary tissues. Nano-based carriers can be designed to deliver drugs directly to the target site in the lungs, allowing for a more focused and efficient treatment approach. This targeted delivery can enhance the therapeutic effect while minimizing systemic side effects.

Nano-drug delivery systems also have the potential to improve drug stability and release profiles. By encapsulating drugs within nanoparticles, their stability can be enhanced, protecting them from degradation or inactivation in the body. Additionally, nanoparticles can be engineered to release drugs in a controlled and sustained manner, ensuring a prolonged therapeutic effect and reducing the frequency of dosing. Furthermore, nano-based carriers can overcome physiological barriers and improve drug penetration into the lungs [177]. For example, nanoparticles can bypass mucus barriers and effectively reach the deep lung regions where the pathology of ARDS occurs. They can also facilitate the transport of drugs across cellular barriers, such as the epithelial and endothelial layers, to reach the target cells within the lungs.

The use of nano-drug delivery systems can also enable stimuli-responsive drug release. These systems can be designed to respond to specific triggers or environmental cues present in the diseased lungs, such as changes in pH, temperature, or enzyme activity. This stimulus-induced drug release allows for on-demand drug delivery and improves the efficiency of therapeutic interventions. Nano-based drug delivery systems hold great potential to enhance ARDS treatment strategies by improving drug delivery, targeting specific tissues, and ensuring sustained and controlled release of bioactive substances. However, further research and clinical studies are needed to evaluate the safety, efficacy, and feasibility of these approaches in the context of ARDS management [177].

Nanotechnology has introduced a promising solution for delivering water-insoluble medications, peptides, and various pharmaceuticals by offering a biocompatible and biodegradable carrier platform. This technology involves the use of nanoparticles, which are tiny particles at the nanoscale, to encapsulate and transport these substances. The nanoparticles provide a protective and controlled environment for the drugs, ensuring their stability and enabling efficient delivery to the target sites in the body. This approach has shown great potential in improving the solubility, bioavailability, and therapeutic efficacy of water-insoluble medicines and enhancing the delivery of peptides and other pharmaceutical compounds [178]. Many critical pharmacological difficulties, such as decreased drug absorption, shorter half-life, poor pharmacokinetics, and so on, are addressed by these delivery strategies. Nanoscale particles have unique physicochemical features that can be utilised to increase medication solubility, selectivity, effectiveness, pharmacokinetics, and toxicity [179]. It also aids in overcoming problems such as long-acting nanocarrier stability, bioavailability, and systemic dispersion [180].

Though pulmonary nanomedicine is a new field, some nano-modified medications have been examined and found to have numerous benefits in the treatment of both chronic and acute lung illnesses [181]. It provides a potential foundation for a variety of drug-delivery vehicles with uniform distribution, sustained drug-release in plasma, and alveolar internalisation [182]. Nanoparticles having a diameter of 5 m have been demonstrated to increase lung deposition and improve the dissolution of poorly water-soluble medications. Nanocarriers have significant advantages in addressing the shortcomings of traditional medication or gene treatment for ARDS. They have shown great promise for critical illness management, with numerous advantages such as improving drug biopharmacokinetics, achieving targeted delivery to increase efficiency while decreasing toxicity, and realising co-delivery of agents for early diagnosis and synergetic treatment [182]. The future of pulmonary medicine is being transformed by advancements in nanomaterials, which utilize various nanoscale delivery systems along with pharmacological elements, peptides, and nucleic acids. This innovative approach has demonstrated considerable promise in effectively addressing lung conditions like ARDS, marking a significant breakthrough in the field [183, 184].

Conclusion and future prospects

Nano-strategies are opening new vistas in the era of personalized medicine and offering novel arena on diagnosis and therapy including drug delivery. The potential and significance of the current usage of theranostic nanoparticles in different lung diseases are covered in this chapter thoroughly. The imaging component using nanoparticles is highly advantageous as it is non-invasive and detectable with high sensitivity. On the other hand, the same nanomaterial being able to deliver the drug locally at the desired site for specific function is of much interest as well as the treatment time can be limited. Such usage of theranostics is also widely recognized in other areas of clinical importance, diagnostics as well as treatment in other diseases like artherosclerosis, cardiovascular problem, rheumatoid arthritis, and other cancers. More studies are needed in the future to generalize and widen the usage of theranostic nanoparticles in various diseases as the future holds great promise for personalized/precision medicine.

References

[1] T. Lammers, S. Aime, W.E. Hennink, G. Storm, F. Kiessling, Theranostic nanomedicine. Acc Chem Res. 44(2011)1029–1038. https://doi.org/10.1021/ar200019c

[2] W. Zhong, X. Zhang, Y. Zeng, D. Lin, J. Wu, Recent applications and strategies in nanotechnology for lung diseases. Nano Res. 14(2021)2067-2089. https://doi.org/10.1007/s12274-020-3180-3

[3] F. Chen, E.B. Ehlerding, W. Cai, Theranostic nanoparticles. J Nucl Med. 55 (2014)1919-22. https://doi.org/10.2967/jnumed.114.146019

[4] J. Xie, S. Lee, X. Chen, Nanoparticle-based theranostic agents. Adv Drug Deliv Rev. 62 (2010)1064–1079. https://doi.org/10.1016/j.addr.2010.07.009

[5] Targeting Cellular Signalling Pathways in Lung Diseases. In Springer eBooks. Springer Nat. (2021). https://doi.org/10.1007/978-981-33-6827-9

[6] M. Doroudian, M. H. Azhdari, N. Goodarzi, D. O'Sullivan, S.C. Donnelly, Smart Nanotherapeutics and Lung Cancer. *Pharma.* 13(2021)1972. https://doi.org/10.3390/pharmaceutics13111972

[7] S. St Claire, H. N. Gouda, K. Schotte, R. Fayokun, D. Fu, C. Varghese, &V. M. Prasad, Lung health, tobacco, and related products: gaps, challenges, new threats, and suggested research. Am. J. Physiol. Lung Cell Mol. 318(2020) L1004–L1007. https://doi.org/10.1152/ajplung.00101.2020

[8] O. K. Kurt, J. Zhang, K. E. Pinkerton, Pulmonary health effects of air pollution. Curr Opin Pulm Med. 22(2016)138–143. https://doi.org/10.1097/mcp.0000000000000248

[9] J. Lelieveld, J. Evans, M. Fnais, D. Giannadaki, & A. Pozzer, The contribution of outdoor air pollution sources to premature mortality on a global scale. Nat. 525(2015)367–371. https://doi.org/10.1038/nature15371

[10] M. Vats, Respiratory Disease and Infection: A New Insight. BoD – Books on Demand. (2013)

[11] M. Obeidat, I. P. Hall, Genetics of complex respiratory diseases: implications for pathophysiology and pharmacology studies. Br. J. Pharmacol. 163(2011)96–105. https://doi.org/10.1111/j.1476-5381.2011.01222.x

[12] K. Vlahovich, A. Sood. A 2019 Update on Occupational Lung Diseases: A Narrative Review. Pulm. Ther, 7(2021)75–87. https://doi.org/10.1007/s41030-020-00143-4

[13] S. Cho, H. W. Stout-Delgado, Aging and Lung Disease. Annual Rev. Psysiol. 82(2020)433–459. https://doi.org/10.1146/annurev-physiol-021119-034610

[14] D. E. Schraufnagel, The world respiratory diseases report [Editorial]. Int. J. Tuberc. Lung Dis., 17(2013)1517. https://doi.org/10.5588/ijtld.13.0743

[15] P. Song, D. Adeloye, H. Salim, J. P. R. D. Santos, H. Campbell, A. Sheikh, I. Rudan, Global, regional, and national prevalence of asthma in 2019: a systematic analysis and modelling study. J. Glob. Health, 12(2022). https://doi.org/10.7189/jogh.12.04052

[16] D. Adeloye, P. Song, Y. Zhu, H. Campbell, A. Sheikh, I. Rudan, Global, regional, and national prevalence of, and risk factors for, chronic obstructive pulmonary disease (COPD) in 2019: a systematic review and modelling analysis. Lancet Respir. Med., 10(2022)447–458. https://doi.org/10.1016/s2213-2600(21)00511-7

[17] F. Prabhu, A. R. Sikes, & I. Sulapas, Pulmonary Infections. Springer eBooks, (2016)1083–1101. https://doi.org/10.1007/978-3-319-04414-9_91

[18] F. Siddiqui, S. Vaqar, A. H. Siddiqui, Lung Cancer. In: StatPearls. Treasure Island (FL): StatPearls Publishing; 2023 https://www.ncbi.nlm.nih.gov/books/NBK482357/

[19] S. D. Shukla, K. S. Vanka, A. Chavelier, M. D. Shastri, M. M. Tambuwala,H. A. Baksh, K. Pabreja, N. M. Ashraf, & R. V. O'Toole, Chronic respiratory diseases: An introduction and need for novel drug delivery approaches. Elsevier eBooks, (2020)1–31. https://doi.org/10.1016/b978-0-12-820658-4.00001-7

[20] R. A. Pleasants, Clinical Pharmacology of Oral Maintenance Therapies for Obstructive Lung Diseases. Respir. Care 63(2018)671–689. https://doi.org/10.4187/respcare.06068

[21] L. C. Lands, S. Stanojevic, Oral non-steroidal anti-inflammatory drug therapy for lung disease in cystic fibrosis. The Cochrane Library, (2019) https://doi.org/10.1002/14651858.cd001505.pub5

[22] A. Kuzmov, T. Minko, Nanotechnology approaches for inhalation treatment of lung diseases. J Controlled Release, 219(2015)500–518. https://doi.org/10.1016/j.jconrel.2015.07.024

[23] Q. Zhou, S. S. Leung, P. A. Tang, T. Parumasivam, Z. H. Loh, & H. Chan. Inhaled formulations and pulmonary drug delivery systems for respiratory infections. Adv. Drug Deliv. Rev., 85(2015)83–99. https://doi.org/10.1016/j.addr.2014.10.022

[24] J. T. Patton, & P. R. Byron. Inhaling medicines: delivering drugs to the body through the lungs. Nat. Rev. Drug Discov., 6(2007)67–74. https://doi.org/10.1038/nrd2153

[25] R. A. Athanazio. Airway disease: similarities and differences between asthma, COPD and bronchiectasis. Clinics, 67(2012)1335–1343. https://doi.org/10.6061/clinics/2012(11)19

[26] M. F. Hashmi, Asthma. StatPearls - NCBI Bookshelf. (2023) https://www.ncbi.nlm.nih.gov/books/NBK430901/

[27] C. Porsbjerg, E. Melén, L. Lehtimäki, D. Shaw. Asthma. The Lancet, 401(2023)858–873. https://doi.org/10.1016/s0140-6736(22)02125-0

[28] J. Quirt, K. J. Hildebrand, J. A. Mazza, F. Noya, & H. Kim. Asthma. AACI, 14 (2018). https://doi.org/10.1186/s13223-018-0279-0

[29] M. Trivedi, E. Denton. Asthma in Children and Adults—What Are the Differences and What Can They Tell us About Asthma? Front Pediatr., 7(2019). https://doi.org/10.3389/fped.2019.00256

[30] T. To, S. Stanojevic, G. Moores, A. S. Gershon, E. D. Bateman, A. A. Cruz, L. Boulet. Global asthma prevalence in adults: findings from the cross-sectional world health survey. BMC Public Health, 12(2012). https://doi.org/10.1186/1471-2458-12-204

[31] J. Greiwe, J. A. Bernstein. Occupational Asthma. Springer eBooks, (2018)1–16. https://doi.org/10.1007/978-3-319-58726-4_16-1

[32] S. T. Holgate, S. E. Wenzel, D. S. Postma, S. T. Weiss, H. Renz, & P.D. Sly. Asthma. Nat. Rev. Dis. Primers, 1(2015). https://doi.org/10.1038/nrdp.2015.25

[33] Z. He, J. Feng, J. Xia, Q. Wu, H. Yang, & Q. Ma. Frequency of Signs and Symptoms in Persons with Asthma. Respir. Care, 65(2020)252–264. https://doi.org/10.4187/respcare.06714

[34] J. L. Mims. Asthma: definitions and pathophysiology. IFAR, 5(2015)S2–S6. https://doi.org/10.1002/alr.21609

[35] H. K. Reddel, L. B. Bacharier, E. D. Bateman, C. E. Brightling, G. Brusselle, R. Buhl, A. A. Cruz, L. Duijts, J. M. Drazen, J. M. FitzGerald, L. Fleming, H. Inoue, F. W. Ko, J. A. Krishnan, M. L. Levy, J. Lin, K. Mortimer, P. M. Pitrez, A. Sheikh, L. Boulet. Global Initiative for Asthma Strategy 2021: Executive Summary and Rationale for Key Changes. Am. J. Respir. Crit. Care Med., 205(2022)17–35. https://doi.org/10.1164/rccm.202109-2205pp

[36] P. G. Gibson, H. Powell, J. Coughlan, A. J. Wilson, M. Abramson, P. Haywood, A. Bauman, M. J. Hensley, E. H. Walters. Self-management education and regular practitioner review for adults with asthma. CDSR, (2003)CD001117. doi: 10.1002/14651858.CD001117.

[37] V. E. Ortega, & M. Izquierdo. Drugs for Preventing and Treating Asthma. MSD Manual Consumer Version. (2022) https://www.msdmanuals.com/en-in/home/lung-and-airway-disorders/asthma/drugs-for-preventing-and-treating-asthma

[38] M. Kupczyk, B. Dahlén, & S. E. Dahlén. Which anti-inflammatory drug should we use in asthma? Pol. Arch. Intern. Med. (2011); https://www.mp.pl/paim/issue/article/1115/

[39] D. M. Williams, &B. K. Rubin. Clinical Pharmacology of Bronchodilator Medications. Respir. Care, 63(2018)641–654. https://doi.org/10.4187/respcare.06051

[40] M. Cazzola, P. Rogliani, & M. G. Matera. Ultra-LABAs for the treatment of asthma. Respir. Med., 156(2019)47–52. https://doi.org/10.1016/j.rmed.2019.08.005

[41] R. Beasley, M. A. Holliday, H. K. Reddel, I. Braithwaite, S. Ebmeier, R. J. Hancox, T. Harrison, C. Houghton, K. Oldfield, A. Papi, I. D. Pavord, M. Williams, & M. Weatherall. Controlled Trial of Budesonide–Formoterol as Needed for Mild Asthma. NEJM 380(2019) 2020–2030. https://doi.org/10.1056/nejmoa1901963

[42] S. Wenzel. Severe asthma: from characteristics to phenotypes to endotypes. Clin. Exp. Allergy, 42(2012)650–658. https://doi.org/10.1111/j.1365-2222.2011.03929.x

[43] H. M. Jin. Biological treatments for severe asthma. J Yeungnam Med Sci., 37(2020)262–268. https://doi.org/10.12701/yujm.2020.00647

[44] I. Agache, I. Eguiluz-Gracia, C. Cojanu, A. Laculiceanu, S. Del Giacco, M. Zemelka-Wiacek, A. Kosowska, C. A. Akdis, & M. Jutel. Advances and highlights in asthma in 2021. ALGY, 76(2021)3390–3407. https://doi.org/10.1111/all.15054

[45] R. E. O'Hehir, N. Varese, K. Deckert, C. Zubrinich, M. C. Van Zelm, J. M. Rolland, M. Hew. Epidemic Thunderstorm Asthma Protection with Five-Grass Pollen Tablet Sublingual Immunotherapy: A Clinical Trial. Am. J. Respir. Crit. Care Med., 198(2018)126–128. https://doi.org/10.1164/rccm.201711-2337le

[46] K. D. G. Van De Kant, L. J. T. M. Van Der Sande, Q. Jöbsis, O. C. P. Van Schayck, E. Dompeling. Clinical use of exhaled volatile organic compounds in pulmonary diseases: a systematic review. Respir. Res., 13(2012)117. https://doi.org/10.1186/1465-9921-13-117

[47] Y. Y. Broza, H. Haick. Nanomaterial-based sensors for detection of disease by volatile organic compounds. Nanomed., 8(2013)785–806. https://doi.org/10.2217/nnm.13.64

[48] B. Pelaz, C. Alexiou, R. A. Alvarez-Puebla, F. Alves, A. M. Andrews, S. Ashraf, L. P. Balogh, L. Ballerini, A. Bestetti, C. Brendel, S. Bosi, M. Carril,W. C. W. Chan, C. Chen, X. D. Chen, X. Chen, Z. Cheng, D. Cui, J. Du, W. J. Parak. Diverse Applications of Nanomedicine. ACS Nano, 11(2017)2313–2381. https://doi.org/10.1021/acsnano.6b06040

[49] T. Keil, D. Feldmann, G. Costabile, Q. Zhong, S. R. P. Da Rocha, & O. M. Merkel. Characterization of spray dried powders with nucleic acid-containing PEI nanoparticles. Eur. J. Pharm. Biopharm., 143(2019)61–69. https://doi.org/10.1016/j.ejpb.2019.08.012

[50] M. Nasr, M. Najlah, A. D'Emanuele, & A. Elhissi. PAMAM dendrimers as aerosol drug nanocarriers for pulmonary delivery via nebulization. Int. J. Pharm. 461(2014)242–250. https://doi.org/10.1016/j.ijpharm.2013.11.023

[51] N. J. Kenyon, J. M. Bratt, J. M. Lee, J. Luo, L. M. Franzi, A. A. Zeki, K. S. Lam. Self-Assembling Nanoparticles Containing Dexamethasone as a Novel Therapy in Allergic Airways Inflammation. PLOS ONE, 8(2013) e77730. https://doi.org/10.1371/journal.pone.0077730

[52] Bhavna, F. J. Ahmad, G. Mittal, G. Jain, G. Malhotra, R. K. Khar, A. Bhatnagar. Nano-salbutamol dry powder inhalation: A new approach for treating broncho-constrictive conditions. Eur. J. Pharm. Biopharm. 71(2009)282–291. https://doi.org/10.1016/j.ejpb.2008.09.018

[53] M. G. Arafa, B. M. Ayoub. Nano-vesicles of salbutamol sulphate in metered dose inhalers: formulation, characterization and in vitro evaluation. Int. J. Appl. Pharm., 9(2017)100. https://doi.org/10.22159/ijap.2017v9i6.22448

[54] K. Wang, Y. Feng, S. Li, W. Li, X. Chen, R. Yi, H. Zhang, & Z. Hong. Oral Delivery of Bavachinin-Loaded PEG-PLGA Nanoparticles for Asthma Treatment in a Murine Model. J. Biomed. Nanotechnol., 14(2018)1806–1815. https://doi.org/10.1166/jbn.2018.2618

[55] S. Chakraborty, I. Ehsan, B. Mukherjee, L. Mondal, S. Roy, K. D. Saha, B. Paul, M. C. Debnath, T. Bera. Therapeutic potential of andrographolide-loaded nanoparticles on a murine asthma model. Nanomed.: Nanotechnol. Biol. Med., 20(2019)102006. https://doi.org/10.1016/j.nano.2019.04.009

[56] V. M. Joshi, A. Adamcakova-Dodd, X. Jing, A. Wongrakpanich, K. N. Gibson-Corley, P. S. Thorne, & A. K. Salem. Development of a Poly (lactic-co-glycolic acid) Particle Vaccine to Protect against House Dust Mite Induced Allergy. Aaps Journal., 16(2014)975–985. https://doi.org/10.1208/s12248-014-9624-5

[57] M. Zedan, A. Darwish, M. E. Wassefy, E. O. Khashaba, E. Osman, A. M. Osman, & N. Ellithy. Association between the chemokine receptor 3 gene polymorphism and clinical asthma phenotypes among Egyptian asthmatic children. Alex J Pediatr., 34(2021)237. https://doi.org/10.4103/1687-9945.337835

[58] M. Kumar, X. Kong, A. K. Behera, G. Hellermann, R. F. Lockey, & S. S. Mohapatra. Chitosan IFN-γ-pDNA Nanoparticle (CIN) Therapy for Allergic Asthma. *Genet. Vaccine ther.*, 1(2003)3. https://doi.org/10.1186/1479-0556-1-3

[59] A. K. Agarwal. Chronic Obstructive Pulmonary Disease. StatPearls - NCBI Bookshelf. (2022); https://www.ncbi.nlm.nih.gov/books/NBK559281/

[60] D. Singh, A. Agusti, A. Anzueto, P. J. Barnes, J. Bourbeau, B. R. Celli, G. J. Criner, P. Frith, D. M. Halpin, M. K. Han, M. Varela, F. J. Martinez, M. M. De Oca, A. Papi, I. D. Pavord, N. Roche, D. D. Sin, R. A. Stockley, J. Vestbo, C. Vogelmeier. Global Strategy for the Diagnosis, Management, and Prevention of Chronic Obstructive Lung Disease: the

GOLD science committee report 2019. Eur. Respir. J., 53(2019)1900164.
https://doi.org/10.1183/13993003.00164-2019

[61] S. Safiri, K. V. Carson, M. Noori, S. A. Nejadghaderi, M. J. Sullman, J. A. Heris, K.
Ansarin, M. A. Mansournia, G. S. Collins, A. Kolahi & J. S. Kaufman. Burden of chronic
obstructive pulmonary disease and its attributable risk factors in 204 countries and territories,
1990-2019: results from the Global Burden of Disease Study 2019. BMJ., (2022) e069679.
https://doi.org/10.1136/bmj-2021-069679

[62] C. J. L. Murray, A. Y. Aravkin, P. Zheng, C. Abbafati, K. Abbas, M. Abbasi-Kangevari, F.
Abd-Allah, A. A. Abdelalim, M. Abdollahi, I. Abdollahpour,K. H. Abegaz, H. Abolhassani,
V. Aboyans, L. G. Abreu, M. R. Abrigo, A. Abualhasan, L. J. Abu-Raddad, A. I. Abushouk,
M. Adabi, S. S. Lim. Global burden of 87 risk factors in 204 countries and territories, 1990–
2019: a systematic analysis for the Global Burden of Disease Study 2019. Lancet.,
396(2020)1223–1249. https://doi.org/10.1016/s0140-6736(20)30752-2

[63] J. R. Hurst, N. Skolnik, G. J. Hansen, A. Anzueto, G. C. Donaldson, M. T. Dransfield, & P.
Varghese. Understanding the impact of chronic obstructive pulmonary disease exacerbations
on patient health and quality of life. Eur. J. Intern. Med., 73(2020)1–6.
https://doi.org/10.1016/j.ejim.2019.12.014

[64] M. A. Changizi, K. Rio. Harnessing color vision for visual oximetry in central cyanosis.
Med. Hypotheses., 74(2010)87–91. https://doi.org/10.1016/j.mehy.2009.07.045

[65] A. Fazleen, & T. Wilkinson. Early COPD: current evidence for diagnosis and management.
Ther. Adv. Respir. Dis., 14(2020)175346662094212.
https://doi.org/10.1177/1753466620942128

[66] T. Nakajima, H. Nakamura, C. A. Owen, S. Yoshida, K. Tsuduki, S. Chubachi, T.
Shirahata, S. Mashimo, M. Nakamura, S. Takahashi, N. Minematsu, H. Tateno, S. Fujishima,
K. Asano, B. R. Celli, & T. Betsuyaku. Plasma Cathepsin S and Cathepsin S/Cystatin C
Ratios Are Potential Biomarkers for COPD. Dis. Markers, (2016)1–9.
https://doi.org/10.1155/2016/4093870

[67] S. Sarkar, P. Bhattacharyya, M. Mitra, & S. Pal. A novel approach towards non-obstructive
detection and classification of COPD using ECG derived respiration. APESM,
42(2019)1011–1024. https://doi.org/10.1007/s13246-019-00800-2

[68] H. U. R. Siddiqui, A. Raza, A. A. Saleem, F. Rustam, I. De La Torre Díez, D. G. Aray, V.
Lipari, I. Ashraf, & S. Dudley. An Approach to Detect Chronic Obstructive Pulmonary
Disease Using UWB Radar-Based Temporal and Spectral Features. Diagn., 13(2023)1096.
https://doi.org/10.3390/diagnostics13061096

[69] E. Ray, D. Culliford, H. Kruk, K. Gillett, M. North, C. Astles, A. Hicks, M. P. Johnson, S.
W. Lin, R. Orlando, M. Thomas, R. Jordan, D. Price, M. Konstantin, & T. Wilkinson.
Specialist respiratory outreach: a case-finding initiative for identifying undiagnosed COPD in
primary care. NPJ Prim. Care Respir. Med., 31(2021). https://doi.org/10.1038/s41533-021-
00219-x

[70] P. D. Scanlon, J. E. Connett, L. A. Waller, M. D. Altose, W. C. Bailey, A. S. Buist & D. P. Tashkin. Smoking Cessation and Lung Function in Mild-to-Moderate Chronic Obstructive Pulmonary Disease. Am. J. Respir. Crit. Care Med. 161(2000)381–390. https://doi.org/10.1164/ajrccm.161.2.9901044

[71] J. A. Wedzicha, P. M. Calverley, & K. F. Rabe. Roflumilast: a review of its use in the treatment of COPD. *Int J Chron Obstruct Pulmon Dis.*, (2016)81. https://doi.org/10.2147/copd.s89849

[72] H. N. Tse, L. Raiteri, Y. Wong, K. S. Yee, N.R. J. Gascoigne, K. Wai, C. K. Loo, & K. M. Chan. High-Dose N-Acetylcysteine in Stable COPD. Chest, 144(2013)106–118. https://doi.org/10.1378/chest.12-2357

[73] A. J. Watson, C. Spalluto, C. McCrae, D. Cellura, H. Burke, D. Cunoosamy, A. Freeman, A. Hicks, M. Huhn, K. Ostridge, K. J. Staples, & O. Vaarala. Dynamics of IFN-β Responses during Respiratory Viral Infection. Insights for Therapeutic Strategies. Am. J. Respir. Crit. Care Med., 201(2020)83–94. https://doi.org/10.1164/rccm.201901-0214oc

[74] N. R. Anthonisen. Effects of Smoking Intervention and the Use of an Inhaled Anticholinergic Bronchodilator on the Rate of Decline of FEV1. JAMA, 272(1994)1497. https://doi.org/10.1001/jama.1994.03520190043033

[75] R. Pauwels, C. Löfdahl, L. A. Laitinen, J. S. A. G. Schouten, D. S. Postma, N. B. Pride, & S. V. Ohlsson Long-Term Treatment with Inhaled Budesonide in Persons with Mild Chronic Obstructive Pulmonary Disease Who Continue Smoking. NEJM, 340(1999)1948–1953. https://doi.org/10.1056/nejm199906243402503

[76] M. Candela, R. Costorella, A. Stassaldi, V. Maestrini, G. Curradi. Treatment of COPD: the simplicity is a resolved complexity. Multidiscip. Respir. Med., 14(2019). https://doi.org/10.1186/s40248-019-0181-8

[77] A. N. Da Silva, F. F. Cruz, P. Pelosi, & M. M. Morales. New perspectives in nanotherapeutics for chronic respiratory diseases. Biophys. Rev., 9(2017)793–803. https://doi.org/10.1007/s12551-017-0319-x

[78] H. Kato, M. Hagihara, J. Hirai, D. Sakanashi, H. Suematsu, N. Nishiyama, H. Mikamo, Y. Yamagishi, & K. Matsuura. Evaluation of Amikacin Pharmacokinetics and Pharmacodynamics for Optimal Initial Dosing Regimen. Drugs R. D, 17(2017)177–187. https://doi.org/10.1007/s40268-016-0165-5

[79] P. Liu, G. Luo, W. Hu, J. Hua, S. Geng, P. K. Chu, J. Zhang, H. Wang, & X. Yu. Mediated Drug Release from Nanovehicles by Black Phosphorus Quantum Dots for Efficient Therapy of Chronic Obstructive Pulmonary Disease. Angew. Chem. 59(2020)20568–20576. https://doi.org/10.1002/anie.202008379

[80] K. Chikuma, K. Arima, Y. Asaba, R. Kubota, S. Asayama, K. Sato, & H. Kawakami. The potential of lipid-polymer nanoparticles as epigenetic and ROS control approaches for COPD. Free Radic. Res. 54(2020)829–840. https://doi.org/10.1080/10715762.2019.1696965

[81] A. M. Mohamed, A. Muth, I. Saleem, & G. A. Hutcheon. Polymeric nanoparticles for the delivery of miRNA to treat Chronic Obstructive Pulmonary Disease (COPD). Eur. J. Pharm. Biopharm., 136(2019)1–8. https://doi.org/10.1016/j.ejpb.2019.01.002

[82] N. Vij, T. Min, R. E. R. Marasigan, C. N. Belcher, S. Mazur, H. Ding, K. Yong, & I. Roy. Development of PEGylated PLGA nanoparticle for controlled and sustained drug delivery in cystic fibrosis. J. Nanobiotechnology., 8(2010)22. https://doi.org/10.1186/1477-3155-8-22

[83] M. Passi, S. Shahid, S. Chockalingam, I. K. Sundar, & P. Gopinath. Conventional and Nanotechnology Based Approaches to Combat Chronic Obstructive Pulmonary Disease: Implications for Chronic Airway Diseases. Int J Nanomedicine., 15(2020)3803–3826. https://doi.org/10.2147/ijn.s242516

[84] F. Mejza, L. Gnatiuc, A. S. Buist, W. M. Vollmer, B. Lamprecht, D. O. Obaseki, P. Nastałek, E. Nizankowska-Mogilnicka, & P. Burney. Prevalence and burden of chronic bronchitis symptoms: results from the BOLD study. Eur. Respir. J., 50(2017)1700621. https://doi.org/10.1183/13993003.00621-2017

[85] A. Widysanto. Chronic Bronchitis. StatPearls - NCBI Bookshelf. (2022); https://www.ncbi.nlm.nih.gov/books/NBK482437/

[86] A. Ferré, C. Fuhrman, M. Zureik, C. Chouaid, A. Vergnenegre, G. Huchon, M. Delmas, & N. Roche. Chronic bronchitis in the general population: Influence of age, gender and socio-economic conditions. Respir. Med., 106(2012)467–471. https://doi.org/10.1016/j.rmed.2011.12.002

[87] V. Kim, & G. J. Criner. Chronic Bronchitis and Chronic Obstructive Pulmonary Disease. Am. J. Respir. Crit. Care Med., 187(2013)228–237. https://doi.org/10.1164/rccm.201210-1843ci

[88] R. Ehrlich, N. J. White, R. J. Norman, R. Laubscher, K. Steyn, C. Lombard, & D. Bradshaw. Predictors of chronic bronchitis in South African adults. IJTLD, 8(2004)369–376. PMID: 15139477.

[89] K. F. Chung, L. McGarvey, W. Song, A. B. Chang, K Lai, B. J. Canning, S. S. Birring, J. A. Smith, & S. B. Mazzone. Cough hypersensitivity and chronic cough. Nat. Rev. Dis. Primers., 8(2022). https://doi.org/10.1038/s41572-022-00370-w

[90] S. Tual, B. Clin, N. Levêque-Morlais, C. Raherison, I. Baldi, & P. Lebailly. Agricultural exposures and chronic bronchitis: findings from the AGRICAN (AGRIculture and CANcer) cohort. Ann. Epidemiol., 23(2013)539–545. https://doi.org/10.1016/j.annepidem.2013.06.005

[91] A. Mamane, I. Baldi, J Tessier, C. Raherison, & G. Bouvier. Occupational exposure to pesticides and respiratory health. ERR, 24(2015)306–319. https://doi.org/10.1183/16000617.00006014

[92] F. J. Kelly, & J. C. Fussell. Air pollution and public health: emerging hazards and improved understanding of risk. Environ. Geochem. Health, 37(2015)631–649. https://doi.org/10.1007/s10653-015-9720-1

[93] M. M. De Oca, R. J. Halbert, M. C. Lopez, R. Pérez-Padilla, C. Tálamo, D. Moreno, A. Muiño, J. R. Jardim, G. Valdivia, J. Pertuzé, & P. C. Hallal. The chronic bronchitis phenotype in subjects with and without COPD: the PLATINO study. Eur. Respir. J., 40(2012)28–36. https://doi.org/10.1183/09031936.00141611

[94] C. A. R. Martinez, V. Kim, Y. Chen, E. A. Kazerooni, S. Murray, G. J. Criner, J. R. Curtis, E. A. Regan, E. S. Wan, C. P. Hersh, E. K. Silverman, J. D. Crapo, F. J. Martinez, & M. K. Han. The clinical impact of non-obstructive chronic bronchitis in current and former smokers. Respir. Med., 108(2014)491–499. https://doi.org/10.1016/j.rmed.2013.11.003

[95] J. Y. Choi, H. K. Yoon, S. Y. Lee, J. Kim, H. J. Choi, Y. Kim, K. S. Jung, K. H. Yoo, W. H. Kim, & C. K. Rhee. Comparison of clinical characteristics between chronic bronchitis and non-chronic bronchitis in patients with chronic obstructive pulmonary disease. BMC Pulm. Med., 22(2022). https://doi.org/10.1186/s12890-022-01854-x

[96] Y. Dotan, J. Y. So, & V. Kim. Chronic Bronchitis: Where Are We Now? COPD, 6(2019)178–192. https://doi.org/10.15326/jcopdf.6.2.2018.0151

[97] P. Sestini, E. Renzoni, S. Robinson, P. Poole, & F. S. F. Ram. Short-acting beta2-agonists for stable chronic obstructive pulmonary disease. The Cochrane Library, (2002); 2010. https://doi.org/10.1002/14651858.cd001495

[98] J. A. Walters, D. J. Tan, C. J. White, P. G. Gibson, R. Wood-Baker, & E. H. Walters. Systemic corticosteroids for acute exacerbations of chronic obstructive pulmonary disease. The Cochrane Library. (2014); https://doi.org/10.1002/14651858.cd001288.pub4

[99] M. K. Han, N. Tayob, S. Murray, M. T. Dransfield, G. R. Washko, P. D. Scanlon, G. J. Criner, R. Casaburi, J. E. Connett, S. C. Lazarus, R. K. Albert, P. G. Woodruff, & F. J. Martinez. Predictors of Chronic Obstructive Pulmonary Disease Exacerbation Reduction in Response to Daily Azithromycin Therapy. Am. J. Respir. Crit. Care Med., 189(2014)1503–1508. https://doi.org/10.1164/rccm.201402-0207oc

[100]R. K. Albert, J. Connett, W. C. Bailey, R. Casaburi, J. A. Cooper, G. J. Jr, Criner, J. L. Curtis, M. T Dransfield, M. K. Han, S. C. Lazarus, B. Make, N. Marchetti, F. J. Martinez, N. E. Madinger, C. McEvoy, D. E. Niewoehner, J. Porsasz, C. S. Price, J. Reilly, P. D. Scanlon, F. C. Sciurba, S. M. Scharf, G. R. Washko, P. G. Woodruff, N. R. Anthonisen, COPD Clinical Research Network Azithromycin for prevention of exacerbations of COPD. N Engl J Med. 365(2011)689–698. doi: 10.1056/NEJMoa1104623.

[101]P. M. Calverley, K. F. Rabe, U. M. Goehring, S. Kristiansen, L. M. Fabbri, F. J Martinez, Roflumilast in symptomatic chronic obstructive pulmonary disease: two randomised clinical trials, Lancet. 374(2009) 685–694. https://doi.org/10.1016/s0140-6736(09)61255-1

[102]F. J. Martinez, P. M. Calverley, U. Goehring, M. Brose, L. M. Fabbri, K. F. Rabe, Effect of roflumilast on exacerbations in patients with severe chronic obstructive pulmonary disease

uncontrolled by combination therapy (REACT): a multicentre randomised controlled trial, Lancet. 385(2015) 857–866. https://doi.org/10.1016/s0140-6736(14)62410-7

[103] F. J. Martinez, K. F. Rabe, S. Sethi, E. Pizzichini, A. McIvor, A. Anzueto, V. Alagappan, S. S. Siddiqui, L. Rekeda, C. J. Miller, S. Zetterstrand, C. Reisner, S. I. Rennard, Effect of Roflumilast and Inhaled Corticosteroid/Long-Acting β_2-Agonist on Chronic Obstructive Pulmonary Disease Exacerbations (RE^{2}SPOND). A Randomized Clinical Trial, Am. J. Respir. Crit. Care Med. 194(2016) 559–567. https://doi.org/10.1164/rccm.201607-1349oc

[104] J Chong, B Leung, P Poole, Phosphodiesterase 4 inhibitors for chronic obstructive pulmonary disease, Cochrane Database Syst Rev. 9(2017) CD002309. https://doi.org/10.1002/14651858.CD002309.pub5

[105] A.Valipour, S. Fernandez-Bussy, A. J. Ing, D. P. Steinfort, G. I. Snell, , J. P. Williamson, T. Saghaie, L. Irving, E. Dabscheck, W. Krimsky, J. R. Waldstreicher, Bronchial Rheoplasty for Treatment of Chronic Bronchitis. Twelve-Month Results from a Multicenter Clinical Trial, Am. J. Respir. Crit. Care Med. 202(2020) 681–689. https://doi.org/10.1164/rccm.201908-1546oc

[106] B. K. Rubin, The pharmacologic approach to airway clearance: mucoactive agents, Respir. Care. 47(2002) 818–822. PMID: 12088552.

[107] M.Cazzola, L. Calzetta, C. P. Page, J. R. Jardim, A. Chuchalin, P.Rogliani, M. G. Matera, Influence of N-acetylcysteine on chronic bronchitis or COPD exacerbations: a meta-analysis, ERR. 24(2015) 451–461. https://doi.org/10.1183/16000617.00002215

[108] Q. Li, S. Huang, H. Wan, H. Wu, T. Zhou, S. D. Minteer, W Deng, Effect of smoking cessation on airway inflammation of rats with chronic bronchitis, Chin. Med. J. 120(2007)1511–1516. https://doi.org/10.1097/00029330-200709010-00009

[109] V. Arkhipov, D. M. Arkhipova, M. Miravitlles, A. Lazarev, E. Y. Stukalina, Characteristics of COPD patients according to GOLD classification and clinical phenotypes in the Russian Federation: the SUPPORT trial, Int J Chron Obstruct Pulmon Dis. 12(2017)3255–3262. https://doi.org/10.2147/copd.s142997

[110] M. Geiser, O. Quaile, A. Wenk, C. Wigge, S. Eigeldinger-Berthou, S. Hirn, M. Schäffler, C. Schleh, W. Möller, M. A. Mall, W. G. Kreyling, Cellular uptake and localization of inhaled gold nanoparticles in lungs of mice with chronic obstructive pulmonary disease, Part. Fibre Toxicol. 10(2013)19. https://doi.org/10.1186/1743-8977-10-19

[111] E. Yu, S. Sharma, Cystic Fibrosis. StatPearls - NCBI Bookshelf. (2022). https://www.ncbi.nlm.nih.gov/books/NBK493206/

[112] W. Poncin, P. Lebecque, L'indice de clairance pulmonaire dans la mucoviscidose. *Rev. des Mal. Respir.* 36(2019) 377–395. https://doi.org/10.1016/j.rmr.2018.03.007

[113] T. M. Endres, M. W. Konstan, What Is Cystic Fibrosis? JAMA, 327 (2022) 191. https://doi.org/10.1001/jama.2021.23280

[114]Q. Chen, Y. Shen, J. Zheng, A review of cystic fibrosis: Basic and clinical aspects. Animal Models and Experimental Medicine, 4 (2021) 220–232. https://doi.org/10.1002/ame2.12180

[115]J. C. Davies, E. W. Alton, A. Bush, Cystic fibrosis, BMJ. 335(2007)1255–1259. https://doi.org/10.1136/bmj.39391.713229.ad

[116]P. H. Gilligan, Infections in Patients with Cystic Fibrosis, Clin. Lab. Med. 34(2014) 197– 217. https://doi.org/10.1016/j.cll.2014.02.001

[117]S. Link, R. P. Nayak, Review of Rapid Advances in Cystic Fibrosis, Mo Med, 117(2020) 548–554. PMID: 33311787; PMCID: PMC7721430.

[118]A. M Coverstone, P. D. Sly, Early Diagnosis and Intervention in Cystic Fibrosis: Imagining the Unimaginable, Front Pediatr. 8 (2021) https://doi.org/10.3389/fped.2020.608821

[119]P. A. Flume, K. A. Robinson, B. O'Sullivan, J. D. Finder, R. L. Vender, D. B.Willey-Courand, T. B. White, B. C. Marshall, Cystic fibrosis pulmonary guidelines: airway clearance therapies, Respir. Care. 4 (2009) 522–537.

[120]M. R.Morrell, J. M. Pilewski, Lung Transplantation for Cystic Fibrosis, Clin. Chest Med. 37(2016)127–138. https://doi.org/10.1016/j.ccm.2015.11.008

[121]C. L. Hardee, L. M. Arevalo-Soliz, B. D. Hornstein, L. Zechiedrich, Advances in Non-Viral DNA Vectors for Gene Therapy, Genes. 8 (2017) 65. https://doi.org/10.3390/genes8020065

[122]M. W. Konstan, P. B. Davis, J. S.Wagener, K. A. Hilliard, R. S. Stern, , L. J. H. Milgram, T. Kowalczyk, S. L. Hyatt, T. L. Fink, C. R. Gedeon, S. M. Oette, J. L. Payne, O. Muhammad, A. G. Ziady, R. C. Moen, M. E. Cooper, Compacted DNA Nanoparticles Administered to the Nasal Mucosa of Cystic Fibrosis Subjects Are Safe and Demonstrate Partial to Complete Cystic Fibrosis Transmembrane Regulator Reconstitution, HGT. 15(2004) 1255–1269. https://doi.org/10.1089/hum.2004.15.1255

[123]E. W. Alton, D. G. Armstrong, D. Ashby, K. J.Bayfield, D. Bilton, E. V. Bloomfield, A. W. Boyd, J. Brand, R. Buchan, R. Calcedo, P. Carvelli, M. Chan, S. H. Cheng, D. Collie, S. Cunningham, H. A. Davidson, G. A. Davies, J. C. Davies, L. A Davies, P. Wolstenholme-Hogg, Repeated nebulisation of non-viral CFTR gene therapy in patients with cystic fibrosis: a randomised, double-blind, placebo-controlled, phase 2b trial, Lancet Respir. Med. 3(2015) 684–691. https://doi.org/10.1016/s2213-2600(15)00245-3

[124]E. Robinson, K. D. MacDonald, K. Slaughter, M. McKinney, S. M. Patel, C. Sun, G. Sahay, Lipid Nanoparticle-Delivered Chemically Modified mRNA Restores Chloride Secretion in Cystic Fibrosis, Mol. Ther. 26(2018) 2034–2046. https://doi.org/10.1016/j.ymthe.2018.05.014

[125]A. K. M. A. Haque, A. Dewerth, J. S. Antony, J. Riethmüller, G. R. Schweizer, P. Weinmann, N. Latifi, H. Yasar, N. Pedemonte, E.Sondo, B.Weidensee, A. Ralhan, J. Laval, P. Schlegel, C. Seitz, B. Loretz, C. Lehr, R. Handgretinger, M. S. D Kormann, Chemically

modified hCFTR mRNAs recuperate lung function in a mouse model of cystic fibrosis. Sci. Rep. 8 (2018). https://doi.org/10.1038/s41598-018-34960-0

[126] A. D. Tagalakis, M. M. Munye, R. Ivanova, H. Chen, C. Smith, A. Aldossary, L. Rosa, D. Moulding, J. Barnes, K. N. Kafetzis, S. E. Jones, D. L. Baines, G. W. J Moss, C.O' Callaghan, R. J. McAnulty, S. D. Hart, Effective silencing of ENaC by siRNA delivered with epithelial-targeted nanocomplexes in human cystic fibrosis cells and in mouse lung, Thorax. 73(2018) 847–856. https://doi.org/10.1136/thoraxjnl-2017-210670

[127] G. Osman, J. D. Rodriguez, S. L. Chan, J. Chisholm, G. A Duncan, N. Kim, A. L. Tatler, K. M. Shakesheff, J. Hanes, J. S. Suk, J. R. Dixon, PEGylated enhanced cell penetrating peptide nanoparticles for lung gene therapy, JCR. 285 (2018) 35–45. https://doi.org/10.1016/j.jconrel.2018.07.001

[128] H. D. C. Smyth, X. Peng, X. Liu, D. Arasappan, D. Wylie, S. Schwartz, J. J. Fullmer, B. C. McWilliams, D. Ghosh, Peptides as surface coatings of nanoparticles that penetrate human cystic fibrosis sputum and uniformly distribute in vivo following pulmonary delivery, JCR. 322, (2020) 457–469. https://doi.org/10.1016/j.jconrel.2020.03.032

[129] O. B.Garbuzenko, N. Kbah, A. Kuzmov, N. Pogrebnyak, V. P. Pozharov, T. Minko, Inhalation treatment of cystic fibrosis with lumacaftor and ivacaftor co-delivered by nanostructured lipid carriers, JCR. 296 (2019) 225–231. https://doi.org/10.1016/j.jconrel.2019.01.025

[130] N. G.Türeli, A. Torge, J. Juntke, B. Schwarz, N. Schneider-Daum, A. E. Türeli, C. Lehr, M. Schneider, Ciprofloxacin-loaded PLGA nanoparticles against cystic fibrosis P. aeruginosa lung infections. Eur. J. Pharm. Biopharm. 117 (2017) 363–371. https://doi.org/10.1016/j.ejpb.2017.04.032

[131] J. Ernst, M. Klinger-Strobel, K. Arnold, J. Thamm, A. Hartung, M.W. Pletz, O. Makarewicz, & D. Fischer, Polyester-based particles to overcome the obstacles of mucus and biofilms in the lung for tobramycin application under static and dynamic fluidic conditions, Eur. J. Pharm. Biopharm. 131(2018)120–129. https://doi.org/10.1016/j.ejpb.2018.07.025

[132] WHO: Pneumonia in children, (2023), Available online. https://www.who.int/news-room/fact-sheets/detail/pneumonia

[133] C.C. Chou, C. F. Shen, S. J. Chen, H. M. Chen, Y. C. Wang, W. S. Chang, Y. T. Chang, W. Y. Chen, C. Y. Huang, C. C. Kuo, et al, Recommendations and guidelines for the treatment of pneumonia in Taiwan, J. Microbiol. Immunol. Infect. 52(2019) 172–199. doi: 10.1016/j.jmii.2018.11.004.

[134] O. Henig, K. S. Kaye, Bacterial Pneumonia in Older Adults, Infect. Dis. Clin. N. Am. 31(2017) 689–713. doi: 10.1016/j.idc.2017.07.015

[135] S. Pakhale, S. Mulpuru, T. J. Verheij, M. M. Kochen, G. G. Rohde, L. M. Bjerre, Antibiotics for community-acquired pneumonia in adult outpatients, Cochrane Database Syst. Rev. 10(2014) CD002109. DOI: 10.1002/14651858.CD002109.pub4.

[136] W. Muhammad, Z. Zhai, S. Wang, C. Gao, Inflammation-modulating nanoparticles for pneumonia therapy, Wiley Interdiscip Rev Nanomed Nanobiotechnol. 14(2022) e1763. doi: 10.1002/wnan.1763.

[137] C. Bao, B. Liu, B. Li, J. Chai, L. Zhang, L. Jiao, D. Li, Z. Yu, F. Ren, X. Shi, et al, Enhanced Transport of Shape and Rigidity-Tuned α-Lactalbumin Nanotubes across Intestinal Mucus and Cellular Barriers, Nano Lett. 20(2020)1352–1361. doi: 10.1021/acs.nanolett.9b04841

[138] R. Lima, F. S. Del Fiol, V. M. Balcão, Prospects for the Use of New Technologies to Combat Multidrug-Resistant Bacteria, Front. Pharmacol. 10(2019)692. https://doi.org/10.3389/fphar.2019.00692.

[139] Z. Huang, S. N. Kłodzińska, F. Wan, H. M. Nielsen, Nanoparticle-mediated pulmonary drug delivery: state of the art towards efficient treatment of recalcitrant respiratory tract bacterial infections, Drug Deliv Transl Res. 11(2021)1634-1654. doi: 10.1007/s13346-021-00954-1.

[140] J. Tang, Q. Ouyang, Y. Li, P. Zhang, W. Jin, S. Qu, F. Yang, Z. He, M. Qin. Nanomaterials for Delivering Antibiotics in the Therapy of Pneumonia, Int. J. Mol. Sci. 23(2022)15738. https://doi.org/10.3390/ijms232415738

[141] V. Kumar, A. K. Abbas, N. Fausto, and R. N. Mitchell, Robbins Basic Pathology, Saunder Elsevier, 8(2007).

[142] A. M. Cooper, Cell-mediated immune responses in tuberculosis, Annu. Rev. Immunol. 27(2009)393–422.

[143] K. Rohde, R. M. Yates, G. E. Purdy, and D. G. Russell, Mycobacterium tuberculosis and the environment within the phagosome, Immunol. Rev. 219(2007) 37–54. https://doi.org/10.1111/j.1600-065X.2007.00547.x

[144] G. R. Stewart, B. D. Robertson, and D. B. Young, Tuberculosis: a problem with persistence, Nat. Rev. Microbiol., 1(2003) 97–105. https://doi.org/10.1038/nrmicro749.

[145] WHO Global Tuberculosis Control Report 2010, Summary, Cent. Eur. J. Public Health. 18(2010) 237.

[146] T. R. Frieden, T. R. Sterling, S. S. Munsiff, C. J. Watt, and C. Dye, Tuberculosis, Lancet, 362(2003)887–899. DOI:https://doi.org/10.1016/j.ccm.2005.02.001

[147] D. J. Morgan, L. McLain, and N. J. Dimmock, WHO declares tuberculosis a global emergency, Soz. Praventivmed. 38(1993)251–252. https://doi.org/10.1007/BF01624546.

[148] D. Bhowmik, R. M. Chiranjib, B. Jayakar, and K. P. S. Kumar, Recent trends of drug used treatment of tuberculosis, J. Chem. Pharm. Res., 1(2009)113– 133.

[149] M. A. Moretton, R. J. Glisoni, D. A. Chiappetta, and A. Sosnik, Molecular implications in the nanoencapsulation of the anti-tuberculosis drug rifampicin within flower-like polymeric micelles, Colloids Surf. B, 79(2010)467–479. https://doi.org/10.1016/j.colsurfb.2010.05.016

[150] Lalloo, U. G., Ambaram, A, New antituberculous drugs in development. Current HIV/AIDS reports, 7 (2010), 143–151. https://doi.org/10.1007/s11904-010-0054-4

[151] B. N. V. Hari, K. P. Chitra, R. Bhimavarapu, P. Karunakaran, N. Muthukrishnan, and B. S. Rani, Novel technologies: a weapon against tuberculosis, Indian J. Pharmacol., 42(2010)338–344. DOI: 10.4103/0253-7613.71887

[152] A.Sosnik, A. M. Carcaboso, R. J. Glisoni, M. A. Moretton, and ´ D. A. Chiappetta, New old challenges in tuberculosis: potentially effective nanotechnologies in drug delivery, Adv. Drug Deliv. Rev. 62(2010) 547–559. https://doi.org/10.1016/j.addr.2009.11.023.

[153] W. A. Ritschel, "Microemulsions for improved peptide absorption from the gastrointestinal tract, Meth. And Find Exp. Clin. Pharamcol.13(1991) 205– 220.

[154] J. M. Sarciaux, L. Acar, and P. A. Sado, Using microemulsion formulations for oral drug delivery of therapeutic peptides, Int. J. Pharm. 120(1995)127– 136. https://doi.org/10.1016/0378-5173(94)00386.

[155] Elalccleston,G. M. Microemulsions, in Encyclopedia of Pharmaceutical Technology, J. Swarbrick and J. C. Boylan, Eds., Marcel Dekker, New York, NY, USA, 9(1992)375–42.

[156] C. Song, D. Yu, Y. Wang, Q. Wang, Z. Guo, J. Huang, S. Li, W. Hu. Dual Primary Cancer Patients with Lung Cancer as a Second Primary Malignancy: A Population-Based Study, Front. Oncol. 10(2020)515606. https://doi.org/10.3389/fonc.2020.515606.

[157] S. Mukherjee, C. R. Patra. Therapeutic application of anti-angiogenic nanomaterials in cancers. Nanoscale. 8(2016)12444–12470. doi: 10.1039/C5NR07887C.

[158] H. J. Burstein, R. S. Schwartz. Molecular origins of cancer. N. Engl. J. Med. 358(2008)527. doi: 10.1056/NEJMe0800065.

[159] M. K. Paul, A. K. Mukhopadhyay. Tyrosine kinase—Role and significance in Cancer. Int. J. Med. Sci. 1(2004)101–115. doi: 10.7150/ijms.1.101.

[160] M. Jamal-Hanjani, G. A. Wilson, N. McGranahan, N. J. Birkbak, T. B. K. Watkins, S. Veeriah, S. Shafi, D. H. Johnson, R. Mitter, R. Rosenthal, et al. Tracking the Evolution of Non-Small-Cell Lung Cancer. N. Engl. J. Med. 376(2017) 2109–2121. doi: 10.1056/NEJMoa1616288.

[161] R. S. Herbst, D. Morgensztern, C. Boshoff. The biology and management of non-small cell lung cancer. Nature, 553(2018) 446–454. https://doi.org/10.1038/nature25183

[162] M. Nicolson. ES05.01 Lung Cancer Survival: Progress and Challenges. J. Thorac. Oncol., 14(2019)S24. DOI:https://doi.org/10.1016/j.jtho.2019.08.087.

[163] R. M. DiSanto, V. Subramanian, Z. Gu. Recent advances in nanotechnology for diabetes treatment. Wiley Interdisasp. Rev. Nanomed. Nanobiotechnol. 7(2015)548–564. doi: 10.1002/wnan.1329.

[164] X. Yue, Z. Dai. Liposomal Nanotechnology for Cancer Theranostics. Curr. Med. Chem. 25(2018) 1397–1408. doi: 10.2174/0929867324666170306105350.

[165] Y. Matsumura, H. Maeda. A New Concept for Macromolecular Therapeutics in Cancer-Chemotherapy—Mechanism of Tumoritropic Accumulation of Proteins and the Antitumor Agent Smancs. Cancer Res. 46(1986) 6387–6392.

[166] A. Mukherjee, M. Paul, S. Mukherjee. Recent Progress in the Theranostics Application of Nanomedicine in Lung Cancer. Cancers (Basel). 11(2019) 597. doi: 10.3390/cancers11050597. PMID: 31035440; PMCID: PMC6562381.

[167] S. Li, S. Xu, X. Liang, Y. Xue, J. Mei, Y. Ma, Y. Liu. Nanotechnology: Breaking the Current Treatment Limits of Lung Cancer. Adv. Health Mater. 10(2021) 2100078. https://doi.org/10.1016/j.apsb.2021.04.023

[168] R. Baskaran, J. Lee, S. G. Yang. Clinical development of photodynamic agents and therapeutic applications. Biomater. Res. 22(2018) 25. doi: 10.1186/s40824-018-0140-z.

[169] H. Kato, T. Horai, K. Furuse, M. Fukuoka, S. Suzuki, Y. Hiki, Y. Ito, S. Mimura, Y. Tenjin, H. Hisazumi, et al. Photodynamic therapy for cancers: A clinical trial of porfimer sodium in Japan. Jpn. J. Cancer Res. 84(1993) 1209–1214. doi: 10.1111/j.1349-7006.1993.tb02823.x.

[170] C. Lin, X. Zhang, H. Chen, Z. Bian, G. Zhang, M. K. Riaz, D. Tyagi, G. Lin, Y. Zhang, J. Wang, et al. Dual-Ligand Modified Liposomes Provide Effective Local Targeted Delivery of Lung-Cancer Drug by Antibody and Tumor Lineage-Homing Cell-Penetrating Peptide, Drug Deliv. 25(2018) 256–266. doi: 10.1080/10717544.2018.1425777.

[171] M. Doroudian, A. Neill, R. Mac Loughlin, A. Prina-Mello, Y. Volkov, & S. C. Donnelly. Nanotechnology in pulmonary medicine. COPHAR, 56(2021) 85–92. https://doi.org/10.1016/j.coph.2020.11.002

[172] Q. Qiao, X. Liu, T. Yang, K. Cui, L. Kong, C. Yang, Z. Zhang, Nanomedicine for acute respiratory distress syndrome: The latest application, targeting strategy, and rational design, APSB B, 11(2021) 3060-3091.

[173] J. Xie, L. Liu, Y. Yang, W. Yu, M. Li, K. Yu, R. Zheng, J. Yan, X. Wang, G. Cai, J. Li, Q. Gu, H. Zhao, X. Mu, X. Ma, H. Qiu, A modified acute respiratory distress syndrome prediction score: a multicenter cohort study in China, J. Thorac. Dis. 10(2018) 5764–5773. doi:10.21037/jtd.2018.09.117.

[174] B. T. Thompson, G. R. Bernard. ARDS network (NHLBI) studies: successes and challenges in ARDS clinical research, Crit. Care Clin. 27(2011) 459–468. doi: 10.1016/j.ccc.2011.05.011.

[175] M. A. Matthay, R. L. Zemans, G. A. Zimmerman, Y. M. Arabi, J. R. Beitler, A. Mercat, M. Herridge, A. G. Randolph, C. S. Calfee, Acute respiratory distress syndrome, Nat. Rev. Dis. Prim. 5(2018) 13. https://doi.org/10.1038/s41572-019-0069-0

[176] T. A. P. F. Doll, R. Dey, P. Burkhard. Design and optimization of peptide nanoparticles, J. Nanobiotechnology. 13(2015). doi: 10.1186/s12951-015-0119-z.

[177] J. S. Brenner. Nanomedicine for the treatment of acute respiratory distress syndrome: the 2016 ATS Bear cage award-winning proposal. Ann. Am. Thorac. Soc. 14(2017) 561–564. doi: 10.1513/AnnalsATS.201701-090PS.

[178] S. Tarvirdipour, C. A. Schoenenberger, Y. Benenson, C. G. Palivan. A self-assembling amphiphilic peptide nanoparticle for the efficient entrapment of DNA cargoes up to 100 nucleotides in length. Soft Matter. 16(2020) 1678–1691. doi: 10.1039/c9sm01990a.

[179] P. N. Navya, H. K. Daima. Rational engineering of physicochemical properties of nanomaterials for biomedical applications with nanotoxicological perspectives, Nano Converg. 3(2016). doi: 10.1186/s40580-016-0064-z.

[180] H. M. Mansour, Y. S. Rhee, X. Wu. Nanomedicine in pulmonary delivery, Int. J. Nanomedicine. 4(2009) 299–319. doi: 10.2147/ijn.s4937.

[181] R. T. Sadikot, Peptide nanomedicines for treatment of acute lung injury, Methods Enzymol. 508(2012) 315–324. doi: 10.1016/B978-0-12-391860-4.00016-1.

[182] R. T. Sadikot, The potential role of nano- and micro-technology in the management of critical illnesses, Adv Drug Deliv Rev. 77(2014) 27-31.

[183] I. Roy, N. Vij. Nanodelivery in airway diseases: challenges and therapeutic applications, Nanomedicine Nanotechnology. Biol. Med. 6(2010) 237–244. doi: 10.1016/j.nano.2009.07.001.

[184] O. S. Thomas, W. Weber, Overcoming physiological barriers to nanoparticle delivery—are we there yet? Front. Bioeng. Biotechnol. 7(2019). doi: 10.3389/fbioe.2019.00415.

Nanoparticles in Healthcare: Applications in Therapy, Diagnosis and Drug Delivery   Materials Research Forum LLC
11Materials Research Foundations 160 (2024) 296-312   https://doi.org/10.21741/9781644902974-11

Chapter 11

Nanoparticles in Glioma Therapy and Drug Delivery Systems

Bhanu Revathi K[1], Hamsaveni S.M[1], Bhavyashree V[1], Rajeswari N[1],
Shinomol George Kunnel[2*]

[1]Department of Biotechnology, Dayananda Sagar College of Engineering, Bangalore, India

[2] Faculty of Life Sciences, Kristu Jayanti College (Autonomous), Bangalore, India

*shinojesu@gmail.com

Abstract

Gliomas are tumours of the glial cells of the brain, and they are highly complicated in terms of treatment, recovery, and diagnosis. Some of them are high grade tumours that require sophisticated imaging tools for diagnosis. Nanoparticles offer many advantages as targeted drug delivery systems causing minimal harm to the surrounding tissues and higher bioavailability at the site of the tumour. This chapter aims in exploring the various treatment methods and the possibility of using cellular macromolecules as targets for drug delivery. Combinatorial approaches along with nanoparticles are required to minimise the recurrence of tumour and to enhance patient recovery in various types of gliomas.

Keywords

Glioma, Glioblastoma, Nanoparticles, Therapy, Drug Delivery

Contents

1.   Introduction

Glioma is a common term used for cancer of the glial cells that surround the nerve endings of the central nervous system. These gliomas progress over time to a subset called glioblastoma which is far more aggressive and recurrent. Gliomas constitute most brain tumours; classified by WHO according to histological progression as grade I, II, III and IV based on characterization of biomarkers [1]. This is crucial for accurate diagnosis and subsequent treatment options. Pilocytic astomas are grade I that are treatable to some extent. The tumours classified as grade II are astrocytomas, oligodendromas and oligoastrocytomas that are aggressive. Grade III gliomas have an intermediate clinical course and constitute anaplastic astrocytomas and anaplastic oligoastrocytomas [2]. The final group is the most aggressive in clinical course. Their diffuse nature and progression over time makes the procedure of surgical removal rather cumbersome. In oligodendromas, codeletions of chromosome arms 1p and 19q bring about sensitivity to chemotherapy.

Molecular alterations in gliomas are important in projecting the response to therapy. In a few patients of lower grade gliomas with Isocitrate dehydrogenases (IDH) and 1p 19q codeletions had greater recoverability. Molecular markers such as loss of or mutations in the EGFR, NF-1, PTEN, TP53, RB 1, TERT promoter and many others were associated with identifications of grade II and III astrocytomas. Similarly, deletions or mutations in the IDH/2 along with TP 53, ATRX, 1p/19q codeletions and amplifications in CDK4/6, PDGFRA and other genes were found to be associated with grade II-III astrocytomas and oligodendromas and sometimes in grade IV glioblastomas [3].

1.1 Conventional treatment strategies

Radiotherapy was combined with chemotherapy aided by drugs such as procarbazine, lomustine and vincristine to treat grade II gliomas and resultant enhancement of patient survival compared to radiotherapy alone. The FDA approved use of Tumour Treating Fields (TTF) along with standard chemoradiotherapy to treat freshly diagnosed glioblastomas is a breakthrough in therapy. TTF s are low intensity, intermediate frequency alternating electric fields.TTF treatment posed no added toxicity other than scalp rashes due to electrode contact. Certain patients with TTF combined with chemoradiation also showed progression free survival with improved quality of life [4].

1.1.1 Radiation therapy

Radiation and subsequent temozolomide and maintenance chemotherapy could enhance survival of patients with grade IV glioblastoma up to 12-18 months only [5]. Combination of dose intense radiotherapy with PCV treatment improved the chances of survival upto15 years in patients with 1p/19q codeletion. A standard dose of 60 gy was prescribed in 30-33 fractions of 1.8-2 gy to treat an infiltrated grade IV glioblastoma also known as Glioblastoma multiforme (GBM). In elderly population 40 gy in 15 fractions may be employed. The margin of tumour defines the intensity and planning even in surgical and radiation therapy. Methylation of O-6 methylguanine DNA methyltransferase (MGMT promoter) was associated with improved survival in patients treated with radiation and temozolomide.

Various image contrast techniques must be used to determine the presence, size, and tissue density of tumours. Apparent Diffusion Coefficient (ADC) values are used to compare tissues around tumour with normal tissues to detect reduced diffusivity as post radiation side effect [6]. Dynamic contrast enhanced and Dynamic susceptibility contrast MRIs are novel imaging techniques that measure rate of diffusion of contrast agents from blood vessels to tumour microenvironment to detect antiangiogenic activity during radiotherapy. Various metabolites in and around the tumour can also be detected to determine tumour activity by Magnetic Resonance Spectroscopy Imaging (MRSI) [7].For the treatment of low-grade gliomas with lesser density and high risk; external beam radiation may be preferred.

1.1.2 Surgical treatment

The diffusivity of tumour and its density determine the complexity of surgical removal. Providing quality life to the patient post operatively is an integral part of planning therapeutic options. Surgical resection forms the initial phase of therapy and pre-operative image plays a major role in the above. False fMRI results were associated with vascular changes in glioblastomas. A novel technique for improved resection is the use of 5- amino levulinic acid: a fluorescent marker that emits fluorescence under blue light. It correlates with the histological grading of the other structures such as blood vessels and normal brain tissues that do not exhibit fluorescence. Diffusion Tumour Imaging Fibre Tracking (DTI-FT) is an anatomic imaging technique for safe resection.it clearly denudates subcortical tracts in white matter which is important to assess respectability and maintaining neurological function after resection. It has improved the clinical outcome of the tumour presurgically [8].

1.1.3 Chemotherapy

High grade gliomas are treated with temozolomide which is an oral chemotherapeutic drug with DNA alkylating cytotoxicity. Combined with radiation, it improves overall survival. High risk patients were administered procarbazine, CCNU and vincristine after radiation. The characterization of molecular markers such as IDH mutations and 1p19q codeletion also helps to assess the effect of chemotherapy on the tissue. Most drugs do not reach the central nervous system with high concentration at the site of tumour. This emphasizes the role of novel and targeted drug delivery systems such as nanoparticles. The main limitations are the obstacles such as the blood brain barrier, diffusion rate, interstitial space, and density of the tumour. Most drugs are injected into the cerebrospinal fluid (CSF) for enhanced bioavailability. Carmustine is a nitrosourea compound that is used to treat most gliomas- nanoparticle mediated delivery systems have been shown to enhance bioavailability of the chemotherapeutic agent up to several weeks [9]. Targeted transfer using retroviral injection into the tumour for transduction of cytosine deaminase gene into the tumour cells to convert flucytosine to its bioavailable for of 5-fluorouracil is under preclinical stage of study [10]. Franz et al. (2013) studied the role of mTOR pathway inhibitors such as Everolimus to inhibit tumour progression in phase II clinical trials [11].

1.1.4 Immunotherapy

Immunosuppression is a major strategy of therapy in brain tumours that includes immunomodulation, cellular augmentation, and vaccination. Angiogenic therapeutics are under extensive study for the treatment of gliomas/ VEGF is a growth factor crucial for tumour vasculature and can be inhibited by humanized monoclonal antibody called Bevacizumab. This was approved by FDA in 2006 due to its miraculous progression free survival for another half a year. Bevacizumab can also be treated in combination with temozolomide and/or radiotherapy. VEGF can also be inhibited by augmenting using proangiogenic myeloid cells such as granulocytes, myocytes, M-2 skewed macrophages and myeloid derived suppressor cells. Combination of Bevacizumab with onartuzumab (humanized antibody that targets MET-R which is upregulated in gliomas) along with TRC-150 (targeting CD105) helps escape angiogenesis in tumours [8].

Infusion of T cells that have affinity towards tumour antigens is an example of cellular augmentation therapy. The cells are irradiated with granulocyte-macrophage colony stimulating factor (GM-CSF); another strategy is infusing ex vivo coculture of tumour cells and autologous T cells. Other procedures also involve transfer of genetically modified T cells as chimeric antigen receptor cells (CAR T cells) with enhanced reactivity to tumour specific antigens, replicative ability, *in-vivo* persistence, and effector activity. Various antigens such as IL13Rα21, Her2, EphA2, EGFRvIII were targeted as clinical trials in CAR T cell therapy and the target was further narrowed down to tumour associated antigens in GBM. These strategies were found to enhance patient survival [12].

On the other hand, vaccines against tumour specific antigens such as EGFRvIII found only in certain subset of glioblastomas have demonstrated success in phase II trials. Heat shock proteins have been used as vaccine candidates to elicit antitumor inflammatory responses. They are important in antigen presentation to dendritic cells which lead to CD4 and CD8 T cell response to receptors such as CD14, CD91, LOX-1 and TLR-2 and TLR-4 [13]. Another example is the success of polyvalent HSP complex -96 (HSPPC-96) developed from patients' tumour associated antigen conjugated to HSP gp-96 in enhancing recovery of the patients after total resection of GBM [14]. Precision medicine is also being used as an alternate method of therapy for targeted treatment from genome profiling, creating autologous tumour vaccines and immunomodulators from patient biopsy samples [15].

2.   Nanoparticles in the treatment of gliomas

The extent of the spread of glioma is difficult to determine due to their ability to invade the neighbouring tissues. Due to this, the traditional methods which are available are not yielding a therapeutic outcome. As an addition to this the physical and the chemical barriers are also preventing the drugs from reaching the site of the tumour and thus resulting in the failure of the traditional approaches. Among these barriers, the most important ones are the blood brain barrier (BBB) and the blood-brain tumour barrier (BBTB) which are stopping the drugs from reaching the brain. Efforts have been made to overcome physical barriers by discovering methods to directly deliver the drugs to the brain; however, most of these methods have failed. Using nanoparticles for drug delivery is one of the most advanced therapeutic strategies. A variety of nanoparticle formulations have been scrutinized for the delivery of chemotherapeutic drugs to the brain. Among these majority of the formulations used polymers which met the stringent regulations required for biological applications [16]. The most common treatments of glioma using nanoparticles is shown below in Figure 1.

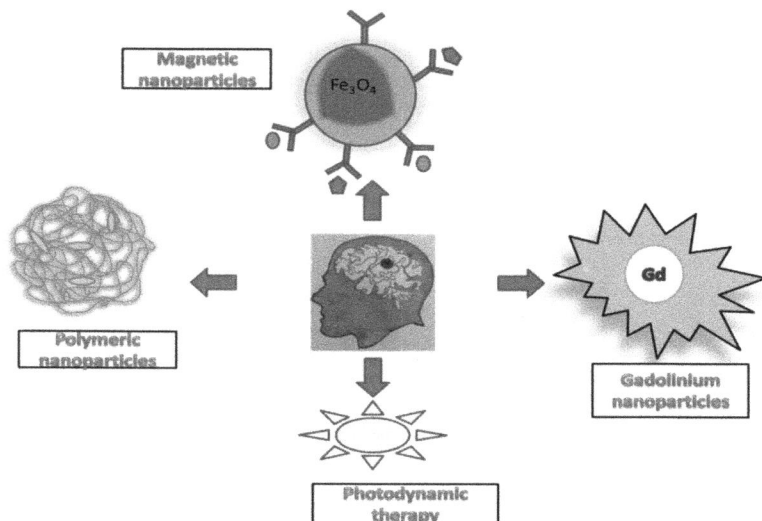

Figure 1. Nanoparticles in the treatment of gliomas.

2.1   Major polymers used in nanoparticle-based gliomas treatment

2.1.1   Polyanhydride

Polyanhydride is formed by the polymerization of carboxylic acid along with anhydride linkages. It is known to be biocompatible as well as biodegradable polymer which can be used in cancer therapy. Before delving into the role of polyanhydride in nanomedicine it is important to understand how it can be used in the treatment of glioma. The Gliadel® wafer which is 1,3-bis(2-chloroethyl)-1-nitrosourea (BCNU) wafer was approved by USFDA in 1996 since it is biodegradable, and a polyanhydride-based implant loaded with carmustine as an alternative to surgery for patients with recurrent GBM. That approval was extended in 2003 to include this treatment as an alternative to surgery and radiation therapy for newly diagnosed patients with high-grade malignant gliomas.

2.1.2   Poly (lactic-co-glycolic acid) (PLGA)

Poly (lactic-co-glycolic acid) (PLGA) is a biodegradable anionic polymer approved by the European Medicines Agency (EMA) as well as FDA. It has broadly been used to capture drugs which are of anti-inflammatory and chemotherapeutic in nature. PLGA can be used to treat many conditions by capturing different proteins and antibiotics on their surface. Maksimenko et al. (2019) reported that PLGA nanoparticles which are doxorubicin loaded and coated with poloxamer 188 which is a copolymer surfactant is known to repair the damaged cells, could cross the physical and chemical barriers and resulting in increasing the efficiency of anti-tumour activity inside the body. There was another study conducted which found that morusin loaded PLGA nanoparticles was bound to a peptide which was specific to chloride channels found in glioma cells named chlorotoxin lead to an increase in the antitumor property in human glioma cell lines *in vitro*.

There are many more polymers which can be used in the treatment of gliomas like Poly(amidoamine) dendrimers, chitosan, Poly (βamino ester), Poly (alkyl cyanoacrylate), Poly (Caprolactone) and others [17]

The broad classification of polymeric nanoparticles is shown in Figure 2.

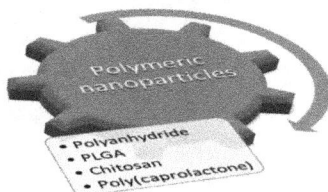

Figure 2. Types of polymeric nanoparticles used in the treatment of gliomas.

2.2   Quantum dot photodynamic therapy

Quantum dots (QDs) are known to be a promising nanomaterial for the bioimaging and destruction of glioma cells due to their unique chemical and optical properties. When compared with organic dyes QDs are comparatively more stable and emissive even after getting exposed to elongated excitation, whereas on the other hand organic dyes are quickly faded. It is possible to change the size and composition of the QDs so that optical tuning can be obtained. By doing this we can ensure that they can emit radiation in the near infrared (NIR) region, which is best for imaging through tissue. QDs have also been used in photodynamic therapy (PDT), a nonintrusive cancer treatment. A photodynamic reaction occurs when a QD is exposed to light and becomes excited. The destruction of cells takes place when an excited QD comes back to its fundamental singlet state and the energy is transferred to the surrounding compounds leading to the production of reactive oxygen species (ROS) which kill cells. Different targeted biological molecules are used in the treatment of glioma. The RGD peptide can bind to integrin v3, which is found in glioma cells. As a targeting ligand, cyclic RGD can be used. The use of NIR QDs for bioimaging and PDT has gained a lot of attention, whereas the use of QD-RGD for gliomas has received little attention [18].

2.3   Magnetic nanoparticles (MNPs) in the treatment of glioma

MNPs are typically made of ferromagnetic iron-oxide (Fe_3O_4). They are typically 1-100 nm in diameter and are invisible to the naked eye. MNPs can be tailored to target cancer cells by attaching a peptide or antibody specific to cancer cells on their surface. They can deliver the drugs to specific areas of the body. It has been proved that MNPs present in a magnetic field have effects on glioma cells that last up to 100 minutes after exposure. In another study, the tumour of squamous cells in rabbits were permanently destroyed when injections consisting of specific MNPs were injected intravenously and then exposed to the external magnetic field. Intravenous mode of administration is not effective when it comes to treatment of brain tumour due to the BBB but can be used for treating tumours in different parts of the body. MNPs can also be used for imaging brain tumours. The MR imaging can be made more accurate by using the ferromagnetic MNPs as contrast agents. MNPs that have been functionalized can be tailored to target brain cancer cells, which can then be detected using MRI [19].

2.4   Gadolinium nanoparticles

GdBNs are gadolinium-based nanoparticles that are helpful in increasing the therapeutic efficiency of the treatment as well as in better performance of MRI. Usually, a good electron emitter is required to improve the radiation effects which is a property of gadolinium due to its atomic mass (Z=64). GdBNs have the ability to improve radiation-induced killing of cells when used in combination with low and high energy X-rays or fast ions. The efficiency of the above particles is so high that they can kill glioma cells taken from a highly aggressive and radiation resistant human tumour. The process of killing starts when the electrons emitted from the irradiated nanoparticles continues to split the water molecule in the presence of radiation which produces free reactive oxygen species (ROS) and when these ROS accumulates in nanoparticles, they cause damages that are dangerous to the cells. By this way NPs amplify the ionizing property and damage in cells while having no effect on macroscopic dose deposition [20] .

2.5   Other treatments

2.5.1   Fluorescent magnetic nanoparticles

Fluorescent magnetic nanoparticles (FMN) are a class of materials that are made up of a magnetic core coated with an inorganic compound or organic polymer that contains a bounded or embedded fluorophore. Fu et al. (2013) reported a novel method for magnetically enhancing cancer targeting *in vivo* using magnetic micromesh and biocompatible FMN in a real-time approach. Using this method, it was possible to magnetically target each carefully administered FMN loaded with one 8 mm super magnetic iron core. Using a mouse as a model to represent human glioma, it was found that NPs can be distributed magnetically in both the tumour neo vasculature and neighbouring tumour tissues.

2.5.2   Ultrasound therapy

In conjunction with the NPs, ultrasound is finding a position in drug administration. Acoustic waves have been linked to the release of pharmacological substances from NPs as well as making cell membranes more permeable, in addition to its non-invasive nature and emphasis on targeted tissues. Ultrasound is utilized to improve the distribution and efficiency of therapeutic agents such as proteins, nucleic acids, and chemotherapeutic medicines in the field of drug delivery.

2.5.3   Cell-mediated nanoparticle delivery

Cell-mediated NP delivery systems have recently evolved as a highly effective technique of delivering medicinal drugs across blood-brain barriers. The cells are used as carriers of nanoparticles which require an accurate determination of loading capacity as well as the identification of intracellular compartments where NPs are transported following endocytosis. This is of tremendous relevance since unique endocytic trafficking routes can eventually influence nanoparticle release mechanism, as well as efficacy and function. Cell selection has been carried out using macrophages, mesenchymal stromal cells, and mesenchymal stem cells.

2.5.4   Thermochemotherapy

When the Medifocus heat treatment was used along with chemotherapy, it increased the tumour shrinkage in the thermochemotherapy by 88.4% while on the other hand when chemotherapy alone was used the tumour shrinkage increased up to 58.8%. Zhao et al. (2013) investigated the magnetic thermochemotherapy effects of unique solar planet structured magnetic nanocomposites (MNCs). Magnetic NPs with amino silane coatings carried magnetic mediated hyperthermia (MMH) and were loaded with Docetaxel. The effects on U251 human glioma cells were studied *in vitro*. All the findings suggested that MNC with a solar-planet configuration was an effective mediator for magnetic thermochemotherapy. MNCs can be used in complete cancer treatment and have a high clinical application potential [21].

3.   Target for nanoparticles gliomas

In response to the tumour microenvironment, glioblastoma cells either overexpress ordinary cell receptors or express unique biomarkers on their membranes in an aberrant manner. As a result, it is easier to construct active targeting techniques based on receptor-mediated endocytosis[22] .The Blood-Brain Barrier (BBB) prevents nearly all large molecules and more than 98% of possible small molecule drugs from entering the brain tissue, which reduces the therapeutic effectiveness of current treatments [23,24].The brain capillary endothelial cells (EC) strongly bound by tight junctions, make up the majority of the BBB(Blood Brain Barrier).One method for drug carriers to cross the BBB is receptor-mediated endocytosis [25].

3.1   LDLR

One of the many receptors that are overexpressed in brain tumours is the LDLR, and as a result, it has been effectively targeted for medication delivery [26]. Since natural LDL is difficult to purify, synthetic versions of LDL have been widely employed in therapeutic methods. The primary ligand for the LDLR (Low Density Lipoprotein Receptor) is Low Density Lipoprotein (LDL), which is also the primary carrier of cholesterol in plasma. LDL is a 22–27 nm particle with a surface coat of unesterified cholesterol, phospholipids and one molecule of apolipoprotein B–100 (apo B). Its core is made of hydrophobic lipids, mostly cholesteryl esters with a little quantity of triglyceride. The binding region for the LDL receptor (LDLR) is nine amino acids (3359–3367) in the 550,000 Da glycoprotein known as apoB-100 [27]. For example, Paclitaxel, a chemotherapeutic drug, has been effectively administered to GBM tumour cells using synthetic LDL-conjugated nanoparticles [28].

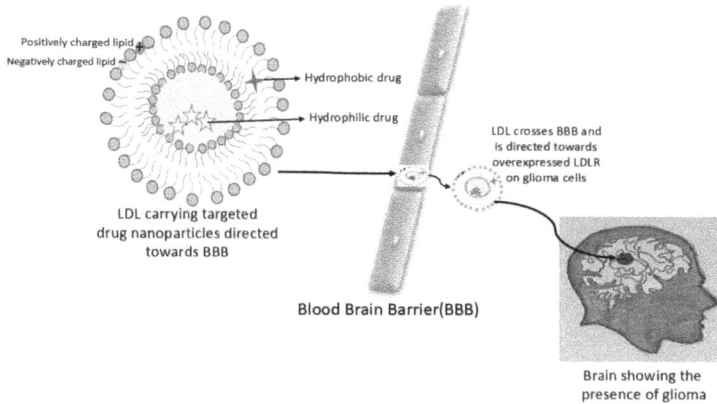

Figure 3. LDLR as target for LDL carrying drug nanoparticles.

3.2   Lactoferrin

Lactoferrin (Lf) belongs to the family of Transferrin (Tf), which may traverse the blood-brain barrier and has 60–80% sequence similarity with Tf. Lf Receptors (LfRs) are also extensively expressed on the cell surface of glioblastomas. In order to circumvent the BBB, Lf conjugated nanoparticles can be used as a brain-targeted nanocarrier [29].The cationic character of Lf contributes to its ability to interact with a variety of cellular anionic ligands, such as glycosaminoglycans, to promote cellular absorption through electrostatic interactions [30].Instead of the more usual passive extravasation transport route through the inter-endothelial gaps, 97% of the NPs may enter the tumours by an active trans-endothelial pathway. This proposed process might explain how Lf-modified nanocarriers enter the brain through the endothelial cells of the brain capillaries. The NPs may bind to endothelial cells before moving through vesicles and transcellular channels into the tumour as shown in figure 4. The concept of Lf-mediated transcytosis and its function in facilitating penetration into deep tumour tissues can thus be further strengthened by this mechanism [31].

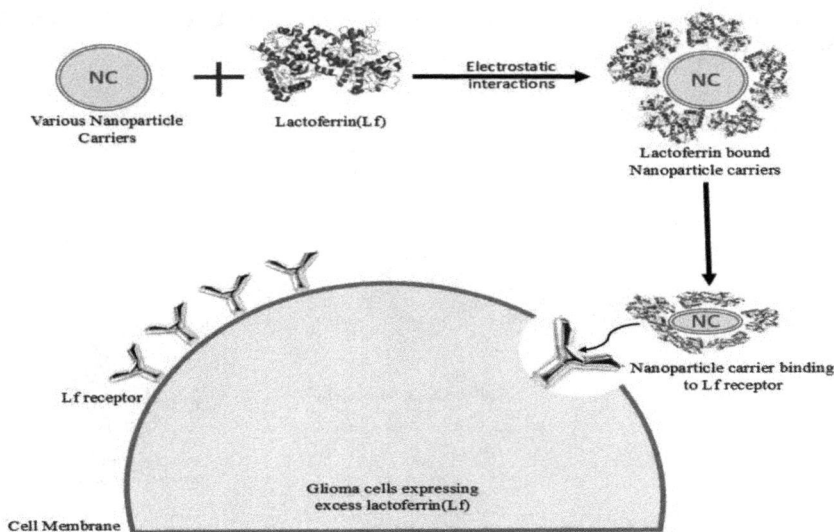

Figure 4. Lactoferrin as a target for nanoparticle carriers

3.3   Transferrin

The transferrin receptor (TfR) is a cell receptor that is overexpressed in both the Glioblastoma cells (GBM) and Blood Brain Barrier (BBB) and is therefore being given increased consideration for dual-targeting techniques. TfR is one of the major proteins expressed on the luminal side of the BBB.As these receptors are in charge of transporting iron into developing cells, GBM cells overexpress them in order to fulfil the heightened need for iron required to maintain the rapid cell division [32].Human Tf (Transferrin) is a 76-kDa blood-plasma glycoprotein that is primarily secreted in the liver and plays a crucial part in iron metabolism as well as being in charge of delivering ferric ion (Fe3+). The oxidation of Fe^{2+} produces Fe^{3+}, the type of iron that binds to Tf, which then attaches to the Tf receptor. The forms of Tf include those that are not iron-bound (apo-Tf), nonferric, or diferric (holo-Tf). Tf removes the harmful iron from the blood and the brain. The conjugated nanoparticles that are to be given to Tfr are included in Tf. Transcytosis, a kind of endocytosis, allows Tf to cross the BBB. Through ferroportin (FPN), Fe^{+2} is exported from the macrophage and binds to Tf, which in turn binds to the TfR on the BBB. Following that, iron is delivered to the brain through the TfR receptor, and the concentration of iron in the brain rises therefore killing the gliomas cells [33]. The mechanism of binding and action of transferrin is depicted in figure 5.

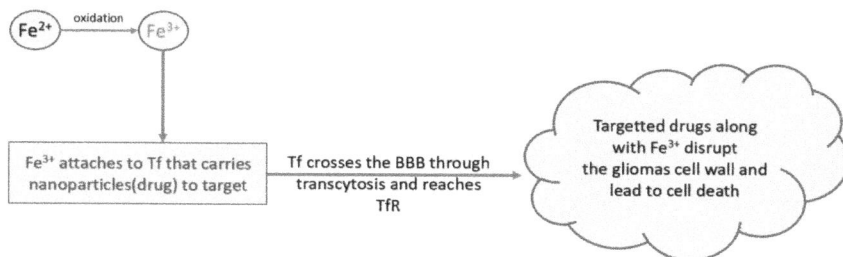

Figure 5. Transferrin as a potential target for the nanoparticles carrying drug to bind and react.

3.4   Matrix metalloproteinases

MMPs are metalloproteinases that may break down a variety of Extracellular Matrix (ECM) macromolecules that contribute to GBM invasion [34]. The active regions of MMPs frequently include Zn2+ atoms, the enzymes need Ca^{2+} ions to function, and the N-terminal propeptide and catalytic domains are largely conserved. MMPs have been classified as gelatinases, collagenases, membrane-type MMPs,stromelysins, etc. based on how they react with different types of substrates [35].One significant cytokine regulator of the MMP-system appears to be the cytokine TGF-ß, which is heavily generated by glioma cells. It increases the production of MMP-2 and MMP-9 while inhibiting TIMP-2, which facilitates the invasion of glioma cells [36]. The mechanism of action of MMP inhibitors is through the inhibition of the proteolytic activity of MMPs, which prevents the degradation of ECM components and reduces glioma invasion. Additionally, the inhibition of MMPs can decrease the production of pro-angiogenic factors, such as vascular endothelial growth factors (VEGF), and inhibit angiogenesis in gliomas [37].

3.5   Others

3.5.1   F3 Peptide

An F3 peptide that binds exclusively to nucleolin, which is extensively expressed on the surface of both glioma cells and endothelial cells of glioma angiogenic blood vessels, is used as a part of a nanoparticulate drug delivery system [38]. The fact that nucleolin is exclusively found in the nucleus of normal cells makes it an appealing target for mediating targeted and effective angiogenic blood vesicles and glioma cells with high cellular internalisation [39]. This results in the dual targeting of glioma cells and neo vasculature as well as effective cellular internalisation [40].

3.5.2   Interleukin receptors

ILRs (Interleukin Receptors) are widely expressed in many tumour cells because they can support the growth and resistance to apoptosis of a variety of cancer forms, including glioma. Interleukin-13 receptor (IL-13R) is one of these ILRs that has been used as a glioma-targeting location for nano-enabled delivery systems. However, IL-13 still has several drawbacks as a targeting ligand, including its high molecular weight and ease of denaturation [41]. Figure 6 shown below summarises all the other potential targets for nanoparticles that are not completely explored.

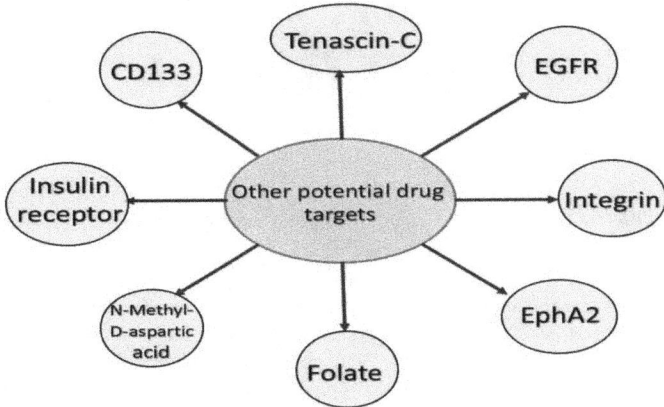

Figure 6. Other potential drug targets for nanoparticles in the treatment of gliomas

References

[1] D.N. Louis, H. Ohgaki, O.D. Wiestler, W.K. Cavenee, P.C. Burger, A. Jouvet, B.W. Scheithauer, P. Kleihues, The 2007 WHO classification of tumours of the central nervous system, *Acta Neuropathol.* 114(2) (2007) 97-109. https://doi.org/10.1007/s00401-007-0243-4

[2] H Suzuki, A. Kosuke, C.Kenichi, S. Yusuke, S. Yusuke, S. Yuichi, S. Teppei et al., Mutational landscape and clonal architecture in grade II and III gliomas. *Nature genetics,* 47, no. 5 (2015) 458-468.

[3] N. A. O. Bush, and B. Nicholas, The effect of molecular diagnostics on the treatment of glioma. *Current oncology reports* 19 (2017) 1-9.

[4] P. Wen, D.Reardon, Progress in glioma diagnosis, classification and treatment, *Nat Rev Neurol* , 12, (2016)69–70. https://doi.org/10.1038/nrneurol.2015.242

[5] P.Y. Wen and S. Kesari. Malignant gliomas in adults. *New England Journal of Medicine* 359, no. 5 (2008) 492-507.

[6] A. Pirzkall,,C. McGue, S. Saraswathy, S. Cha, R. Liu, S. Vandenberg, K. R. Lamborn, M. S. Berger, S. M. Chang, and S. J. Nelson, Tumor regrowth between surgery and initiation of adjuvant therapy in patients with newly diagnosed glioblastoma. *Neuro-oncology* 11, no. 6 (2009) 842-852.

[7] J. Kalpathy-Cramer, E. R. Gerstner, K. E. Emblem, O. C. Andronesi, and B Rosen. Advanced magnetic resonance imaging of the physical processes in human glioblastoma. *Cancer research* 74, no. 17 (2014) 4622-4637.

[8] M. R. Gilbert, J. J. Dignam, T. S. Armstrong, J. S. Wefel, D.T. Blumenthal, M.A. Vogelbaum, H. Colman, A. Chakravarti, S. Pugh, M. Won, R. Jeraj, P. D. Brown, K. A. Jaeckle, D. Schiff, V. W. Stieber, D. G. Brachman, M. Werner-Wasik, I. W. Tremont-Lukats, E. P. Sulman, K. D. Aldape, W. J. Neurosurg Rev Curran, M. P Mehta. A randomized trial of bevacizumab for newly diagnosed glioblastoma. 370; *N Engl J Med* (2014) 699–708:. https://doi.org/10.1056/NEJMoa1308573

[9] M. Westphal, D.C. Hilt, E. Bortey, P. Delavault, R. Olivares, P. C. Warnke, I. R. J. Whittle, Jääskeläinen, Z. Ram, A phase 3 trial of local chemotherapy with biodegradable carmustine (BCNU) wafers (Gliadel wafers) in patients with primary malignant glioma. 5;. *NeuroOncol* (2003):79–88. https://doi.org/10.1215/S1522-8517-02-00023-6

[10] C. Alifieris, D. T. Trafalis. Glioblastoma multiforme: pathogenesis and treatment. 152; *Pharmacol Ther* (2015),63–82. https://doi.org/10.1016/j. pharmthera.2015.05.005

[11] D. N. Franz, E. Belousova, S. Sparagana, E. M. Bebin, M. Frost, R. Kuperman, O. Witt, M. H. Kohrman, J. R. Flamini, J. Y. Wu, P. Curatolo, P. J. de Vries, V. H. Whittemore, E. A. Thiele, J. P. Ford, G. Shah, H. Cauwel, D. Lebwohl, T. Sahmoud, S. Jozwiak. Efficacy and safety of everolimus for subependymal giant cell astrocytomas associated with tuberous sclerosis complex (EXIST-1): a multicentre, randomised, placebocontrolled phase 3 trial. 381; *Lancet Lond Engl* (2013).125–132. https://doi.org/10.1016/S0140-6736(12)61134-9

[12] S. Phuphanich, C.J. Wheeler, J. D. Rudnick, M. Mazer, H. Wang, M. A. Nuño, J. E. Richardson, X. Fan, J. Ji, R. M. Chu, J. G. Bender, E. S. Hawkins, C. G. Patil, K. L. Black, J. S. Yu. Phase I trial of a multi-epitopepulsed dendritic cell vaccine for patients with newly diagnosed glioblastoma. CII 62; *Cancer Immunol Immunother* (2013).125–135. https://doi.org/10.1007/s00262-012-1319-0

[13] E. T. Sayegh, T. Oh, S. Fakurnejad, O. Bloch, A. T. Parsa. Vaccine therapies for patients with glioblastoma. 119 *J Neurooncol* (2014).531– 546. https://doi.org/10.1007/s11060-014-1502-6 79.

[14] O. Bloch, C. A. Crane, Y. Fuks, R. Kaur, M. K. Aghi, M. S. Berger, N. A. Butowski, S. M. Chang, J. L. Clarke, M. W. McDermott, M. D. Prados, A. E. Sloan, J. N. Bruce, A. T. Parsa. Heat-shock protein peptide complex-96 vaccination for recurrent glioblastoma: a phase II, single-arm trial. 16; *Neuro-Oncol (*2014):274–279. https://doi.org/10.1093/neuonc/ not203

[15] M.D. Prados, S. A. Byron, N. L. Tran, J. J. Phillips, A. M. Molinaro, K. L. Ligon, P. Y. Wen, J. G. Kuhn, I. K. Mellinghoff, J. F de Groot, H. Colman, T. F. Cloughesy, S. M. Chang, T. C. Ryken, W. D. Tembe, J. A. Kiefer, M. E. Berens, D. W. Craig, J. D. Carpten, J. M. Trent. Toward precision medicine in glioblastoma: the promise and the challenges. 17; Neuro-Oncol (2014):1051–1063. https://doi.org/10.1093/neuonc/nov031

[16] B. S. Mahmoud, A. H. AlAmri, and C. McConville. Polymeric nanoparticles for the treatment of malignant gliomas. *Cancers* 12, no. 1 (2020). 175. https://doi.org/10.3390/cancers12010175

[17] C. A. Caraway, H. Gaitsch, E. E. Wicks, A. Kalluri, N. Kunadi, and B. M. Tyler. Polymeric Nanoparticles in Brain Cancer Therapy: A Review of Current Approaches, *Polymers* 14,(2022.) no. 14: 2963. https://doi.org/10.3390/polym14142963

[18] M. X. Liu, J. Zhong, N.N. Dou, M. Visocchi, and G. Gao. One-pot aqueous synthesization of near-infrared quantum dots for bioimaging and photodynamic therapy of gliomas. *Trends in Reconstructive Neurosurgery: Neurorehabilitation, Restoration and Reconstruction* (2017). 303-308. https://doi.org/10.1007/978-3-319-39546-3_44

[19] K. Mahmoudi and C. G. Hadjipanayis. The application of magnetic nanoparticles for the treatment of brain tumors. *Frontiers in chemistry* 2 (2014). 109. https://doi.org/10.3389/fchem.2014.00109

[20] L. Štefančíková, S. Lacombe, D. Salado, E. Porcel, E. Pagáčová, O. Tillement, F. Lux, D. Depeš, S. Kozubek, and M. Falk. "Effect of gadolinium-based nanoparticles on nuclear DNA damage and repair in glioblastoma tumor cells." *Journal of nanobiotechnology* 14, no. 1 (2016): 1-15. https://doi.org/10.1186/s12951-016-0215-8

[21] M. Li, H. Deng, H. Peng, Q. Wang. Functional nanoparticles in targeting glioma diagnosis and therapies. *Journal of nanoscience and nanotechnology* 14, no. 1 (2014). 415-432.

[22] H. E. Marei. Multimodal targeting of glioma with functionalized nanoparticles. *Cancer Cell International* 22, no. 1 (2022). 265. https://doi.org/10.1186/s12935-022-02687-8

[23] H. Xin, X. Jiang, J. Gu, X. Sha, L. Chen, K. Law, Y. Chen, X. Wang, Y. Jiang, and X. Fang. Angiopep-conjugated poly (ethylene glycol)-co-poly (ε-caprolactone) nanoparticles as dual-targeting drug delivery system for brain glioma. Biomaterials 32, no. 18 (2011). 4293-4305. https://doi.org/10.1016/j.biomaterials.2011.02.044

[24] B. Zhang, X. Sun, H. Mei, Y. Wang, Z. Liao, J. Chen, Q. Zhang, Y. Hu, Z. Pang, and X. Jiang. LDLR-mediated peptide-22-conjugated nanoparticles for dual-targeting therapy of brain glioma. Biomaterials 34, no. 36 (2013). 9171-9182. https://doi.org/10.1016/j.biomaterials.2013.08.039

[25] P. P. Di Mauro, A. Cascante, P. B. Vilà, V. Gómez-Vallejo, J. Llop, and S. Borrós. Peptide-functionalized and high drug loaded novel nanoparticles as dual-targeting drug delivery system for modulated and controlled release of paclitaxel to brain glioma. International Journal of Pharmaceutics 553, no. 1-2 (2018). 169-185. https://doi.org/10.1016/j.ijpharm.2018.10.022

[26] S. Pawar, Shreya, T. Koneru, E. McCord, K. Tatiparti, S. Sau, and A. K. Iyer. LDL receptors and their role in targeted therapy for glioma: a review. Drug Discovery Today 26, no. 5 (2021). 1212-1225. https://doi.org/10.1016/j.drudis.2021.02.008

[27] M. Nikanjam, Mina, E. A. Blakely, K. A. Bjornstad, X. Shu, T. F. Budinger, and T. M. Forte. Synthetic nano-low density lipoprotein as targeted drug delivery vehicle for glioblastoma multiforme. International journal of pharmaceutics 328, no. 1 (2007). 86-94. https://doi.org/10.1016/j.ijpharm.2006.07.046

[28] F. Pourgholi, J.N. Farhad, H. S. Kafil, and M.Yousefi. Nanoparticles: Novel vehicles in treatment of Glioblastoma. Biomedicine & Pharmacotherapy 77 (2016). 98-107. https://doi.org/10.1016/j.biopha.2015.12.014

[29] M. M. Song, H.L. Xu, J.X. Liang, H. H. Xiang, R. Liu, and Y. X. Shen. Lactoferrin modified graphene oxide iron oxide nanocomposite for glioma-targeted drug delivery. Materials Science and Engineering: C 77 (2017). 904-911. https://doi.org/10.1016/j.msec.2017.03.309

[30] M. M. Agwa and S. Sabra. Lactoferrin coated or conjugated nanomaterials as an active targeting approach in nanomedicine. International Journal of Biological Macromolecules 167 (2021), 1527-1543. https://doi.org/10.1016/j.ijbiomac.2020.11.107

[31] A. O. Elzoghby, M. A. Abdelmoneem, I. A. Hassanin, M. M. Abd Elwakil, M. A. Elnaggar, S. Mokhtar, J. Y. Fang, and K. A. Elkhodairy. Lactoferrin, a multi-functional glycoprotein: Active therapeutic, drug nanocarrier & targeting ligand. Biomaterials 263 (2020). 120355. https://doi.org/10.1016/j.biomaterials.2020.120355

[32] M. J. Ramalho, J. A. Loureiro, M. A. N. Coelho, and M. C. Pereira. Transferrin receptor-targeted nanocarriers: overcoming barriers to treat glioblastoma. Pharmaceutics 14, no. 2 (2022). 279. https://doi.org/10.3390/pharmaceutics14020279

[33] T. Koneru, E. McCord, S. Pawar, K. Tatiparti, S. Sau, and A. K. Iyer. Transferrin: biology and use in receptor-targeted nanotherapy of gliomas. ACS omega 6, no. 13 (2021). 8727-8733. https://doi.org/10.1021/acsomega.0c05848

[34] M. Li, H. Deng, H. Peng, and Q. Wang. Functional nanoparticles in targeting glioma diagnosis and therapies. Journal of nanoscience and nanotechnology 14, no. 1 (2014). 415-432. https://doi.org/10.1166/jnn.2014.8757

[35] G. A. Cabral-Pacheco, I. Garza-Veloz, C. Castruita-De la Rosa, J. M. Ramirez-Acuna, B. A. Perez-Romero, J. F. Guerrero-Rodriguez, N. Martinez-Avila, and M. L. Martinez-Fierro. The roles of matrix metalloproteinases and their inhibitors in human diseases. International journal of molecular sciences 21, no. 24 (2020), 9739. https://doi.org/10.3390/ijms21249739

[36] A. O. Elzoghby, M. A. Abdelmoneem, I. A. Hassanin, M. M. Abd Elwakil, M. A. Elnaggar, S. Mokhtar, J. Y. Fang, and K. A. Elkhodairy. Lactoferrin, a multi-functional glycoprotein: Active therapeutic, drug nanocarrier & targeting ligand. Biomaterials 263 (2020). 120355. https://doi.org/10.1016/j.biomaterials.2020.120355

[37] G. A. Cabral-Pacheco, I. Garza-Veloz, C. Castruita-De la Rosa, J. M. Ramirez-Acuna, B. A. Perez-Romero, J. F. Guerrero-Rodriguez, N. Martinez-Avila, and M. L. Martinez-Fierro. The roles of matrix metalloproteinases and their inhibitors in human diseases. International journal of molecular sciences 21, no. 24 (2020). 9739. https://doi.org/10.3390/ijms21249739

[38] Q. Hu, G. Gu, Z. Liu, M. Jiang, T. Kang, D. Miao, Y. Tu, Z. Pang, Q. Song, L. Yao, and H. Xia, F3 peptide-functionalized PEG-PLA nanoparticles co-administrated with tLyp-1 peptide for anti-glioma drug delivery. Biomaterials 34, no. 4 (2013): 1135-1145. https://doi.org/10.1016/j.biomaterials.2012.10.048

[39] M. Li, H. Deng, H. Peng, and Q. Wang. Functional nanoparticles in targeting glioma diagnosis and therapies. Journal of nanoscience and nanotechnology 14, no. 1 (2014). 415-432. https://doi.org/10.1166/jnn.2014.8757

[40] Q. Hu, G. Gu, Z. Liu, M. Jiang, T. Kang, D. Miao, Y. Tu Z. Pang, Q. Song, L. Yao, H. Xia. F3 peptide-functionalized PEG-PLA nanoparticles co-administrated with tLyp-1 peptide for anti-glioma drug delivery. Biomaterials 34, no. 4 (2013). 1135-1145. https://doi.org/10.1016/j.biomaterials.2012.10.048

[41] S. Wang, Y. Meng, C. Li, M. Qian, and R. Huang. Receptor-mediated drug delivery systems targeting to glioma. Nanomaterials 6, no. 1 (2015). 3. https://doi.org/10.3390/nano6010003

Keyword Index

About the Editors

Dr. Dileep Francis, an Assistant Professor in the Department of Life Sciences at Kristu Jayanti College, Autonomous, Bengaluru, holds a Ph.D. in Life Sciences from Kannur University, Kerala, India. He achieved an impressive all India rank of 60 in the CSIR-NET exam and received junior and senior research fellowships from the Council for Scientific and Industrial Research (CSIR), Government of India, to support his doctoral research. During his doctoral studies, he successfully identified a vaccine candidate against Methicillin resistant *Staphylococcus aureus* infections. His current research is focused on interventions for diagnosing and treating infectious diseases, with recent forays into material sciences for biomedical applications. Dr. Dileep has a robust publication record, including high-impact journal articles, edited books, conference proceedings, book chapters, and popular science articles. Additionally, he serves as the editor of the Kristu Jayanti Journal of Core and Applied Biology, a biannual journal published by the Department of Life Sciences at Kristu Jayanti College, Autonomous. Dr. Dileep is a life member of the Kerala Academy of Sciences and a member of the research advisory committee at Kristu Jayanti College, Autonomous. Furthermore, he coordinates the Star College Scheme of the Department of Biotechnology (DBT), Government of India, at Kristu Jayanti College, Autonomous.

Dr. Manoj Balachandran is a Professor in the Department of Physics and Electronics at CHRIST (Deemed to be University), Bengaluru, Karnataka, India. He received his/her Ph.D. Degree in Physics from Bharathiar University in 2013.He is having 24 Years of Experience in the field of Functional nanomaterials and Energy Science. His area of Expertise includes: Low dimensional and layered materials for Energy Harvesting-Third generation solar cells, Energy Storage-supercapacitor, Water purification, Antibacterial & antifouling studies, membrane filtering, Bioimaging, Soft robotics and Photocatalysis. He is the author of more than 130 national-international publication and 22 patent publication, recipient of DST Grants for development of third generation solar cells, layered materials for supercapacitors etc.

Dr. Elcey C Daniel is a Professor and the Head of the Department of Life Sciences at Kristu Jayanti College, Autonomous, Bengaluru, India. She earned her Ph.D. degree from Banaras Hindu University in 1992 and continued her research by completing a CSIR Research Associateship in the Department of Food Microbiology at the Central Food Technological Research Institute, focusing on Environmental Microbiology. Additionally, she received a Research Fellowship from the Ministry of International Trade and Industry in Japan and completed her tenure at ONRI in Osaka (AIST Kansai, Ikeda Osaka). Following her research endeavors, she has dedicated 20 years as a faculty member at Kristu Jayanti College, Autonomous. Dr. Elcey C Daniel has accumulated 35 years of experience in the fields of Teaching and Research. Her areas of expertise encompass Microbial Technology, Microbial Biodegradation, and Nano-biotechnology. She is the author of numerous international peer-reviewed research articles, along with a book and a book chapter. Moreover, Dr. Elcey has supervised a Ph.D. scholar and guided many postgraduate dissertations. She has received industrial and institutional research grants and has also served as a consultant on industrial projects.

Dr. Shinomol George Kunnel is an Assistant Professor in the Department of Life Sciences at Kristu Jayanti College, an autonomous institution in Bengaluru. She completed her BSc in Zoology, Botany, and Chemistry at Mahatma Gandhi University, Kottayam, India, and her MSc in Biochemistry from the School of Biosciences at the same university. Furthermore, she obtained her Ph.D. in Biochemistry from CSIR-Central Food Technological Research Institute, University of Mysore, Mysore. Following her Ph.D., she conducted Post-Doctoral Research as a DST-Women Scientist at the National Institute of Mental Health and Neurosciences (NIMHANS) in Bengaluru, specializing in the field of neurotoxicology. Dr. Shinomol George K has secured and continues to work on grants from various national and state agencies as the Principal Investigator. She is an approved Ph.D. guide at Visvesvaraya Technological University, Belagavi, and has ongoing Ph.D. students under her mentorship. She has a substantial publication record in journals of international repute and has contributed to several book chapters. Furthermore, she has been the recipient of multiple international and national travel grants and has served as a Ph.D. Examiner for institutions such as Osmania University, Hyderabad, India; National Institute of Pharmaceutical Education and Research, Guwahati, India; and the University of Mysore, Mysore. Dr. Shinomol George K also serves as an ad hoc reviewer for various scientific journals and is a life member of several professional societies. Her research interests span a wide range of topics, including neurotoxicology related to Parkinson's, Alzheimer's, and Huntington's disease-causing toxins, neuropharmacology involving the use of Ayurvedic plants to ameliorate neuronal mitochondrial dysfunctions, phytochemistry, mosquito repellent and bionanoparticle studies of ethnopharmacological plants, environmental toxicology of heavy metals, and the potential of Tardigrades as effective survival models under extreme environmental conditions.

www.ingramcontent.com/pod-product-compliance
Lightning Source LLC
Chambersburg PA
CBHW071326210326
41597CB00015B/1368